THE PRINCIPLES OF ELECTRONIC AND ELECTROMECHANIC POWER CONVERSION

THE PRINCIPLES OF ELECTRONIC AND ELECTROMECHANIC POWER CONVERSION

A Systems Approach

BRAHAM FERREIRA
Delft University of Technology

WIM VAN DER MERWE
ABB Corporate Research

Published by John Wiley & Sons, Inc., Hoboken, New Jersey
Published simultaneously in Canada

Library of Congress Cataloging-in-Publication Data:

Ferreira, Braham, 1958-
The principles of electronic and electromechanic power conversion : a systems approach / Braham Ferreira, Wim van der Merwe.
1 online resource.
Includes bibliographical references and index.
Description based on print version record and CIP data provided by publisher; resource not viewed.
ISBN 978-1-118-79884-3 – ISBN 978-1-118-79885-0 (ePub) – ISBN 978-1-118-65609-9 (cloth)
1. Power electronics. 2. Electric generators. I. Van der Merwe, Wim, 1977- II. Title.
TK7881.15
621.31′7–dc23

2013029833

10 9 8 7 6 5 4 3 2 1

CONTENTS

PREFACE

During the past forty years, electrical power conversion systems have evolved considerably. Some technologies have matured, while others have undergone substantial new development. Conversion systems have become more diverse and sophisticated, and feature complex architectures that have improved performance. Central and embedded controllers are now playing a prominent role in managing power flow. Power is applied to loads with high precision, and energy conversion efficiency is optimized at the same time. The volume of power conversion knowledge has increased, and the average electrical engineer deals increasingly with system aspects and less with component details.

The typical undergraduate student needs to study fundamentals as well as the various disciplines of electrical and electronics engineering during a three or four year study program. It is a full program, and it is not possible to introduce more course material in a power conversion course on system-related issues without condensing the content on components. In this book, we have strived to strike a new balance between systems and components in a one semester course that is intended for a bachelors undergraduate program. Alternatively, the book may be used as a self-study book on the principles of power conversion by a professional in science and engineering, who has sufficient mathematical background.

This textbook is a result of many discussions with colleagues. Our observation was that available undergraduate textbooks at this level follow a bottom-up approach that is forty years old. Issues such as magnetic circuits and dc machine field winding configurations receive a substantial amount of attention, while few modern engineers need this knowledge. Power electronics is frequently treated in a 1970s style, describing circuit topologies that are now being phased out and using old-fashioned power devices. In this book, a compact treatise on magnetic circuits is presented where the principles are introduced to gain a good qualitative understanding, leaving practical magnetic circuit analysis and design for follow-up courses. A modern simple approach to machines that makes the principles of field-oriented control and space vector theory approachable to undergraduate students is introduced. In power electronics, the focus is on topologies that use a series transistor and diode combination that is connected to a dc source because this has become a standard building block. The trend is that semiconductor switches are becoming ever better and, therefore, this treatise uses ideal switches.

A top-down approach is followed in this textbook, where the role and system context of power conversion functions are first introduced. The building blocks of the system are defined, and the theory of how they exchange power with each other is described. Then the building blocks are opened, and principles of static and electromechanical power conversion are discussed. Chapter 1 introduces the system architectures used in modern power conversion systems and defines the functions of the building blocks. The first components in a power conversion chain are the sources. As the theory of most energy sources falls outside the scope of the book, the treatise in Chapter 2 is limited and focuses only on a few key concepts. Chapters 3–5 cover electric, magnetic and mechanical power transfer. In Chapter 3, electrical power theory is expanded to include periodic nonsinusoidal voltage and current waveforms because shapes of waveforms other than dc and sinusoidal ac now exist in systems; this phenomena is increasingly due to the prevalence of power electronics in the modern grid. Magnetic coupling is addressed in Chapter 4, and it is not only relevant for coupling different electrical circuits but also plays a role in electromechanical conversion. Chapter 5 is a short chapter on mechanical components and was included because of the important role that mechanical inertias play in the dynamics of systems and because mechanical coupling and electrical coupling are often interchangeable in system architectures. Chapter 6 introduces power electronics at the hands of switchmode dc–dc converters. The main focus is on the quasi-steady-state circuit analysis and the basic conversion functions where step-up, step-down, and conversion involving magnetic coupling is introduced. Chapter 7 introduces electromechanics at the hand of the Lorenz force and evolves the principle from linear motion through rotational motion to a practical dc machine. The concept of orthogonality between torque and speed control in separately excited machines receives attention because of its important role in high-performance electrical drive systems. In Chapter 8, the source of variable frequency, variable amplitude ac, namely a power electronics inverter, is first introduced followed by a treatise on the ac machines stator as a device that can be connected to a three-phase ac source. Two types of rotors then yield either a synchronous or an asynchronous machine. The loss mechanisms are superficially discussed and the calculation of power conversion efficiency is also treated in the two chapters on electrical machines.

During the development of the text, we have placed considerable emphasis on keeping the descriptions and notations mathematically correct and consistent. Although for simplification it might be possible to rewrite the text using different notation methods and conventions for different parts of the book and to shy away from the more complicated mathematical derivations, it is our view that such an approach would be to the eventual detriment of the student. It is our sincere hope that this text strikes the balance of both making the contents accessible to a wide range of undergraduate students, not least of all the students who will eventually specialize in a different engineering field, and, at the same time, to lay a solid and mathematically sound foundation for those students who wish to specialize in power conversion systems either as a career or through postgraduate studies.

We are grateful to colleagues and students for constructive discussions and advice to improve the content of the book. In particular, we want to acknowledge the effort and support of Sjoerd de Haan and Fatih Çağlayan.

Delft, the Netherlands
Baden, Switzerland
August 2013

BRAHAM FERREIRA
WIM VAN DER MERWE

CHAPTER 1

INTRODUCTION TO ELECTRICAL SYSTEMS AND POWER CONVERSION

1.1 ELECTRICITY AS AN ENERGY CARRIER

All through human existence, some of the greatest advancements in the standard of living came about by learning how to convert energy from one form to another. The first controlled and intentional fire (converting chemical energy to heat energy) was the first step in this process. Man employed the energy from biomass for heating and cooking by burning dried leaves, wood and animal waste. The chemical energy trapped in the organic material was released and converted into heat and light energy.

Later, humans discovered that wind could be used for transportation on water. By using sails, the energy present in moving air can be used to propel a ship in water. Wind energy is a form of kinetic energy that is used to overcome the resistance of water and make the ship move.

The generation of mechanical energy to replace human or animal power came later in human history with the development of simple devices to harness the energy of flowing water and wind. The earliest machines built were waterwheels used initially for grinding grain but later adopted for various other functions such as for driving saw mills and pumps. The oldest reference to a water mill dates to about 85 BC, appearing in a poem by an early Greek writer. The source of energy harnessed here is the potential energy of water, which flows from high areas along rivers to the sea.

The Principles of Electronic and Electromechanic Power Conversion: A Systems Approach, First Edition. Braham Ferreira and Wim van der Merwe.

Windmills, like waterwheels, were among the original machines that replaced animal muscle as a source of power. They were used for centuries in various parts of the world, converting the energy of the wind into mechanical energy for grinding grain, pumping water and draining lowland areas.

The rapid growth of industry from the mid-18th century created a need for new sources of motive power, particularly solutions that are independent of geographic location and weather conditions. This situation, together with certain other factors, set the stage for the development and widespread use of the steam engine, the first practical device for converting thermal energy into mechanical energy. In 1765, James Watt, a Scottish instrument maker and inventor, made important modifications to the steam engine which resulted in a fuel cost reduction of about 75%. This was the first modern breakthrough in improving the efficiency of a machine that converts energy. The unit for electrical power, the rate of energy flow or energy conversion, was named after James Watt.

Electrical energy conversion emerged during the 19th century when the English physicist and chemist Michael Faraday discovered a means by which to convert mechanical energy into electricity. This set the scene for the use of electricity to provide light. Lighting was an important driver for the large scale use of electrical energy because incandescent lamps were much easier and safer to use than oil lamps.

The conversion of mechanical energy into electrical energy can also be reversed; the same generator can be operated as an electrical motor, which converts electrical energy into mechanical energy. The advantage of electrical motors in factories was soon realised. Instead of using a complex system of belts and pulleys to distribute power to the various work stations in the factory, copper cables and electrical motors could be used. In this way, the first small electrical grid came into being with a steam engine driving a generator that supplied the electrical energy for lamps and electrical motors.

An advantage of electrical energy is that it can be easily controlled, by using, for example, switches. An oil lamp can be ignited with a match, while it is much simpler to push the button of a switch to turn on a light. Today, you do not even have to get out of the chair to turn on the light or television; you can use a remote control that is carried in your pocket.

An alternative source of electrical energy is batteries in which chemical energy is directly converted to electrical current. Electrical automobiles using batteries succeeded steam-powered self-propelled vehicles but were later displaced by combustion engine-powered automobiles because the energy density of gasoline is much higher than that of a battery.

Electrical energy can, in general, be converted into other forms of energy with very high efficiency. It is also possible to convert many other forms of energy into electrical energy. The introduction of electrical machines set the first step towards making electrical energy the best universal energy carrier. Semiconductor-based energy converters made it possible to accurately control the flow of electrical energy. The era of electronics began in the 1950s with rapid advances in the design and construction of semiconductor diodes and transistors. When thyristors

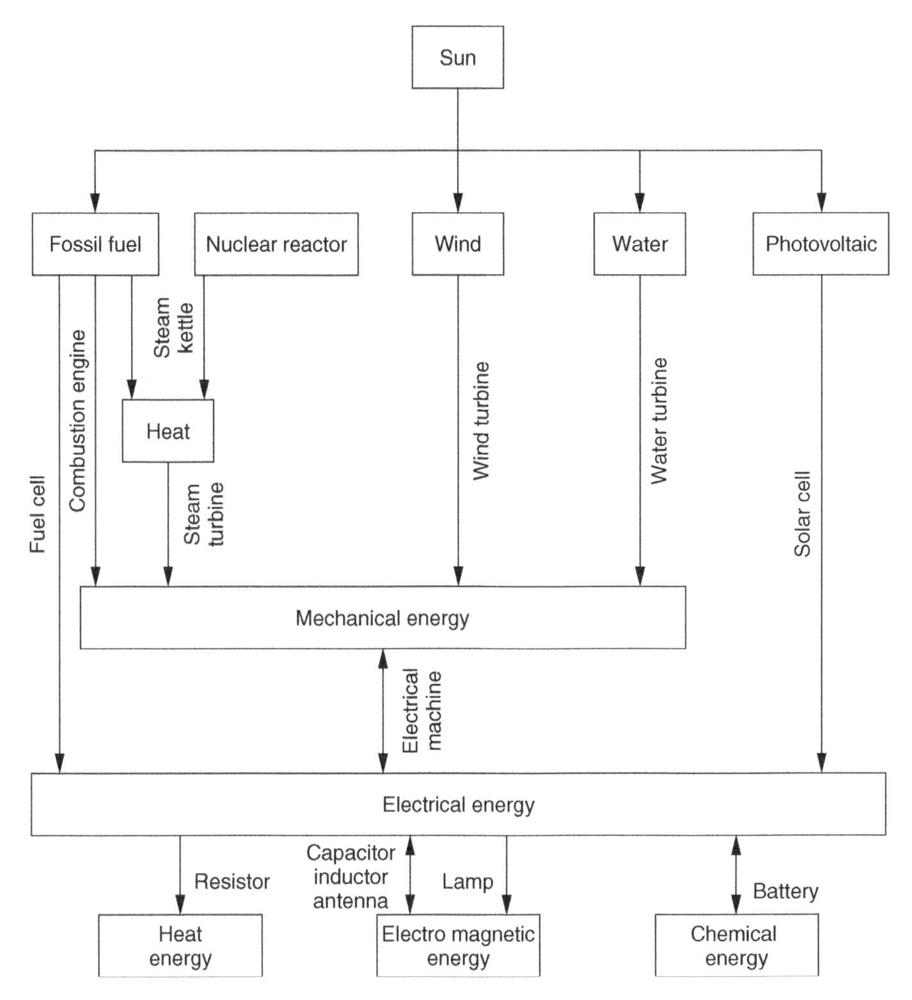

Figure 1.1 The interaction between different forms of energy.

appeared a few years later in 1957, it became possible to convert energy using electronic circuits.

Figure 1.1 illustrates the interaction between different forms of energy. All energy comes from nuclear reaction, taking place either in the sun or on the earth. Energy from water, wind and solar radiation is called *renewable* because it cannot be depleted. Fossil fuels, on the other hand, while a cheap and easily available source of energy, can only be consumed once and, as a result, these sources of energy are being depleted.

Most energy is first converted into mechanical energy as an intermediate step before it is converted into electrical energy. By using power electronics, electrical energy is converted into heat, light, mechanical energy and chemical energy for homes, offices and factories. Note that all of these energy flows can be bidirectional, with the exception of heating.

One of the major problems with electrical energy is the willing buyer/willing seller principle. It is impossible to generate electrical energy if there is no consumer of that form of energy. Electrical energy can only be used as a carrier to transport energy from one place (of generation) to another (of consumption). Although it is possible, as we will discuss later in this book, to store limited amounts of electrical energy, we can only generate electrical energy if we have a consumer, even if the storage device is the consumer for that period in time. One of the big challenges to be solved in future energy systems is the fact that the electrical grid will be supplied by a growing amount of renewable energy, which fluctuates between day and night and when wind conditions vary. This will imply that energy may not be available when it is needed by the consumer or the energy could be in abundance it is not needed. Large energy storage systems can solve this problem but it is easier said than done. This puzzle will need to be solved with the Smart Grid in the future.

1.2 DEVELOPMENT OF ELECTRICAL ENERGY CONVERSION SYSTEMS

The process of energy conversion needs to be controlled. The early humans quickly learned how much wood was required to prepare food for an evening meal. Then they had to feed fuel to the fire at a specific rate to ensure that the flame was large enough, but not too big, to cook food or to roast meat. The same process of controlling the application of energy was used when coal was shovelled into the furnace chamber built into the boiler of a steam locomotive, see Fig. 1.2. Generating the right amount of steam for propulsion and controlling the pressure in the boiler was complicated, and driving a steam engine requires years of practice and apprenticeship training.

In a steam locomotive, a complicated system of regulators and throttle valves is employed to control the steam pressure and the flow of energy to the pistons that are directly mounted on the driving wheels of the locomotive. The steam engine

Figure 1.2 A steam locomotive.

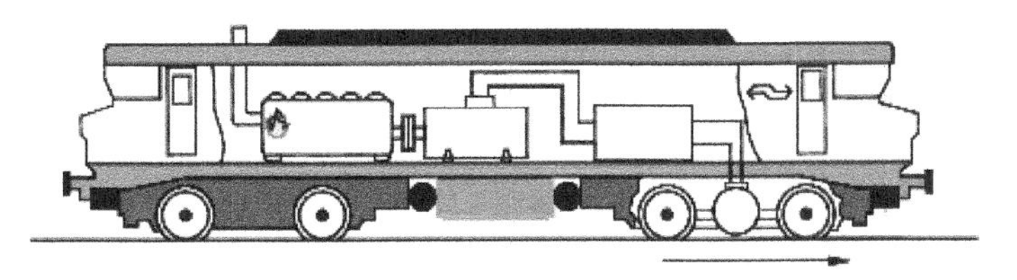

Figure 1.3 A diesel-electrical locomotive.

was improved by feeding the steam back through boiler tubes to superheat (to heat something to beyond its boiling point) the steam to increase the engine efficiency and power. The modification of the steam engine set the trend to apply advanced technology to improve the efficiency of energy conversion systems. As a result, modern energy conversion systems are complex and have many conversion stages.

Continuing with the example of the train locomotive, we observe in Fig. 1.3 how the distance between the furnace or chamber where the combustion takes place and mechanical energy conversion has increased. In a diesel-electrical locomotive, the combustion takes place in a diesel engine that is connected to an electrical generator. The energy required by the train is then supplied, in electrical form, to an electronic power converter that controls both the speed and traction power delivered to the wheels by controlling the flow of electrical energy delivered to the electrical motors. This power electronic converter fulfils the same function as that of the gears and throttle in an automobile. In the final stage of the conversion chain, the electrical energy is converted to mechanical energy by an electrical motor that is directly connected to the axle.

In the case of an electrical locomotive, Fig. 1.4, the combustion of fuel takes place at a power station that feeds the electrical grid. The electrical energy flows hundreds of kilometres and goes through many conversion stages before it reaches the locomotive. The electrical locomotive has to share its source of energy with homes, offices, factories and other systems that are connected to the grid. Once again a power electronic converter in the locomotive controls the flow of electrical

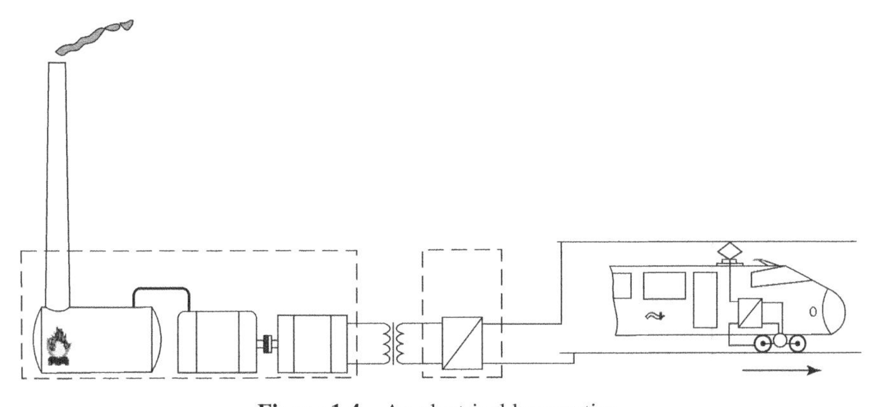

Figure 1.4 An electrical locomotive.

energy from the grid to the driving electrical motors to control the speed and the delivered traction power.

1.3 SYSTEM BUILDING BLOCKS

Look around you and notice all the advanced power conversion systems that have become part of modern life. We briefly discussed the evolution of the train, but similar developments have taken place in automobiles, ships and aircraft. The trend is to have more power conversion stages and to increase the levels of electrification in systems. Growing complexity, improved performance and energy efficiency can also be observed if we study the evolution of telecommunication and computer systems. For example, the original personal computer of 1980 had one power supply and a single microprocessor, while any modern cell phone, game console, laptop or desktop computer today has multiple processors and power supplies that are all specialised for their specific functions.

A systems engineer puts a system together from various building blocks. He needs to consider many issues and has to meet certain performance criteria. For example, if it is a power conversion system, then energy efficiency is an important parameter. The system building blocks are often physical building blocks. Consider, as an example, the PVS300 converter shown in Fig. 1.5. This converter can connect up to 8 kW of solar panels to the electrical grid. Although this sounds simple, this process implies that firstly the flow of dc power from the solar panels must be controlled to make it possible to extract maximum power from the panels. Then, for the second step, the low dc voltage delivered by the panels must be transformed into a higher dc voltage from where an ac voltage waveform can be created which can

Figure 1.5 ABB PVS300 solar power converter.

Figure 1.6 An induction machine with integrated drive electronics.

be connected to the electrical grid. A power converter such as this can easily consist of more than ten different smaller power supply units, several micro controllers and a multitude of other component systems such as sensors.

Another example is the induction machine that is shipped with all the drive circuitry included in one simple housing, as shown in Fig. 1.6. We still have one set of electrical terminals to connect wires that provide the electrical power. However, instead of an electrical output, we now have a rotating shaft that is connected to a mechanical system.

Although it is possible to design and build many systems by using various building blocks and by reading the instructions carefully, a successful engineer will always need a fair amount of knowledge of the inner working of each of these blocks. In this book, we will 'open' each of these system blocks to discuss their individual operation.

1.4 GUIDE TO THE BOOK

The purpose of the book is to provide a hands-on understanding of the fundamentals of electrical power processing. The examples and exercises are designed to help students apply their knowledge to practical situations. The power conversion systems that were discussed earlier can be broken down into system building blocks that convert energy from one form to another, convert electrical energy into mechanical energy, change mechanical energy from one form into another or convert electrical energy into a form where the energy can be stored as depicted in Fig. 1.7. Using these blocks it is, in principle, possible to build a variety of modern power conversion systems ranging from microgrids to vehicle drivetrains.

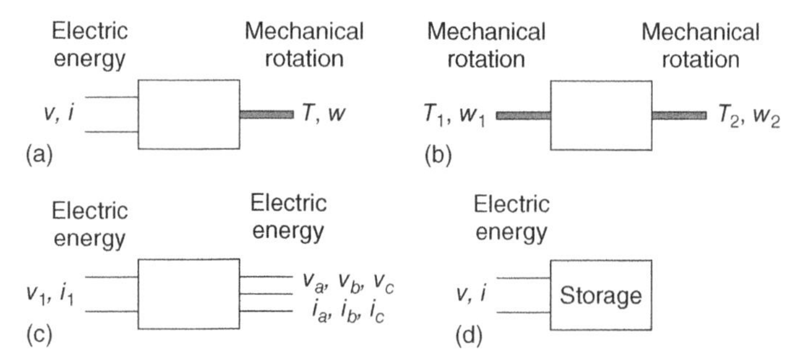

Figure 1.7 The energy conversion system blocks of interest: (a) electrical machines, (b) mechanical transmission systems, (c) electrical energy converters, and (d) electrical energy storage systems.

1.4.1 Generation, Storage and Consumption of Electricity

Chapter 2 serves as an introduction to power generation and energy storage components. Because the theory of most sources falls outside the scope of the book, the focus is on just a few key concepts including power balancing, energy housekeeping and efficiency calculations. Chapters 7 and 8 discuss conversion of electrical power to and from mechanical power, which is important from a system viewpoint because it deals with energy that enters or leaves the electrical power system. Most of the energy handled by the electrical grid comes from electromechanical power conversion.

1.4.2 Power Transfer and Matching of Loads and Sources

In addition to balancing power and energy, the interaction between the different building blocks in the system must also be optimised from a performance point of view. Drawing up a specification of how the building blocks must behave is usually the task of the engineer in charge of integration of the system. Some parameters play an important role in determining the performance of the system and the effectiveness of operation.

Chapter 3 discusses the theory of how electrical power is transferred between the building blocks by the voltage and current waveforms on the electrical conductors and introduces figures of merit that describe their effectiveness.

Since power is defined as the product of two terms, for example voltage times current or torque times speed, converter building blocks can be used to better match the characteristics of a source and a load to get a good power transfer. Techniques that are used to adjust the ratio of the two terms in the power equation are

1. changing the ratio of windings (Chapter 4);
2. the diameter of gears and pulleys (Chapter 5);
3. the duty cycle of power electronic converters (Chapter 6) and
4. the pole pairs in electrical machines (Chapters 7 and 8).

1.4.3 Electromechanics

The conversion of electrical power into mechanical power has its roots in physics. For this reason, Ampere's circuital law and Faraday's law are introduced in Chapter 4, followed by the Lorentz force equation in Chapter 7. Starting with very simple electromechanical devices, the theory for practical machines is derived in Chapters 7 and 8. Taking the viewpoint that power electronics-generated ac will dominate in future, thanks to variable speed drives replacing fixed speed motors, synchronous and asynchronous ac machines are introduced by assuming that they are fed by variable frequency and variable amplitude ac voltage.

1.4.4 Power Electronics

Power electronics is introduced in Chapter 6 where the application of the switching approach to state space theory is discussed without going into too much detail. As the next step, a simplified analysis method is introduced to analyse the buck and boost converter. It is then shown that the buck and boost converter can be combined to create a phase arm, which is the power electronics building block that is needed to construct converters that can drive electrical machines. In Chapters 7 and 8, this basic power electronics converter then becomes the power source for electrical machines.

For a study program that would include dc power supplies, isolated converters are dealt with in the second part of Chapter 6. Using knowledge on transformers and inductors that was covered in Chapter 4, the flyback and forward converter are introduced and analysed.

PROBLEMS

This introductory chapter has not discussed much theory. An important observation has been that modern energy systems contain many power conversion steps. The efficiency can easily be calculated by tracing the power flow through the various conversion stages. The overall system power efficiency is obtained by simply multiplying the conversion efficiencies of the system building blocks with each other.

1.1 Energy conversion chains have become longer despite the fact that the system has become more complex. Explain the improvement in terms of energy efficiency and effective application of power for the following two applications.

1. A steam train of 1900 compared to an electrical train of 2010.
2. The transmission system automobile of 1980 (using gearboxes with three or four gear ratios compared to a 2012 automobile (using gearboxes with six to nine gear ratios or a hybrid electrical drive train).

1.2 Two options of providing power to several subsystems from a single 18 V supply are shown in Fig. 1.8. Each of the blocks represents a linear regulator.

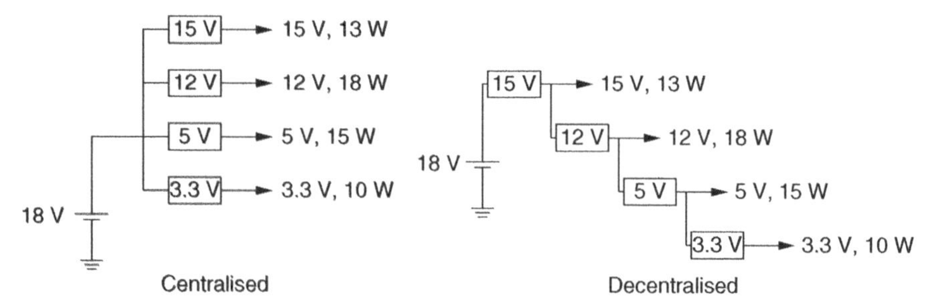

Figure 1.8 Problem 1.2: A centralised and a decentralised power delivery system.

Remembering that the losses of a linear regulator is $P_l = \Delta V\, I$ where ΔV is the voltage difference between the input and output and I is the current through the regulator.

1. Compare the power efficiency of centralised and decentralised systems.
2. What will the power efficiency be if switchmode converters with an efficiency of 96% are used in stead of linear regulators?

1.3 The power required by an electrical boat, in kilowatts, at a specific speed (in kilometres per hour) can be found by the relationship

$$p(v) = 0.0236v^2.$$

The boat uses a 12 V lead-acid battery with a rating of 100 Ah. The electrical motor is 95% efficient. The electrical motor is coupled to the battery with a converter rated 5 kW maximum. If α is the proportion of maximum power delivered by the converter, then the efficiency of the converter is determined as

$$\eta(\alpha) = 0.85 + 0.1\sin(\pi\alpha).$$

Calculate the maximum range at

1. 5 km/h,
2. the maximum speed.

CHAPTER 2

ELECTRICAL POWER SOURCES AND ENERGY STORAGE

2.1 INTRODUCTION

In this chapter, we focus on the different energy sources that are normally associated with our modern society. As we are predominantly interested in the conversion of electrical energy, this chapter focuses on the sources normally associated with electrical systems.

In general, we can identify two types of energy sources: primary and secondary sources. The distinction is that primary sources are one-directional sources where energy is transferred from one state to another without a viable option of reversing the process. On the other hand, secondary sources are, in essence, storage sources where we first store the energy we want to use and then extract it again at a later (and hopefully more convenient) time.

In modern systems, energy sources are connected to the grid using power electronic converters as intermediate coupling interfaces. These converters are required since the grid frequency is 50 or 60 Hz, and when the energy source is dc or works at a different ac frequency, a method of converting the power to the form required by the grid is needed. Because primary sources are unidirectional, the converter needs to handle power in one direction. In contrast, a power electronic converter that exchanges electrical energy between a system and the grid in both directions is needed for a secondary electrical energy source. One example of such a converter is

The Principles of Electronic and Electromechanic Power Conversion: A Systems Approach, First Edition.
Braham Ferreira and Wim van der Merwe.

the drive unit of an electrical locomotive. When the train is accelerating or running at a constant speed, the converter is taking energy from the grid and providing the train with propulsion. However, when the train is decelerating, the converter effectively converts the excess unwanted mechanical energy into electrical energy and therefore the power flow through the converter reverses, and the converter can now feed the excess energy back in the grid. It is, however, true that these converters are often found operating in tandem with other energy storage devices, for example, in a hybrid electrical vehicle (EV), or even in a full EV, in which the excess kinetic energy of the vehicle during braking is stored, using the converter, in the vehicle's battery pack.

2.2 PRIMARY SOURCES

Primary sources are, in essence, unidirectional and are used to supply electrical energy to the grid or even to isolated systems. We can divide the discussion on primary sources by looking at the difference between centralised sources and distributed sources.

2.2.1 Centralised Sources

Centralised power sources are normally large power stations that deliver energy to the grid. These power stations are located in areas where one form of energy that can be exploited cheaply is abundant; once this energy is converted into electricity, it can then be easily used in faraway locations where no such primary energy source is available. Of all the centralised power station types, the thermal station has been the most prevalent in the past.

2.2.1.1 Thermal Power Plant The basic operation of a thermal power plant is shown in Fig. 2.1. The plant operates by using heat to generate steam at very high pressure. This mechanical pressure is translated into mechanical rotation through the use of a steam turbine. In the final step, the rotational mechanical energy is converted into electricity by means of a generator, usually a synchronous machine.

Different sources of heat can be used as the primary energy source of a thermal power station. In large parts of the world, coal-fired power stations provide

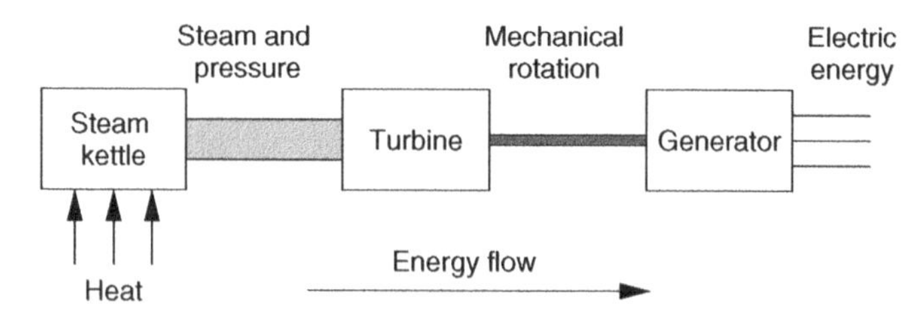

Figure 2.1 A thermal power station.

the bulk of the electrical energy. Although coal power stations are a major contributor to unwanted emissions, the relatively cheap cost of coal as fuel makes it relatively unlikely that the existing coal stations will be decommissioned in the near future. Actually, in most countries, the majority of planned new power stations are still coal fired; fortunately, with newer technologies, it is possible to remove about 90–99% of unwanted emissions such as sulphur dioxide, nitrogen oxide, mercury and particulate matter (ash and other particles).

Nuclear energy is also often used as a source of thermal energy. The basic operation of a nuclear station is similar to that of a coal-fired station, as depicted in Fig. 2.1; however, a controlled nuclear reaction is used as the source of heat. Although nuclear stations are very expensive to design and build, the low cost of fuel and other operational costs make nuclear power stations a financially viable option. It is also true that no harmful emissions are associated with nuclear stations. However, the possible consequences when things go wrong are all too clear as the events at Three Mile Island (1979), Chernobyl (1988) and Fukushima (2011) remind us. There is also the small issue of storing the harmful nuclear waste to consider. That said, the new generation of designs has dramatically improved safety features, and the technology is still very attractive for new stations primarily because of the relatively low generation cost and zero emissions.

There are also some renewable options for use as the source of heat energy. In areas with an abundant supply of wood, such as Scandinavia, biomass can be used, while in areas with geothermal activity, such as Iceland and the Western United States, the heat of the earth's mantle can be used. In some cases, the heat is extracted from hot water or steam that is forced from the lower layers of rock, so-called *natural geysers*, while in many other instances, water is pumped into an underground reservoir to be heated to a sufficiently high temperature.

Figure 2.2 A thermal solar power heat collector system using a parabolic trough.

In the last decade or two with newer technologies, it has also become viable to use the sun's energy as a heat source in a commercial-scale thermal power plant. One example of this technology is a concentrated solar tower that uses many mirrors around a collector to focus the sun's energy on to a specific point where an energy carrier is heated (Fig. 2.2). From this point onwards, the station functions in a similar manner to other thermal power plants. However, to enable the plant to produce power for the hours after sunset, molten salt is often used as the energy carrier in an effort to store energy that could be used at this time. The heat is extracted from the molten salt to create steam.

There are considerable losses involved in conversion process from heat to steam to rotational energy to electricity. Let us take the modern coal-fired station as an example. One kilogram of high quality coal can contain up to 30 MJ of latent energy; however, not all this energy can be converted to electricity. Let us assume that we start with 1 kg coal with a heating value of 20 MJ (a reasonable assumption as not all coal is exactly A-grade), this is equivalent to 5.56 kWh. Not all the energy in the coal can be released as heat when the coal is burned because of factors such as incomplete combustion, and the presence of small amounts of moisture in the coal that take a certain amount of energy to heat. Furthermore, not all the heat energy can be transferred to the water mainly because some of the heat is lost to the atmosphere. The total efficiency of the transfer of the energy from coal to water is around 90%. The maximum losses in the power plant occur during the thermodynamic energy conversion stage where this heat energy in the water is converted into rotational mechanical energy. This process is described by the Rankine cycle, which falls outside our scope but which describes why the efficiency of this process is limited to about 43% for most thermal plants. It is, however, true that some modern power plants can, mainly by increasing the temperature of the steam, operate with a thermodynamic efficiency of about 47%. Finally, the conversion from mechanical energy to electricity in the generator is around 95% efficient, while stepping up the generated voltage to a high voltage suitable for transmission over long distances is about 97% efficient. This brings the efficiency of the total process for converting the energy from coal to transmission voltage electrical energy to about 35.7%. This means that from the possible 5.56 kWh in the coal, we can only expect around 2 kWh to leave the station.

Example 2.1 How much coal is burned and CO_2 (expressed in kilograms) is emitted into the atmosphere to fill a bath with 80 l of water heated to 55 °C. Assume the efficiencies mentioned above and use the heating value of coal to be 20 MJ/kg and that the emissions of the power plant amounts to 1.1 tons/MWh (about the average for a coal-fired power plant built in the 1980s). Furthermore, we make the assumption that the water is heated using electricity and that the water entering the system is at 5 °C and that the combination of the water kettle and the transmission system is 91% efficient.

Solution

The energy required to heat 80 l of water by 50 °C

$$W_{req} = 4181.3 \times 80 \times 50 = 16.73\,\text{MJ}$$

This amounts to 4.65 kWh. Using the fact, as outlined above that 1 kg of coal can deliver 2 kWh of energy and that the efficiency is 91%, we know that we will need

$$W_{gen} = \frac{4.65}{0.91} = 5.1\,\text{kWh}$$

at the point of generation, which translates to about 2.55 kg coal! Furthermore, since 1 MWh translates to 1100 kg of CO_2 emissions, the generation of 5.1 kWh will imply that around 5.6 kg of CO_2 will be emitted into the atmosphere.

2.2.1.2 Kinetic Energy Power Plant It is also possible to bypass the Rankine cycle if a suitable source of kinetic energy is available. The energy source most often used for these classes of power stations is hydroelectrical energy. A small hydroelectrical plant is shown in Fig. 2.3.

In a hydroelectric plant, large amounts of water are allowed to flow downhill and the resulting kinetic energy is converted into rotational energy and finally into electrical energy using a generator. An overview of such a power plant is included in Fig. 2.4. A well-designed water turbine can extract as much as 90% of the kinetic energy from the water and convert it into rotational mechanical energy. This brings the overall efficiency of a hydroelectric power station to around 83%.

Figure 2.3 A small hydroelectrical station using a Pelton wheel as turbine.

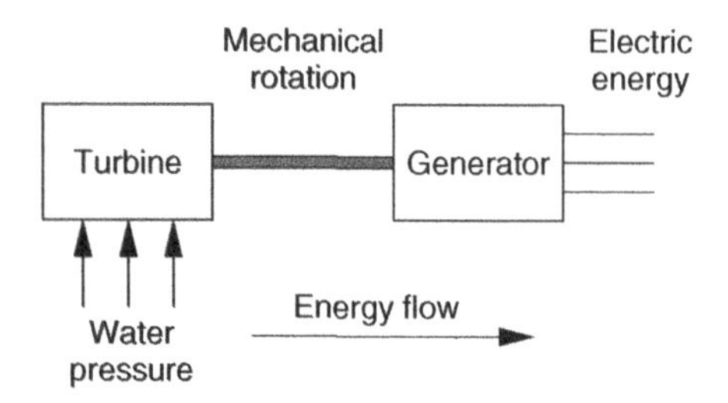

Figure 2.4 A hydropower station.

Example 2.2 How much water is needed in a hydropower station with a head (the vertical distance through which water moves) of 10 m to heat the water in Example 2.1. Assume the same efficiency of the kettle and transmission network and that the hydroplant is 80% efficient.

Solution

As found in Example 2.1, the energy required at the point of generation is 5.1 kWh. This implies that

$$W_{\text{water}} = \frac{5.1}{0.8} 60^2 = 22.95\,\text{MJ}$$

of kinetic water energy is required. With a head of 10 m, this translates to

$$V = \frac{22.95 \times 1000}{10 \times 9.81} = 234\ \text{kl}.$$

There are also other sources of mechanical energy that can be used. For many applications, it is convenient to have a controllable power source to provide power on demand and which can be brought online relatively quickly. The most common plants of this type are internal combustion engines (most often diesel engines) and open-cycle gas turbines. An open-cycle gas turbine is similar to an aircraft engine but adapted for use in a stationary environment. The hot exhaust gases of the turbine are often also used for residential heating and other similar purposes. The turbine is called a *combined heat and power plant*. Nowadays, the open-cycle gas turbine running with natural gas is preferred for new power plants. This is mainly due to the fact that even if the thermal efficiency of the gas turbine is only around 35%, when hot exhaust gasses are used to heat water to create steam to operate the thermal power plant in addition to the kinetic energy plant, then the overall efficiency can increase to as much as 58%. If the power plant is built close to a residential area, which is possible as natural gas is relatively clean burning, and the waste heat is used for heating purposes, the efficiency of the system can approach the 80% mark.

One renewable source of mechanical energy is wind power. A wind turbine harvests the kinetic energy of air molecules by converting it to mechanical energy when the air passes over the blades of the wind turbine. The basic operation of a wind power plant is similar to that of Fig. 2.4, however, the rotational speed of a

Figure 2.5 An off-shore wind farm off the coast of Denmark.

wind turbine is limited and also varies with the wind speed. An electrical generator is often connected to the wind turbine via a mechanical gearbox to increase the rotational speed. The variable wind speed is, however, a large problem because the frequency of the grid is regulated to a fixed frequency, 50 Hz for most of Europe, and any source connected to the grid must operate at the same frequency. Power electronic converters are often used in wind power plants to convert the generated electrical energy to the required frequency. The power coming from all the wind turbines in the park is collected and then the voltage level is adjusted to the right value by using a transformer, before the power is fed into the grid. The seabed provide ample space for building wind farms (Fig. 2.5) and an overview of how a wind power generator works, is shown in Fig. 2.6.

2.2.2 Decentralised Sources

Decentralised sources are normally situated quite close to the point where the energy is to be used. Some of the well-known versions of decentralised power sources are in fact scaled down versions of the power plants that we have discussed. Think, for argument's sake, of a small petroleum power generator sometimes used to provide power at a remote location or even a small wind turbine used to provide power on a sailboat.

There are three commonly used decentralised power sources that we have not discussed thus far. The first is the photovoltaic solar cell. In a photovoltaic cell, the energy of a photon entering the cell is transferred to an electron, which is then liberated from its valence band. A voltage and, if the circuit is closed, a current, can therefore be formed by this supply of liberated electrons. Since photon energy is converted into electrical energy, the amount of electrical power generated is directly proportional to the amount of incoming irradiation (up to a certain

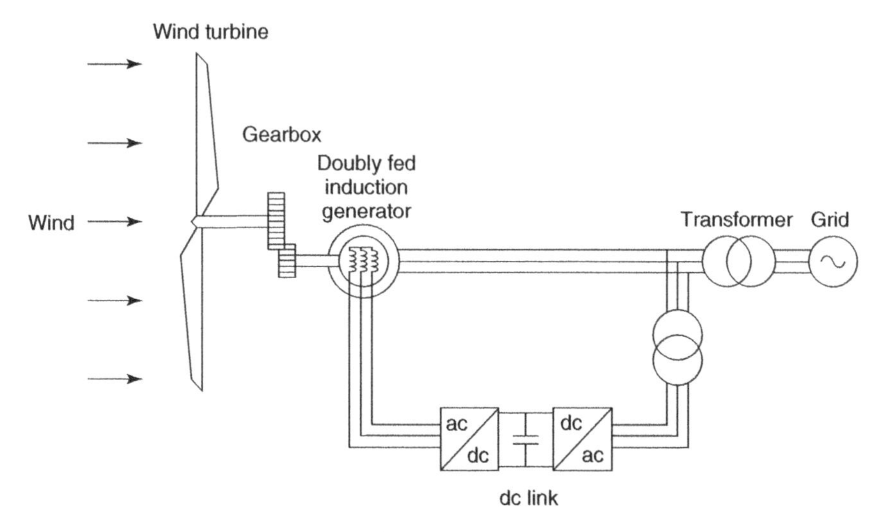

Figure 2.6 A wind turbine.

point). This implies that the voltage of the solar cell and the current delivered by the cell vary in proportion to the amount of irradiation and also the power delivered by the cell. A single solar cell delivers electrical power, of about 3 W maximum, at a dc voltage between 0 and 0.6 V. By connecting many of the cells together (both in series and in parallel) to form a solar panel, a panel output of around 100 W at 48 V is quite possible. The output voltage of the panel will, however, also vary according to the amount of power extracted from the panel and the incoming irradiation. A power electronic converter is often used to either store the excess energy from the solar panel in a storage device (normally a battery) or convert it into a usable form such as 230 V at 50 Hz as used by most household appliances in Europe. A typical photovoltaic power generation system is shown in Fig. 2.7.

The final two distributed primary sources are the disposable battery and the unidirectional power electronic converter. The disposable battery operates by converting chemical energy into electrical energy. However, unlike a rechargeable battery, once the chemical reaction has run its course and all the energy is extracted from the battery, it cannot be reused. The energy flow is distinctly in one direction. Additionally the unidirectional power electronic converter is often found in household appliances; think, for instance, of cellphone charger circuits or power supply modules to appliances such as televisions or computers. These converters convert electrical power from one voltage level and frequency to another. Again the energy flow is only in one direction and therefore these systems can also be considered as primary sources.

As a final note, the distinction between centralised and decentralised sources can be blurred at times. Consider the photovoltaic solar farm shown in Fig. 2.8. In this case, although photovoltaic panels are used, the installation is probably closer to a centralised source. The exact point where a primary source moves from being a

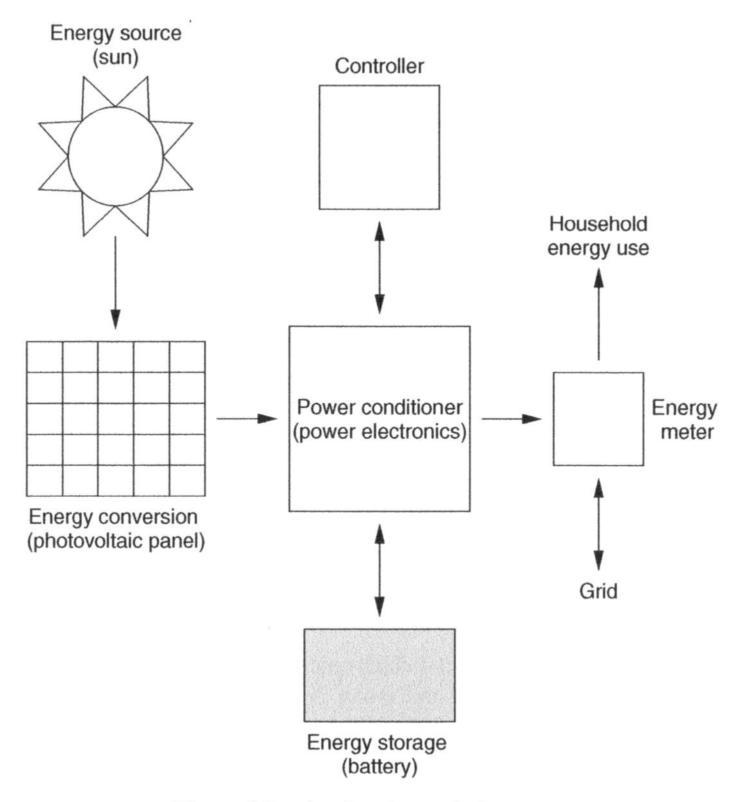

Figure 2.7 A solar photovoltaic system.

Figure 2.8 A photovoltaic solar farm.

decentralised source to a centralised primary source is not so clear, however, good common sense often brings us to the right conclusion.

2.3 SECONDARY SOURCES

A secondary source of electrical energy is a source where electrical energy is stored to be used later. The main difference between a primary source and a secondary source is the fact that the energy flow in the source is bidirectional. It is also true that a secondary source does not convert energy from an abundant source of energy, such as coal or wind power, into electricity. In fact, to extract electrical energy from a secondary source, one must first store the electrical energy in the source.

An in-depth discussion of secondary sources is outside the scope of this text. However, much can be learned by treating the secondary source as a pure energy storage device. In the remainder of this chapter, some terminology associated with storage devices will be discussed together with the basic operation of the most prevalent storage mechanisms.

2.3.1 Basic Concepts

During the last decade, the personal electronics market, most prominently mobile phones, personal audio devices, and tablet and laptop computers, has made us accustomed to electrical energy storage devices. We are used to thinking of an energy storage device as a vessel that needs to be filled up. Although there is much more to energy storage devices, this view suffices for most applications.

2.3.1.1 Definitions Before we discuss electrical energy storage devices, we must define the following:

Capacity The rated amount of energy that can be stored in a device. Although it is possible to overcharge some devices, the rated capacity is the maximum amount of energy that can be used if the device is charged to its rated capacity. For many devices, some energy still remains in the device if all the usable energy had been used after it was charged to full capacity. This is due to the fact that the last bit of energy cannot be extracted from the device in an efficient manner. Therefore the storage device capacity cannot be defined as the total amount of energy stored in a device but merely the total amount of usable energy stored in a device.

Active Shelf Life The period of time (under specified conditions) that a charged storage device can be stored before its capacity falls to an unusable level. This can vary considerably from device to device; for example, non-rechargeable alkaline batteries have a shelf life of five to ten years, while rechargeable nickel-metal hydride loses up to 40% of its stored energy in a month if left on a shelf after being fully charged. To complicate matters further the shelf life varies quite a bit if the storage conditions change.

Ampere-Hour The unit used to describe the energy storage capacity. Abbreviated as Ah, the ampere-hour is well suited for energy storage devices, such as batteries, which deliver energy at a constant (or rated) voltage. Some other types of energy storage devices can also be rated using the ampere-hour as measurement if used in conjunction with a converter that keeps the output voltage constant.

Cycle Life The number of cycles to a specified depth-of-discharge (DOD) a storage device can undergo before failing to meet capacity or efficiency criteria. The cycle life of some storage devices is a strong function of usage criteria such as operating temperature, DOD and discharge and/or charge rate. For this reason, the cycle-life specification can only be used as a guideline as real-life operating conditions rarely match those used in deriving the cycle-life specification.

Life The usable life of an energy storage device, specified in cycles or years. The distinction between life and cycle-life hinges on the fact that some storage technologies do not exhibit a cycle life but rather a maximum lifetime expressed in years. Should a storage technology have a cycle life, then the life and the cycle life are the same.

Charge The process of storing energy in a storage device. Sometimes the term *charge* is also used to describe the energy stored in a device (especially when speaking about batteries). However, to avoid confusion, it is advisable to, in this case, rather use the term *stored energy*.

Discharge The process of removing energy from a storage device.

Charge Rate The maximum rate at which a device can be charged.

Cycle The cycle from fully charged, the subsequent discharge of the usable capacity of a device, followed by a recharge to full capacity. Although it is convenient to work in full discharge cycles, most cycles are not full cycles; consider how and when you recharge your mobile phone battery, only rarely will you wait until it is completely flat before recharging it.

Depth-of-Discharge, DOD The amount of energy removed from a fully charged device. This can be measured in percentage or directly in terms of energy measured in ampere-hours or joules. When the storage device is charged to full capacity, the DOD is 0% while the DOD is 100% when it is completely discharged.

State-of-Charge, SOC An alternate method of describing the amount of usable energy left in a device. The SOC is closely related to the DOD. One can think of the SOC as being similar to the fuel gauge in a car; 100% implies that the storage device is fully charged and consequently 0% is empty.

Self-Discharge Rate The rate at which a storage device will discharge if left fully charged and not in use. For most energy storage technologies, the self-discharge rate is a function of external factors such as temperature.

Round-Trip Efficiency A measure of the efficiency of the storage device. This can be calculated by dividing the total energy recuperated from the device by the total amount of energy stored in the device. Naturally, the round-trip efficiency cannot be greater than 100%; if it is, then we are dealing with an energy source and not an energy storage device. The round-trip efficiency is a function of many factors such as the (dis)charge rate, the temperature and the length of time that the energy is stored. For this reason, round-trip efficiency is often specified for specific operating conditions.

Specific Energy The amount of energy stored per weight. Specific energy is normally expressed in terms of kilojoules per kilogram. The specific energy is important for mobility applications where a light weight storage device would be preferable to a heavier device, provided that the energy stored is the same.

Energy Density The amount of energy stored per volume, normally measured in terms of kilojoules per cubic metre or even kilojoules per litre. Although this measure is closely related to specific energy, a high energy density does not imply a high specific energy. A simple example is that of hydrogen. As it is very light, the specific energy is high but it has a low energy density. One of the major drawbacks of a hydrogen-powered car is the large volumetric requirement of the hydrogen storage tank in comparison with a petrol or diesel tank.

Specific Power The output power capability of the storage device by weight, measured in kilowatts per kilogram. Specific power is related to the specific energy in that it is also a measure that is normalised by weight.

Power Density The output power capability by volume, measured in kilowatts per cubic metre or kilowatts per litre.

2.3.1.2 Non-Electrical Energy Storage

In our discussion of electrical energy storage, we will only discuss methods of storing electrical energy. However, there are other methods of energy storage, not discussed in this book, which could influence electrical systems. The best example is probably the storage of thermal energy, especially when used in conjunction with an electrical HVAC (heating ventilation and air-conditioning) unit. As electricity prices are cheap during off-peak periods in the early morning, this cheap energy can be used to make ice that can be used during the day to help cool the building down. Likewise, in winter, cheap electrical energy can be used to heat oil or other fluids, especially molten salts. This heat energy is then used during the day to heat the building.

Using cheap energy during off-peak times to store thermal energy for use during peak times can in one way be considered to be electrical energy storage. Let us consider it in this way: without the thermal energy storage we would use a certain amount of electrical energy during the day. However, because we have stored thermal energy we are using less, but we have used this portion of electrical energy earlier to store the thermal energy. Therefore, we have stored electrical energy for use later. However, for the purposes of this book, we will tighten the definition of electrical energy storage to some extent. Let us define electrical energy storage as the process of storing electrical energy in another form which can be recovered as electrical energy for use at a later point in time.

2.3.2 Storage as Chemical Energy—Hydrogen

Because of its low atomic weight hydrogen has a very high specific energy. Hydrogen does not occur naturally in large quantities and therefore unlike oil or natural gas cannot be considered a primary energy source. That said, it is possible to generate hydrogen using other forms of energy, thereby using it as an energy carrier or equally as an energy storage device. This process is commonly referred to as the *hydrogen economy*.

In terms of an electrical energy storage device, hydrogen can be generated from electricity using electrolysis. Although the efficiency of this process has improved during the last thirty years, especially when combined with an external heat source, such as waste heat from a nuclear reactor, the efficiency of this process is relatively low, in the range of 50–80%. The resulting hydrogen can be recovered in electrical form through the use of a fuel cell. Fuel cells convert the chemical energy of hydrogen into electricity through a chemical reaction with oxygen. The efficiency of this process is also relatively low, between 40 and 60%. Although these values sound low, compared with an internal combustion engine, with a chemical energy to mechanical energy efficiency of about 25%, it is about twice as efficient.

The round-trip efficiency of hydrogen as an electrical energy storage device can be calculated as

$$0.5 \times 0.4 = 0.2 \leq \eta_{\mathrm{rt}} \leq 0.48 = 0.8 \times 0.6. \tag{2.1}$$

Owing to this low round-trip efficiency, hydrogen is not often used as an electrical storage device. The exception to this rule seems to be in the electrical automotive industry where the high specific energy of hydrogen is seen as a huge advantage. In some instances, the benefit of not carrying heavy battery packs is seen as a benefit enough, even considering the low round-trip efficiency.

2.3.3 Storage as Electrochemical Energy

Batteries are the most commonly used electrical energy storage devices, especially in lower power (<1 kW) applications. The recent penetration of personal electronic equipment such as laptops, personal music players and mobile phones has resulted in the general public interacting with several batteries during the course of a day. It is therefore not surprising that annual sales of batteries are estimated at as high as US$ 50 billion.

For the purposes of this discussion, we will make a distinction between small batteries and utility-scale batteries—the distinction being that utility-scale batteries are rated for storage capabilities greater than 1 GWh and power levels in excess of 400 kW.

Because we are interested in electrical energy storage, primary cells (or non-rechargeable) batteries are not discussed here. For all practical purposes, a primary cell is simply an energy source, but as excess electrical energy cannot be stored in the cell for later use, it cannot be considered an energy storage device.

Several types of batteries exist. The types that are most prevalently used are lead-acid, lithium-ion, nickel-metal hydride and nickel–cadmium batteries. Although a detailed discussion of the chemical processes of each battery type is outside our scope, they will briefly be introduced.

2.3.3.1 Lead-Acid Battery Almost every automobile has a lead-acid battery under its bonnet. Its function is to ignite the combustion engine and, together with the alternator, power the electrical systems of the vehicle. Sometimes these batteries are found in small electric vehicles (EVs) such as golf carts or small boats. Deep-cycle lead-acid batteries are sometimes used in uninterruptable power supply (UPS) applications. The batteries are cheap to manufacture and the technology is very well developed. The cycle life is a strong function of the DOD. Even if the battery is fully discharged only once, there is a noticeable decrease in cycle life. Compared to other batteries, lead-acid batteries have relatively low power and energy densities.

2.3.3.2 Lithium-Ion Batteries Nowadays these batteries are very common and are found in low power handheld equipment such as mobile phones and laptops. Lithium-ion batteries have a high specific power, and their specific energy makes them attractive for EV applications. The batteries have high energy and power densities, a low self-discharge rate and a long cycle life. One word of caution though—lithium-ion batteries must not be overcharged because they can catch fire or even explode.

2.3.3.3 Nickel-Metal Hydride Batteries Nickel-metal hydride batteries are relatively cheap and pose no immediate threat to the environment as the raw elements used are not inherently poisonous. These batteries are often used in consumer electronics such as cordless phones. It has been the favoured battery technology for hybrid EV applications such as the Toyota Prius and the Ford Escape. The battery can handle many cycles but has a very high self-discharge rate. A nickel-metal hydride cell can lose between 5 and 10% of its charge on the first day and thereafter about 1% per day.

2.3.3.4 Nickel–Cadmium Batteries These batteries use cadmium, which is harmful to the environment and must be recycled. Recently, they have been replaced by nickel-metal hydride batteries in portable electronics and toys. They are, however, still used in equipment requiring high surge currents such as remote-controlled aeroplanes and handheld power tools. These batteries may suffer from the memory effect, also sometimes called the *lazy battery effect*. This is, however, not a concern in most applications.

2.3.4 Storage as Electrical Energy

Although energy can be stored in electrical forms using inductors or capacitors, the storage capacity of these devices is so small that they cannot be considered as storage devices in the true sense of the word. However, two alternatives exist—the supercapacitor and the superconducting inductor.

2.3.4.1 Supercapacitors Supercapacitors are essentially capacitors with a very large capacity. This increase in capacity is achieved by using plates with an extremely large surface area, typically activated carbon. As a result, a supercapacitor might have, for the same size, a capacitance in excess of 100 times greater than a comparable capacitor. However, because of the construction of the capacitor, the voltage rating is not very high, typically less than 10 V.

The voltage-stored energy relationship differs in a significant manner from a battery in the sense that the terminal voltage of a battery remains relatively constant during the full discharge cycle. The stored energy in a supercapacitor can be found as

$$W = \frac{1}{2}CV^2 \tag{2.2}$$

implying that the voltage is a strong function of the stored energy.

The specific energy of a supercapacitor is much higher than an ordinary capacitor but still about thirty times lower than that of lithium-ion batteries. For that reason supercapacitors are rarely used as bulk energy storage devices. However, that said, the supercapacitor has a very high specific power, up to ten times larger than that of a lithium-ion battery and a very large cycle life of up to a million cycles. This makes the supercapacitor very attractive for power-smoothing applications where it is used in conjunction with a traditional battery. Short pulses of energy are typically stored in the supercapacitor, while the battery provides the bulk power. By utilising the supercapacitor in parallel with the battery, it is also possible to reduce the number of cycles the battery experiences, thereby increasing the battery life. This combination is sometimes referred as a *hybrid battery* or more correctly a *hybrid storage device*.

2.3.4.2 Superconducting Magnetic Energy Storage Inductors store energy in the form of magnetic fields. The energy contained in an inductor is

$$W = \frac{1}{2}LI^2. \tag{2.3}$$

Theoretically, the energy stored in an inductor can be increased to any value by either increasing the current or by increasing the inductance by adding more turns. The problem with this approach can be expressed in one word—*resistance*. Firstly, any increase in the number of turns must necessarily increase the resistance. Secondly, this increase in resistance is a problem in terms of the standing losses experienced by the storage device. The standing losses can be found as

$$P_{sl} = I^2 r \tag{2.4}$$

where r is the resistance of the coil.

The high standing losses can be minimised through the use of a superconductor. A superconductor is a metal or ceramic which has a special property that its resistance falls to zero when cooled below a critical temperature. For most known materials, the critical temperature is in the region of absolute zero, that is, 0 K or −273 °C. However, some materials exist with critical temperatures as high as 125 K or −148 °C. This is significant as it implies that it can be cooled with liquid nitrogen (with boiling point 77.4 K or −195.6 °C).

When a superconducting magnetic energy storage (SMES) is operating correctly, that is, the current is within limits and the conductors are cooled to below the critical temperature, the standing losses due to resistive losses are nearly zero. However, the energy requirements of the cooling system are of concern. Some systems, especially systems using a conductor with a very low critical temperature, will require a significant amount of cooling energy.

As the cooling requirements of the system remain constant and the resistive losses of the system are negligible, one might be tempted to think that the current can be increased to any value, thus increasing the energy storage capacity of the device. However, the capacity of SMES is governed by two important design considerations. Firstly, the high currents and resulting high magnetic fields exert very large forces on the mechanical structure. The magnetic field tries to pull the whole structure together, almost to implode the structure. There is thus a very specific design limit to the forces that the structure can withstand and therefore to the maximum current that the SMES can carry. Secondly, superconducting materials have a property whereby the material characteristics change if the current exceeds a specific level. This current is called the *critical current*. If the current exceeds the critical current, the material becomes resistive and the superconducting properties are lost. These phenomena constitute the second limitation on the storage capacity of the SMES.

2.3.5 Storage as Mechanical Energy

Energy can be stored in mechanical systems in the form of potential or kinetic energy. However, not all types of mechanical energy are equally well suited for the storage of electrical energy. Three methods that are used in practice are discussed here: using a flywheel, pumped hydro and compressed air.

2.3.5.1 Flywheel The energy contained in a spinning mass can be expressed as

$$W = \frac{1}{2} J \omega^2 \tag{2.5}$$

where J is the moment of inertia and ω is the rotational speed. It is difficult to store high values of energy at low rotational speeds, for example, a 1000 kg cylinder with a 1 m radius spinning at 1000 rpm stores 'only' 2.7 MJ of energy. Although this sounds like quite a lot, it is only the equivalent of the energy stored in 75 ml

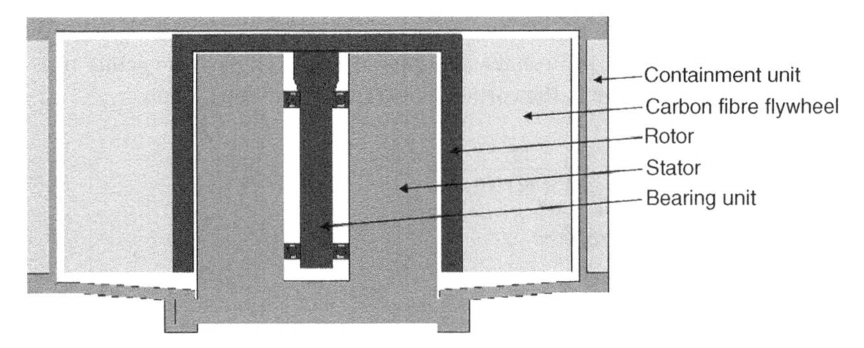

Figure 2.9 Cross section of a flywheel energy storage system.

diesel fuel. However, the efficiency of a diesel engine is only about 30%, so about 250 ml will be needed to create the same mechanical energy using a diesel engine. The energy capacity of flywheel systems is normally not increased through an increase in the inertia of the rotor but through an increase in the rotational speed. Consider a rotor with a 1 m radius weighing 100 kg spinning at 45 000 rpm. This rotor stores 555 MJ, 205 times more than the 1000 kg rotor of the previous example. A discussion of the calculation of the moment of inertia is included in Chapter 5.

The high rotational speed of flywheel systems has two results that must be given due consideration. Firstly, the large amount of energy stored as kinetic energy can have disastrous consequences during failure. Should the rotor suffer structural damage during operation and disintegrate, the fragments will fly away from the structure with very high velocity and with tremendous force. For this reason, flywheel systems are often buried in an effort to contain the energy in the event of a failure and to preserve human life and equipment in the vicinity. Secondly, the high rotational speed can cause high standing losses. All rotational systems experience some sort of rotational friction, both through air resistance and through frictional resistance in the bearings. If the total rotational friction coefficient is given as B, then the frictional torque experienced is

$$T_f = B\omega \tag{2.6}$$

and the standing power loss is

$$P_{sl} = B\omega^2. \tag{2.7}$$

To minimise the standing losses, flywheel systems are often designed to have a vacuum chamber that contains the rotor. This is done to eliminate the frictional force of the air resistance. Furthermore, magnetic levitation bearings are used to minimise the friction of the bearings (Fig. 2.9).

Example 2.3 How much energy is stored in the rotation of the earth about its axis? The earth's moment of inertia is $J = 8.04 \times 10^{37}$ kg m^2

Solution

Let us assume that the earth rotates about its axis in 24 hours (the actual rotation period is slightly less due to the earth's motion about the sun). Then

$$\omega = \frac{2\pi}{24 \times 60 \times 60} = 72.72 \times 10^{-6}\ \text{rad/s}.$$

The stored energy is therefore

$$E = \frac{1}{2} J\,\omega^2 = 212.6 \times 10^{27}!$$

To give an idea of how much energy this is, consider the fact that the total power delivery of all the world's power plants combined is approximately 4500 GW. Therefore, to generate the same energy as the energy contained in the rotation of the earth will take

$$\frac{212.6 \times 10^{27}}{4500 \times 10^{9}} = 47.24 \times 10^{15}\ \text{s}$$

or about one and a half million millennia!

2.3.5.2 Hydroelectrical Energy Storage For very high storage capacity systems, energy can be stored using hydroelectrical energy. Two dams are constructed at two different elevations, the larger the difference in elevation the larger the storage capacity per cubic metre of water. When energy is stored in the system, water is pumped from the low lying dam into the uppermost dam (Fig. 2.10). When energy is extracted from the system, hydroelectricity is generated by allowing water to flow from the top dam to the dam at the bottom.

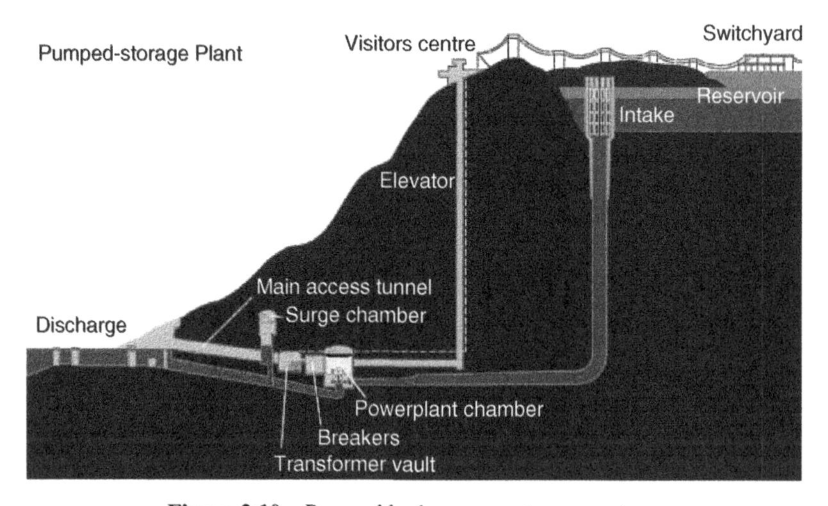

Figure 2.10 Pumped hydroenergy storage system.

Owing to the sheer size and cost of a pumped hydro storage system, it is mostly used at utility scale. At utility scale, the storage of energy is of primary importance. Traditionally, grid energy storage was used to balance the high energy requirements during the day with the surplus of generated power during the night (or off-peak). In the modern grid, energy storage is even more important because of the penetration of renewable resources. Take the high installed wind generation base in North-Western Europe as an example. During periods with high, good and consistent wind, much energy is produced and in the event of excess energy production, the excess energy should be stored.

Pumped hydro schemes are associated with high capacity and high power capabilities. These schemes are also very capital intensive and require specific sites with space for two large dams and a difference in height.

2.3.5.3 Compressed Air Energy Storage Large amounts of energy can also be stored using compressed air. Although compressed air in gas cylinders can be used to store small amounts of energy for long periods of time, bulk energy storage is usually done in underground reservoirs. Large amounts of energy can be stored in this way. The round-trip efficiency of compressed air systems is, however, of concern. The temperature of air increases when it is compressed and when it is stored for long periods (in compressed form), this heat energy is lost to the surrounding area. If the compressed air is used shortly after compression, the round-trip efficiency can approach 99%; however this falls quickly if the heat energy is lost.

2.4 HIGHLIGHTS

- Electrical energy sources are generally grouped in two categories:
  - Primary sources are defined as sources that can only deliver energy. These are unidirectional sources.
  - Secondary sources are defined as bidirectional sources where energy must first be stored in the device before it can be used at a later time.
- Electrical energy is often used as an energy carrier, that is to say that electricity is used to transport energy from one location to another.
- As electricity is only a carrier, it cannot exist in its own right. The reality is that electrical energy can only be generated when a user of the energy is available.
- Centralised sources are large generators of electrical energy, typically located at geographic locations where one form of energy is abundant.
- Naturally abundant sources of energy that are often exploited in centralised power plants are
  - Coal
  - Nuclear
  - Water
  - Wind power

  - Solar energy
  - Geothermal energy
- Decentralised sources are smaller sources that are connected very close to the point of use. These types of sources vary from household solar panels to small petroleum-powered generators to batteries.
- On a utility scale, electrical energy can be stored by converting it to another form of energy such as kinetic energy, potential energy or chemical energy.
- No energy storage device is without flaw and often the best device depends on the specific application at hand.

PROBLEMS

2.1 What are the differences between the following concepts, when applied to energy storage devices?

1. Ampere-hour, charge and SOC.
2. Specific energy, energy density and power density.

2.2 The initial SOC of a 12 V battery with a rated capacity of 90 Ah is 60%. The maximum allowable DOD is 40%. During daytime, the battery is charged from a 60 W solar panel for 3 hours. What is the SOC after charging? For how long can the battery be used to supply 100 W for a TV set, before a DOD of 65% is reached?

2.3 A 12 V battery with a rated capacity of 40 Ah is initially charged to 90% SOC. The battery is subjected to the energy usage profile of Fig. 2.11. Subsequent levels are +1000, 0, −1500, +1500 W; time in seconds). Draw graphs of the current, voltage, SOC and DOD versus time. (Positive power is energy extracted, while negative energy constitutes charging.)

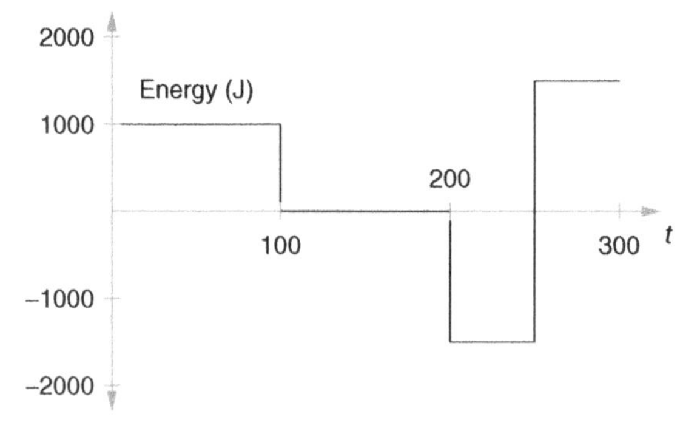

Figure 2.11 Energy usage profile.

2.4 A 40 F supercapacitor in a remote-controlled car rated for 8 V is initially charged to 60% SOC. During braking, 1500 W is delivered to the battery in 1.2 s. Calculate the SOC and the voltage after braking.

2.5 The energy consumption of a household can be approximated by the curve

$$p(t) = 1 - \frac{1}{2}\cos\left(\frac{2\pi t}{24}\right)\,\text{kW},$$

where t is in hours. In an effort to decrease energy costs, ten solar panels are installed. The energy supplied by the solar panels is indicated in Fig. 2.12.

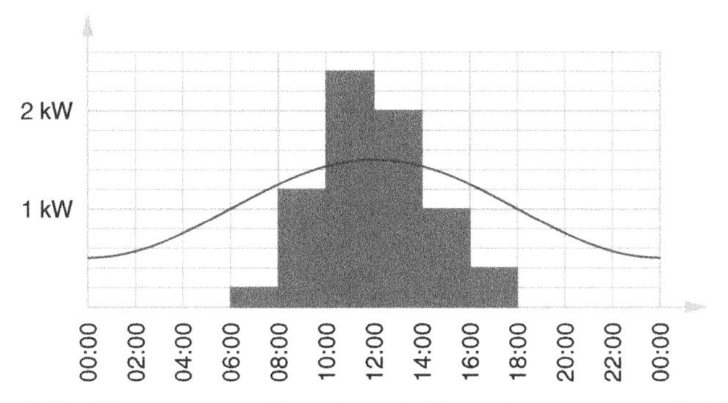

Figure 2.12 The energy used by a household and the energy supplied by ten solar panels.

1. Without energy storage, what proportion of the total energy used can come from the solar panels?
2. The owner wants to use all the energy supplied from the solar panels. How large a storage battery would he need to install?
3. What should the maximum power rating of this battery be?

2.6 A typical 12 V lead-acid battery is rated for 125 Ah. To increase the lifetime of the battery, it should not be used to a DOD greater than 60%. If you have two large water containers and are able to place one 5 m higher than the second, how much water should you place in the containers to make a pumped hydro storage system with the same capacity as one battery? Assume the efficiency of the pump-turbine to be 97%.

2.7 Sometimes we do not appreciate quite how much energy is contained in fuel and the challenges EVs face. Consider two vehicles, one is a diesel vehicle with a tank-to-wheel efficiency of 26%, the other is an EV with a battery-to-wheel efficiency of 92%. Assume that the battery pack of the EV operates at 400 V. The energy contained in 1 l diesel is 35.86 MJ/l.
The diesel vehicle can be filled with 50 l in 2 minutes at a gas station. How much current must a charger deliver to the EV to charge the battery of the vehicle (also in 2 minutes) to enable the vehicle to have the same amount of energy available at the wheels?

2.8 A 500 F supercapacitor rated for 8 V is initially charged to 90% SOC. The capacitor is subjected to the energy usage profile of Fig. 2.13. Draw graphs of the current, voltage, SOC and DOD versus time. (Positive power is the energy extracted, while negative energy constitutes charging.)

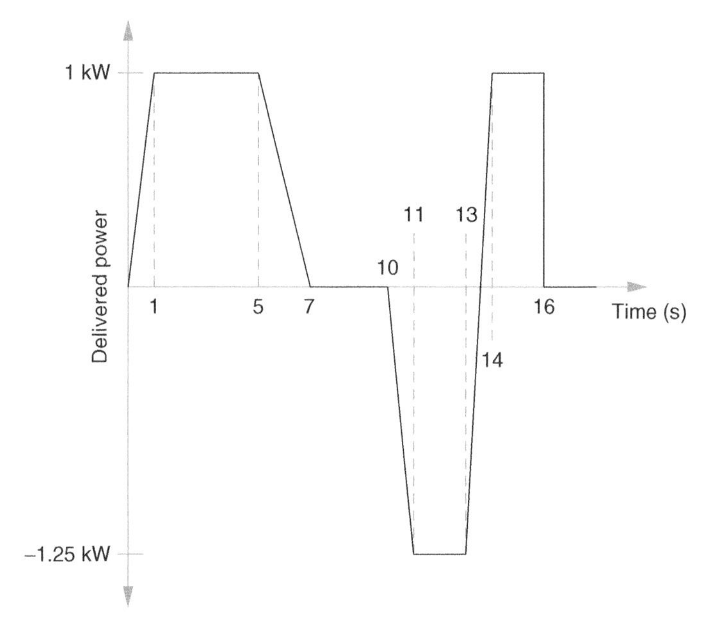

Figure 2.13 Energy usage profile.

2.9 Consider a supercapacitor that delivers power described by the function

$$p(t) = 750\sin(0.2\pi t)W.$$

The capacitor is rated at 200 F and 6 V. At time zero, the SOC is 85%. Find expressions for the voltage, SOC and current.

CHAPTER 3

POWER, REACTIVE POWER AND POWER FACTOR

3.1 INTRODUCTION

Electricity is the best way to transport energy. Electrons are amazingly good at supporting power flow. Power can be transferred very effectively over distances of thousands of kilometres by simply connecting two copper wires between the source and the load. For electrical power, we need both voltage and current; additionally, the direction of power flow can be reversed, in which case, the role of source and load is reversed.

The last hundred years have been dominated by ac power. Most of the world chose 50 Hz as the carrier frequency, while America mostly uses 60 Hz. More recently time functions other than sinusoidal waveforms are being applied more and more. Nowadays, many loads and sources are dc, for example, computers, TVs and solar cells. Many electrical motors in systems such as pumps and vehicles vary in their speed, needing adjustable frequencies rather than the fixed 50 or 60 Hz. Not all ac waveforms are sinusoidal. In fact, the voltages on many modern machines, including wind turbines, are square waves. If you measure the voltages and currents on the wires connecting electrical equipment, you find that it is still mostly sinusoidal, but you will also encounter a surprising diversity of other waveforms.

How do we transfer power with voltage and current, and how does the shape of the waveform affect the transmission of power? This is the topic of this chapter.

The Principles of Electronic and Electromechanic Power Conversion: A Systems Approach, First Edition.
Braham Ferreira and Wim van der Merwe.

3.2 POWER IN DC CIRCUITS

If the instantaneous value of the voltage, $v(t)$, and the current, $i(t)$, are known, then the value of the power in the circuit at that point in time can be calculated as

$$p(t) = v(t)\, i(t). \tag{3.1}$$

In dc circuits, the current and the voltage do not vary with time and are therefore, by definition, always in phase with each other. In dc circuits, the time dependency of the voltage and current, as shown in (3.1), is probably a little redundant as these quantities are constant. That said, it is true that in *real life* values change, for example, the terminal voltage of batteries decreases as they discharge and the resistance of elements is temperature dependent. Although, in theory at least, it is possible to model systems taking all these variations into consideration, in most cases, it adds too much complexity, and systems are modelled using assumptions of the steady state of the circuit. That is, the circuit is assumed to operate with ideal components with fixed values.

Consider the circuit in Fig. 3.1. It is clear that

$$I = \frac{V_{\mathrm{dc}}}{R} \tag{3.2}$$

$$V_{\mathrm{r}} = V_{\mathrm{dc}} \tag{3.3}$$

$$P = IV_{\mathrm{r}} = \frac{V_{\mathrm{r}}^2}{R} = \frac{V_{\mathrm{dc}}^2}{R} = I^2R. \tag{3.4}$$

Notation Note 3.1

Variable quantities are denoted by small letters, for example, $v(t)$. Constant quantities are denoted by capital letters, for example, V.

This result is hardly surprising. However, consider the following hypothetical experiment: this circuit can exist somewhere without the influence of any other circuits or systems, say, the circuit exists in a separate universe consisting *only* of this circuit, and *nothing* else. Because we know that energy cannot be created or

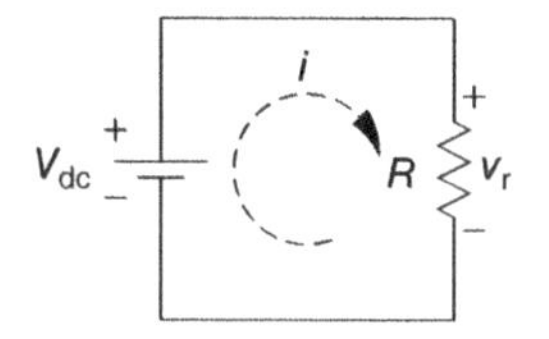

Figure 3.1 A simple dc circuit.

destroyed, it is true that all the energy dissipated in the resistor must be coming from somewhere inside 'our universe' and the only other element in 'our universe' (remember that our wire connections are ideal) is the battery source. Therefore, the power dissipated in the resistor must be supplied by the source!

Inspection of Fig. 3.1 gives the following insight: the current is flowing into the positive terminal of the resistor while it is flowing into the negative terminal of the battery. Therefore, calculating the power of the source, P_s, and the power of the resistor, P_r, yields

$$P_r = IV_{dc} \tag{3.5}$$

$$P_s = I(-V_{dc}) = -IV_{dc} = -P_r. \tag{3.6}$$

It is clear that the nett power in the circuit is $P_s + P_r = 0$.

Definition 3.1

The sign of the power indicates whether an element is delivering or dissipating power. We will use the passive sign convention, which states that negative power flow indicates the delivery of power, while positive power flow indicates the dissipation of power.

Example 3.1 In most cases (with dc sources), it is easy to see where the power is coming from and where it is going. However, sometimes it is not quite so simple. Consider the circuit in Fig. 3.2. How much power is dissipated and where?

Solution

Using Kirchhoff's first law (or Kirchhoff's current law), it is possible to calculate the voltage v_a as 9.813 V. The currents can then be found as

$$i_1 = -906.4\,\text{mA} \quad i_2 = 729\,\text{mA}$$
$$i_3 = 187.2\,\text{mA} \quad i_4 = 9.8\,\text{mA}.$$

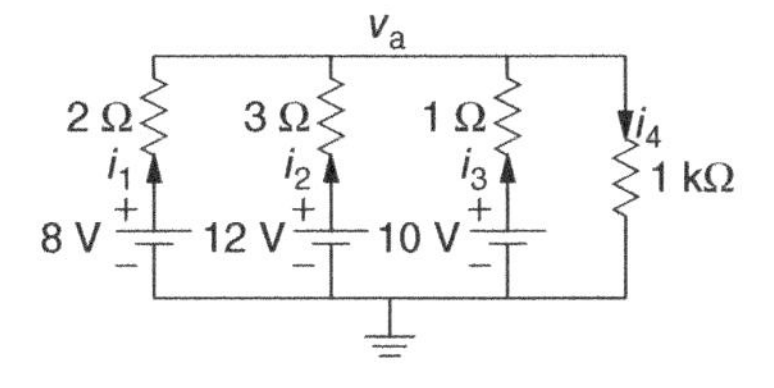

Figure 3.2 Power Flow Example.

The 12 V source delivers 8.75 W, but the 1 kΩ resistor dissipates 96.3 mW! When the power flow is analysed completely, it is seen that the 8 V source absorbs 7.25 W. The 10 V source delivers 1.87 W and the balance of the power is dissipated in the series resistances.

3.3 POWER IN RESISTIVE AC CIRCUITS

Consider a circuit as shown in Fig. 3.3. The source is a sinusoidal voltage source and the voltage and current in the circuit can be described as

$$v_{ac}(t) = v_r(t) = V_m \sin(\omega t) \tag{3.7}$$

$$i(t) = I_m \sin(\omega t) = \frac{V_m}{R} \sin(\omega t). \tag{3.8}$$

The instantaneous power in the resistor can now be calculated as

$$p(t) = V_m I_m \sin^2(\omega t). \tag{3.9}$$

Using the trigonometric identity

$$\sin^2(\theta) = \frac{1 - \cos(2\theta)}{2}$$

the instantaneous power can be then rewritten as

$$p(t) = \frac{V_m I_m}{2}(1 - \cos(2\omega t)). \tag{3.10}$$

It is clear that power delivery varies with time at double the frequency of the voltage and current. The voltage, current and instantaneous power are shown in Fig. 3.4 for $V_m = 1$ and $I_m = \frac{1}{2}$. The average power can be found as follows, where $T = \frac{2\pi}{\omega}$:

$$\begin{aligned} P &= \frac{1}{T}\int_{t_0}^{t_0+T} \frac{V_m I_m}{2}(1 - \cos(2\omega t))dt \\ &= \frac{V_m I_m}{2}. \end{aligned} \tag{3.11}$$

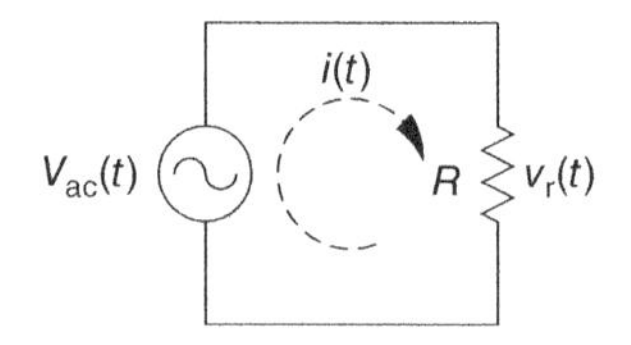

Figure 3.3 A simple ac circuit.

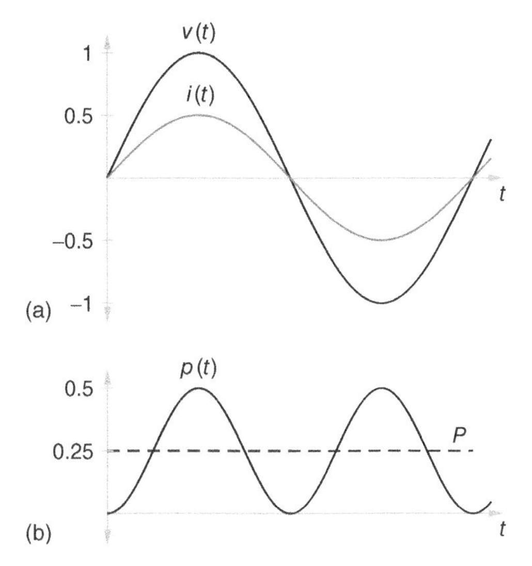

Figure 3.4 Voltage, current and instantaneous power in an ac circuit consisting of only a resistor.

3.4 EFFECTIVE OR RMS VALUES

In dc circuits, all values are constant, and it is easy to calculate the power dissipated in a resistor using I^2R, $\frac{V^2}{R}$ or VI. However, in ac circuits, the average power equation is not quite as clear-cut, see (3.11). For this circuit, the average power can be written as

$$P = \frac{V_m I_m}{2} = \frac{V_m^2}{2R} = \frac{I_m^2 R}{2}. \tag{3.12}$$

For sinusoidal sources, it is clear that the power relationships developed for dc circuits still hold; however, there is a factor of 2 present in the equations. If we knew we would always work only with pure sinusoidal signals, we could probably live with the introduction of an extra '2' into our power equations. However, the question remains: what would the power equations look like for signals that are not purely sinusoidal?

To simplify the discussion of periodic time-variant signals, we define an *effective value of a periodic waveform*. This effective value can be used to represent either currents or voltages. The effective value of a waveform is similar to the average of a waveform as it is a constant value associated with the waveform. Consider the average of a periodic waveform, although the instantaneous value of the waveform changes with time, the average does not; it is a constant value.

Let us now define the effective value of a periodic signal. To simplify our discussion, let us consider a periodic voltage waveform applied across a resistor. We know that this waveform will dissipate a certain amount of power in the resistor, and although this power dissipation will be a function of time, we can calculate the average power dissipation using (3.11). We want the effective value of this periodic voltage waveform to be equal to the dc voltage that should be applied to the resistor and to yield the same average power. Let V_{eff} be the effective dc voltage across a resistor, R. The average power can now be found as

$$P = \frac{V_{\text{eff}}^2}{R}. \tag{3.13}$$

If the same power is delivered to the resistor by a periodic voltage waveform, $v(t)$, with period, T, then it must be true that

$$P = \frac{1}{RT}\int_{t_0}^{t_0+T} v^2(t)dt. \tag{3.14}$$

Equating these expressions yields

$$\boxed{V_{\text{eff}} = \sqrt{\frac{1}{T}\int_{t_0}^{t_0+T} v^2(t)dt}.} \tag{3.15}$$

Investigation of (3.15) reveals that finding the effective value amounts to finding the *mean* value of the *square* of the current and then taking the square *root* of this value. The effective value of the current is consequently often referred to as the *root mean square*, or simply *rms*, value of the current.

Example 3.2 Find the rms value of the sinusoidal waveform $i(t) = I_m \sin(\omega t)$.

Solution

The rms value of the waveform can be found as

$$\begin{aligned} I_{\text{rms}} &= \sqrt{\frac{1}{T}\int_0^T I_m^2 \sin^2(\omega t)\,dt} \\ &= \sqrt{\frac{\omega}{2\pi}\int_0^{\frac{2\pi}{\omega}} \frac{I_m^2}{2}(1 - \cos(2\omega t))\,dt}. \end{aligned}$$

As the integral of the sinusoid over a period (or two periods as in this case) is zero, the equation simplifies to

$$\begin{aligned} I_{\text{rms}} &= \sqrt{\frac{\omega}{2\pi}\int_0^{\frac{2\pi}{\omega}} \frac{I_m^2}{2}\,dt} \\ &= \sqrt{\frac{I_m^2 \omega}{2\pi}\frac{t}{2}\bigg|_0^{\frac{2\pi}{\omega}}} \\ &= \frac{I_m}{\sqrt{2}} \end{aligned} \tag{3.16}$$

Notation Note 3.2

As the power dissipation and the rms value of a periodic signal are closely related, the rms value of a signal will most often be discussed. The rms value is therefore denoted without any subscript, that is to say V will be used rather than V_{rms}. The peak value will be denoted by V_m. The following is therefore true: $v(t) = V_m \sin(\omega t) = \sqrt{2}V \sin(\omega t)$.

3.5 PHASOR REPRESENTATION

Although it is possible to analyse ac circuits using the time domain approach, as we have done up to this point, it is convenient to introduce a new method of looking at things to simplify our analysis. Consider the sinusoidal signal

$$v(t) = V_m \cos(\omega t + \theta_v). \tag{3.17}$$

It is possible to describe this signal to a third person by using only three descriptors, the amplitude (V_m), the frequency (ω) and the phase shift (θ_v). In an ac circuit analysis, the frequency is normally implicit to the problem at hand which does not vary. Therefore, several signals in the same system can be described uniquely using only the amplitude and the phase shift as descriptors.

Although it is convenient to think about the description of sinusoidal signals in this manner, we have to find some mathematical basis for this. Luckily, it is not too difficult. Recall Euler's equation, which states

$$\boxed{e^{j\omega t} = \cos(\omega t) + j\sin(\omega t).} \tag{3.18}$$

If we select the voltage across a resistor as the non-realisable function

$$v(t) = V_m e^{j\theta_v} e^{j\omega t} = V_m \cos(\omega t + \theta_v) + jV_m \sin(\omega t + \theta_v) \tag{3.19}$$

then the current can be written as

$$i(t) = \frac{V_m}{R} e^{j\theta_v} e^{j\omega t}. \tag{3.20}$$

This is of course true because the systems in question are linear and the principle of superposition holds. The fact that the driving function cannot be realised in real life is, however, a slight stumbling block.

Let us look at this driving function, taking two different approaches. First, we take the more classical mathematical approach; an alternative definition for $\cos(\theta)$

can be found using the addition of two complex conjugate functions:

$$\boxed{\cos(\theta) = \frac{1}{2}e^{j\theta} + \frac{1}{2}e^{-j\theta}} \tag{3.21}$$

Let us now consider the voltage across a resistor

$$v(t) = V_m \cos(\omega t + \theta_v) \tag{3.22}$$

and rewrite the function using this alternate definition as

$$v(t) = \frac{1}{2}V_m(e^{j\theta_v}e^{j\omega t} + e^{-j\theta_v}e^{-j\omega t}). \tag{3.23}$$

The current through the resistor can now be written as

$$i(t) = \frac{1}{2R}V_m(e^{j\theta_v}e^{j\omega t} + e^{-j\theta_v}e^{-j\omega t}). \tag{3.24}$$

Although this approach is mathematically correct and we still need only the amplitude and phase angle (remember that the frequency is implicit to the system) to uniquely describe a sinusoid: this approach is a bit long winded.

Another approach is to 'ignore' the imaginary part of the complex driving function. It is possible to write the voltage across a resistor as

$$\boxed{v(t) = V_m \cos(\omega t + \theta_v) = \text{Re}\left\{V_m e^{j\theta_v} e^{j\omega t}\right\}} \tag{3.25}$$

where $\text{Re}\{z\}$ is the real part of the complex number z. Therefore, if $z = x + jy$ then $\text{Re}\{z\} = x$. The current through the resistor now simply becomes

$$i(t) = \text{Re}\left\{\frac{V_m}{R}e^{j\theta_v}e^{j\omega t}\right\} = \frac{V_m}{R}\text{Re}\left\{e^{j\theta_v}e^{j\omega t}\right\}. \tag{3.26}$$

Therefore, by taking the frequency of the system as constant and implicit to the circuit at hand, it is possible to describe any sinusoidal signal using only its amplitude and phase as descriptions. This description is called a *phasor* and takes the shorthand form of $A\angle\theta$. Formally, this shorthand form is called the *Steinmetz notation*, named after the German-American engineer who devised this method of solving ac circuits using complex numbers. Although his concept is widely used in electrical engineering, in most cases, the phasor is not completely defined. The definition used in this book is as follows:

Definition 3.2

A phasor is a complex quantity associated with a phase-shifted cosine wave such that, if the phasor is in polar form, its magnitude is the effective (rms) value of the voltage or current and its angle is the phase angle of the phase-shifted cosine wave. Therefore, $A\angle\theta = \sqrt{2}A\cos(\omega t + \theta)$.

As a side note to this definition, for obvious reasons it is equally possible to define the phasor to imply peak value rather than the rms value and also to use a sine wave rather than a cosine wave, or even any combination of these. No consensus exists about this definition in the electrical engineering field, and the definitions are sometimes used interchangeably. However, the cosine wave choice is made easier if we realise that by using the sine wave definition, we would have to work with the imaginary part of the rotating phasor, not an impossibility but definitely a more daunting proposition than working with the real part. Likewise, all power calculations are done with rms values, so it makes more sense to define the phasor using rms values and therefore to do the conversion between peak and rms right at the start. Also keep in mind that all voltage and current measurement equipment yields an rms value and not a peak value. It is therefore easier to verify calculations done using the rms notation.

Notation Note 3.3

As the phasor consists of both a magnitude and a phase shift, the phasor will be denoted using boldface variables. Furthermore, because the phasor has a constant value, it will be denoted using capital letters. Therefore, according to the definition, it is correct to write $v(t) = V_m\cos(\omega t + \theta_v)$ or, equivalently, $\mathbf{V} = V\angle\theta_v$ where $\sqrt{2}V = V_m$.

The use of the phasor to analyse circuits also yields another benefit, a quick and easy way to visualise the operation of the circuit. The benefits can best be described using an example.

Example 3.3 Consider the circuit in Fig. 3.5.

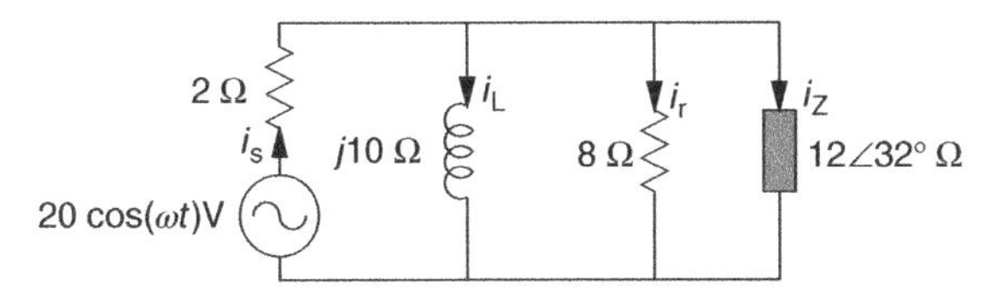

Figure 3.5 Phasor diagram example circuit.

The voltage source can be represented as $\mathbf{V} = 14.14\angle 0°$ V. The parallel combination of the inductor, the resistor and the general impedance is

$$\begin{aligned}\mathbf{Z}_{\text{eq}} &= \left(\frac{1}{10\angle 90°} + \frac{1}{8\angle 0°} + \frac{1}{12\angle 32°}\right)^{-1} \\ &= 4.11\angle 36.38°\ \Omega.\end{aligned} \tag{3.27}$$

The source current is therefore

$$\mathbf{I}_{\text{s}} = \frac{\mathbf{V}}{2\angle 0° + \mathbf{Z}_{\text{eq}}} = 2.42\angle - 24.67°\ \text{A} \tag{3.28}$$

and the voltage across the parallel combination of elements is

$$\mathbf{V}_{\mathbf{p}} = \mathbf{V} - \mathbf{I}(2\angle 0°) = 9.95\angle 11.7°\ \text{V}. \tag{3.29}$$

Therefore, the currents can be found as

$$\mathbf{I}_{\text{L}} = 1\angle - 78.3°\ \text{A} \tag{3.30}$$

$$\mathbf{I}_{\text{r}} = 1.24\angle 11.7°\ \text{A} \tag{3.31}$$

$$\mathbf{I}_{\text{z}} = 0.83\angle - 20.29°\ \text{A} \tag{3.32}$$

It is true that the currents can be visualised by drawing a graph as shown in Fig. 3.6. However, this does not give us much information and is rather confusing. It is, for example, very difficult from this representation to see that $i_{\text{s}}(t) = i_{\text{L}}(t) + i_{\text{r}}(t) + i_{\text{z}}(t)$.

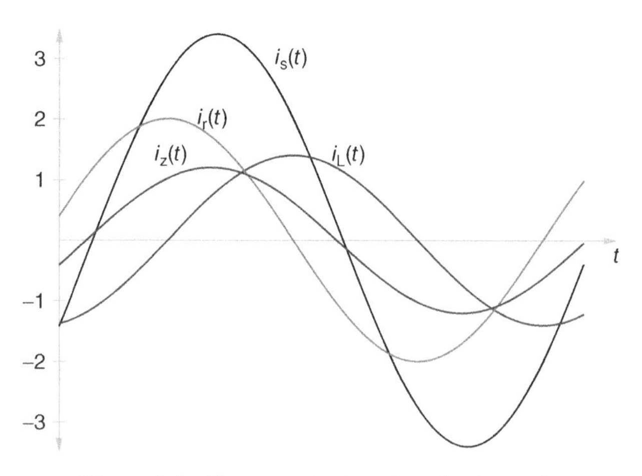

Figure 3.6 The currents plotted on the time axis.

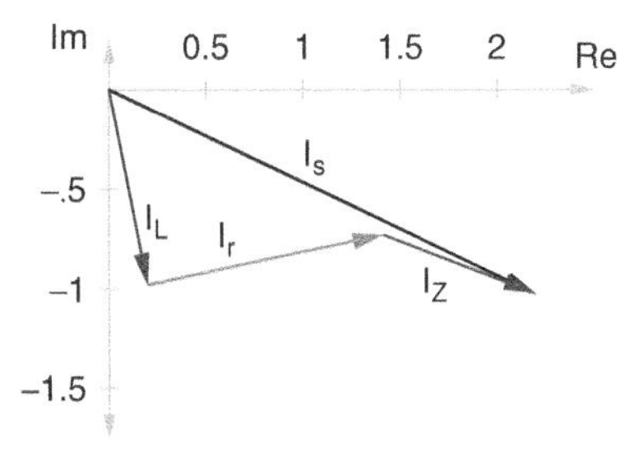

Figure 3.7 The currents plotted as a phasor diagram.

There is, however, a simpler way that gives the same basic information but conveys much more. If the phasors rather than the time domain waveforms are visualised, the system operation is clearer. It is also easy to verify that $\mathbf{I}_s = \mathbf{I}_L + \mathbf{I}_r + \mathbf{I}_Z$. This phasor diagram is shown in Fig. 3.7.

3.6 POWER IN AC CIRCUITS

When an ac circuit consists of only a resistor, as in Fig. 3.3, the current and the voltage will always be in phase. Most circuits, however, do not consist of only a resistive element. Consider the circuit in Fig. 3.8; Z can be any impedance.

The steady-state voltage and current for this circuit can be written as

$$v(t) = V_m \sin(\omega t + \theta_v) \tag{3.33}$$

$$i(t) = I_m \sin(\omega t + \theta_i). \tag{3.34}$$

Following the same argument as with ac circuits consisting only of a resistor, the instantaneous power is then

$$p(t) = V_m I_m \sin(\omega t + \theta_v) \sin(\omega t + \theta_i). \tag{3.35}$$

Using the trigonometric identity that

$$\sin(\alpha)\sin(\beta) = \frac{\cos(\alpha - \beta) - \cos(\alpha + \beta)}{2}$$

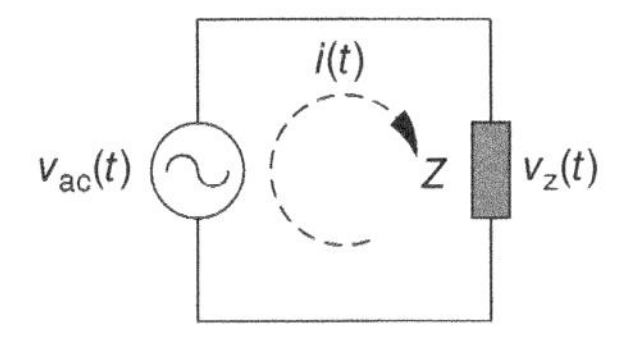

Figure 3.8 A simple ac circuit.

this expression can be rewritten as

$$p(t) = \underbrace{\frac{V_m I_m}{2}\cos(\theta_v - \theta_i)}_{\text{Constant}} - \underbrace{\frac{V_m I_m}{2}\cos(2\omega t + \theta_v + \theta_i)}_{\text{Time varying}}. \tag{3.36}$$

The average power can again be found as (realising that the integral of a sinusoid over a period is equal to zero)

$$\begin{aligned} P &= \frac{1}{T}\int_{t_0}^{t_0+T} p(t)\,dt \\ &= \frac{1}{T}\int_{t_0}^{t_0+T} \frac{V_m I_m}{2}\cos(\theta_v - \theta_i) - \frac{V_m I_m}{2}\cos(2\omega t + \theta_v + \theta_i)\,dt \\ &= \frac{V_m I_m}{2}\cos(\theta_v - \theta_i) \end{aligned} \tag{3.37}$$

which we can rewrite in terms of rms values as

$$P = VI\cos(\theta_v - \theta_i). \tag{3.38}$$

The voltage, current and instantaneous power waveforms for a phase shift of 40° between the voltage and current is shown in Fig. 3.9. Although the current and voltage magnitude are the same as in Fig. 3.4, the average power is lower. This is not the only strange thing that happens, we also see that the power becomes negative at times. According to the definition of power, as discussed in Section 3.2, this means that the load is delivering power to the source. Is this possible?

3.6.1 Power in a Capacitive Circuit

Consider the circuit in Fig. 3.10. The voltage and current are

$$v_c(t) = \sqrt{2}V_c \sin(\omega t) \tag{3.39}$$

$$i(t) = \sqrt{2}I \sin\left(\omega t + \frac{\pi}{2}\right) \tag{3.40}$$

of course

$$I = \omega C V_c. \tag{3.41}$$

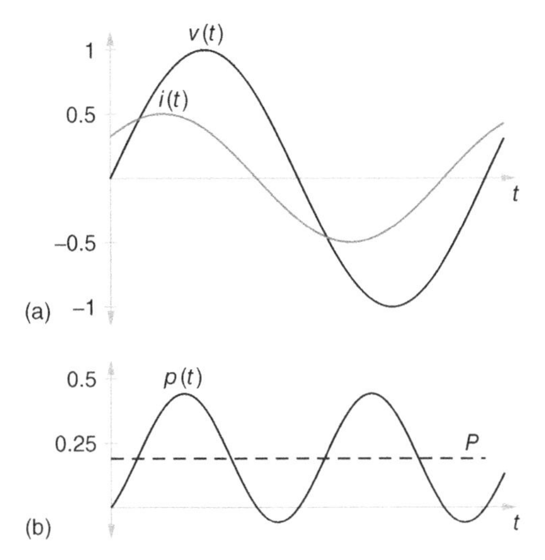

Figure 3.9 Voltage, current and instantaneous power in an ac circuit with an arbitrary impedance as load.

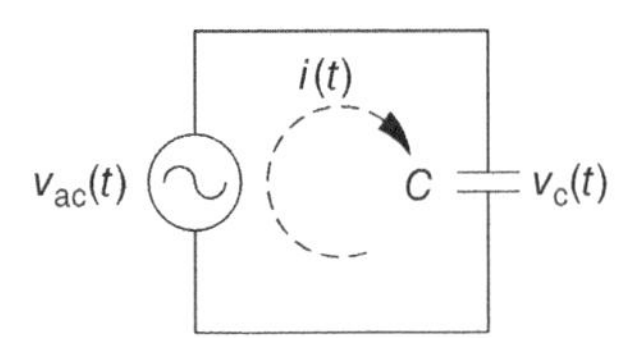

Figure 3.10 A simple ac circuit with a capacitor.

Now, writing the instantaneous power in the same way as in (3.36) yields

$$\begin{aligned} p(t) &= \sqrt{2}V_c\sqrt{2}I\sin(\omega t)\sin\left(\omega t+\frac{\pi}{2}\right) \\ &= \underbrace{V_cI\cos\left(-\frac{\pi}{2}\right)}_{=0} - V_cI\cos\left(2\omega t+\frac{\pi}{2}\right). \end{aligned} \tag{3.42}$$

The voltage, current and instantaneous power waveforms are shown in Fig. 3.11. From this figure and from (3.42), it is clear that the average power is equal to zero. In fact, the capacitor receives energy during half a cycle and then gives the energy back during the next half cycle.

From classical physics, the work, W, is defined as the exchange of energy from one system to another, and if heat flow is ignored (which is a sensible assumption for electrical systems), then

$$W = \Delta E. \tag{3.43}$$

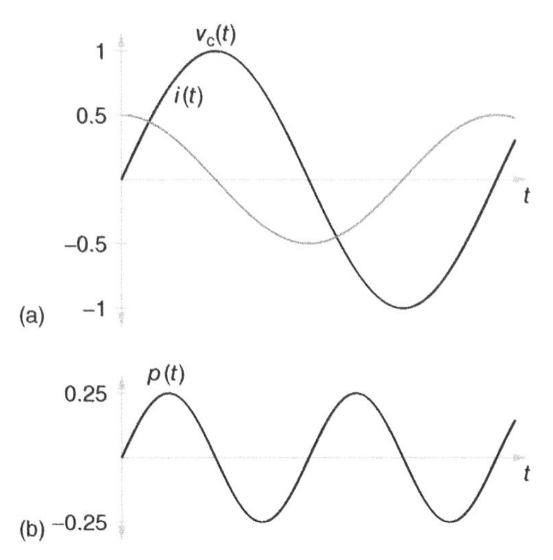

Figure 3.11 Voltage, current and instantaneous power in an ac circuit consisting of only a capacitor.

Power is defined as the amount of work completed in a period of time, or

$$P = \frac{\Delta W}{\Delta t}. \tag{3.44}$$

Instantaneous power is the change in work at any point in time, therefore

$$p(t) = \frac{dW(t)}{dt}. \tag{3.45}$$

Now, let us consider the energy stored in the capacitor, given as

$$E = \frac{1}{2}CV^2. \tag{3.46}$$

The capacitor stores electrical energy in the electrostatic field that develops across the capacitor's dielectric (the insulating material between the capacitor's terminals). Therefore, the energy in the capacitor as a function of time becomes (remembering that V_c is an rms value)

$$\begin{aligned} e(t) &= \frac{1}{2}C\left(\sqrt{2}V_c \sin(\omega t)\right)^2 \\ &= CV_c^2 \sin^2(\omega t) \end{aligned} \tag{3.47}$$

which, using the fact that

$$\sin^2(\theta) = \frac{1 - \cos(2\theta)}{2}$$

becomes

$$e(t) = \frac{1}{2} C V_c^2 (1 - \cos(2\omega t)). \tag{3.48}$$

The instantaneous power can be calculated by taking the first derivative of the energy in the capacitor,

$$\begin{aligned} p(t) &= \frac{de(t)}{dt} \\ &= \frac{d}{dt}\frac{1}{2} C V_c^2 (1 - \cos(2\omega t)) \\ &= V_c^2\, C\omega \sin(2\omega t) \\ &= V_c\, I\, \sin(2\omega t). \end{aligned} \tag{3.49}$$

This is the same result as in (3.42) because

$$\cos\left(2\theta + \frac{\pi}{2}\right) = \sin(2\theta). \tag{3.50}$$

It is important to note from (3.48) that the capacitor never contains negative energy; that would, of course, be a violation of the laws of nature. The capacitor simply stores a certain amount of energy as the voltage increases and delivers the same amount of energy back to the source as the voltage decreases again. As the capacitor returns all the energy back to the source, no average power is delivered although some power flow is observed. We call this temporary flow of energy (that is first stored and returned later) the *reactive power flow*.

3.7 APPARENT POWER, REAL POWER AND POWER FACTOR

When an electrical circuit is investigated, it is normally easy to determine the rms current and voltage of the system. In dc circuits, this is convenient because the average power can be determined from these values. However, in ac circuits, it is not so simple. Let us define a new quantity called *apparent power*, denoted as $|\mathbf{S}|$.[1] This is the power that appears to be in an ac circuit and is defined as

$$\boxed{|\mathbf{S}| = VI.} \tag{3.51}$$

This definition of apparent power is useful as it gives a feel of the size and power handling capability of electrical equipment. Take, for example, a capacitor; although the capacitor is not delivering any real power, it must still be rated for the applied voltage and for the current.

[1] It will become clear in Section 3.8 why a boldface letter is used and not $|S|$.

With this definition of apparent power, the expression for average power in(3.38) takes on a new meaning. The equation can be rewritten as

$$P = VI\cos(\theta_v - \theta_i) = |\mathbf{S}|\cos(\theta_v - \theta_i). \tag{3.52}$$

It is clear that the average power is the apparent power multiplied by a factor that depends on the phase shift between the current and the voltage. Recall from the discussion in Section 3.3 that when the current and the voltage are in phase, the average power is high and no energy is transferred back to the source. Conversely, in the case where the current and the voltage are 90° out of phase, the average power delivery is zero, as discussed in Section 3.6.1.

This factor describes the portion or factor of the apparent power that is converted into real or average power; it is therefore called the *power factor*, denoted by pf. The definition of the power factor follows from (3.52) as

$$\boxed{\text{pf} = \cos(\theta_v - \theta_i)} \tag{3.53}$$

and therefore

$$\boxed{P = |\mathbf{S}|\text{pf}.} \tag{3.54}$$

3.8 COMPLEX POWER

The definition of instantaneous power (3.36) can be simplified by using the newly defined phasor. Assume that the voltage and current through a specific element is given by

$$\mathbf{V} = V\angle\theta_v \text{V} \tag{3.55}$$

$$\mathbf{I} = I\angle\theta_i \text{A}. \tag{3.56}$$

In Section 3.7, apparent power was defined as

$$|\mathbf{S}| = VI \tag{3.57}$$

and average power as

$$P = |\mathbf{S}|\text{pf}, \tag{3.58}$$

where

$$\text{pf} = \cos(\theta_v - \theta_i). \tag{3.59}$$

Now let us define a new term *complex power* with the following definition

$$\boxed{\mathbf{S} = \mathbf{VI}^*,} \tag{3.60}$$

where $\mathbf{Z}^*$ is the complex conjugate of the complex number $\mathbf{Z}$. Therefore, if $\mathbf{Z} = A\angle\phi$ then $\mathbf{Z}^* = A\angle - \phi$. The unit of $\mathbf{S}$ is (VA) or volt-ampere. If this definition of complex power is expanded, we see the following

$$\begin{aligned}\mathbf{S} &= \mathbf{V}\mathbf{I}^* \\ &= VI\angle(\theta_v - \theta_i) \\ &= \underbrace{\overbrace{VI}^{|\mathbf{S}|}\ \overbrace{\cos(\theta_v - \theta_i)}^{\text{pf}}}_{P} + jVI\sin(\theta_v - \theta_i).\end{aligned} \tag{3.61}$$

We can see that the real part of the expression is simply the average power. The imaginary part of the expression is maybe a bit more tricky. Recall from Section 3.6.1 that we called the energy that is stored in a capacitor (or an inductor) during half a cycle only to be delivered back to the rest of the circuit during the next half cycle the reactive power flow. Furthermore, it is true that the reactive power flow is equal to zero when the current and the voltage are in phase. This is perhaps better explained by expanding the definition of instantaneous power. Let $v(t) = \sqrt{2}V\cos(\omega t + \theta_v)$ and $i(t) = \sqrt{2}I\cos(\omega t + \theta_i)$, then using the fact that

$$\cos(\alpha + \beta) = \cos(\alpha)\cos(\beta) - \sin(\alpha)\sin(\beta)$$

this yields

$$\begin{aligned}p(t) &= VI\cos(\theta_v - \theta_i) + VI\cos(2\omega t + \theta_v + \theta_i) \\ &= VI\cos(\theta_v - \theta_i) + VI\cos(\theta_v - \theta_i)\cos(2\omega t) - VI\sin(\theta_v - \theta_i)\sin(2\omega t) \\ &= P + P\cos(2\omega t) - Q\sin(2\omega t),\end{aligned} \tag{3.62}$$

where

$$\boxed{Q = VI\sin(\theta_v - \theta_i).} \tag{3.63}$$

Here, we see a formal definition for Q, the reactive power. From the definition, it is true that the basic unit for Q will also be (VA) or volt-ampere, however, this becomes a little confusing because the unit for $\mathbf{S}$ is also (VA). To avoid confusion, the Romanian electrical engineer Constantin Budeanu suggested a new unit (var), which stands for volt-ampere-reactive. This unit was accepted by the International Electrotechnical Commission (IEC) as the standard unit in 1930. It is worth mentioning that although (var), (VAR) and even (VARs) are used in some texts the generally accepted unit is (var)[2].

[2]Both the IEC and the IEEE (Institute for Electrical and Electronic Engineers) states (var) as the correct unit.

When the definition of the average power (3.60) is compared to the definition of reactive power (3.63), it makes sense to define a new concept: the reactive factor or rf. The reactive factor is defined as

$$\boxed{\text{rf} = \sin(\theta_v - \theta_i).} \tag{3.64}$$

Using the reactive factor, the definitions for P and Q can be written in a similar manner

$$P = |\mathbf{S}|\text{pf}$$
$$Q = |\mathbf{S}|\text{rf}.$$

Apart from bringing consistency in the definitions of P and Q, the reactive factor tells us something that the power factor alone cannot: whether the circuit is absorbing or delivering reactive power. As $\cos(\alpha) = \cos(-\alpha)$, the power factor is the same irrespective of whether the current is leading or lagging the voltage. However, the sign of the reactive factor will show this as $\sin(\alpha) = -\sin(-\alpha)$. That said, in most cases, it is enough to know whether the current is leading or lagging as this would give the same information as the knowledge of the reactive factor.

With the definition of Q covered let us revisit the definition of $\mathbf{S}$ in (3.61), which can now be rewritten as

$$\boxed{\mathbf{S} = P + jQ.} \tag{3.65}$$

This would of course mean that

$$|\mathbf{S}|^2 = P^2 + Q^2 \tag{3.66}$$

a fact that is easily verified using (3.51), (3.60) and (3.63) to show that

$$\begin{aligned} |\mathbf{S}|^2 &= P^2 + Q^2 \\ V^2I^2 &= (VI\cos(\theta_v - \theta_i)^2) + (VI\sin(\theta_v - \theta_i)^2) \\ &= V^2I^2\underbrace{(\cos^2(\theta_v - \theta_i) + \sin^2(\theta_v - \theta_i))}_{=1}. \end{aligned}$$

3.9 ELECTRICAL ENERGY COST AND POWER FACTOR CORRECTION

Electricity is a very good and convenient way of transporting energy from its source of generation to where it is needed. When using electricity, we should therefore think of it as energy usage, the same as when using gas, coal or even an ox turning a wheel. With this viewpoint, it makes sense that the supplier bills the customer for the energy used in a billing period, normally a calendar month. Recall that the energy used in a given period (the same as the work done in the period) is calculated as

$$W = \int_T p(t)\,dt \tag{3.67}$$

with unit (J) or joule, which is essentially one watt-second. However, the joule is a rather small unit, which is for that reason not well suited for energy billing. Consider the fact that you need 355 410.5 J to bring 1 l of water to boil from 15°C.[3] To counteract this problem, electrical engineers have adopted the kilowatt-hour as the preferred unit of energy. The unit of the kilowatt-hour is (kWh) sometimes written as (kW · h). The kilowatt-hour is defined as the energy dissipated in a 1 kW load in 1 h, therefore

$$1\ \text{kWh} = 1000 \cdot 60 \cdot 60 = 3.6\ \text{MJ}. \tag{3.68}$$

The kilowatt-hour is well dimensioned for the monthly energy usage billing purposes of residential and small business consumers and general discussions of energy usage. We can express the amount of energy used to boil water, as calculated above, using the kilowatt-hour as about 0.1 kWh. It is however worthwhile mentioning that large customers and energy suppliers generally also use megawatt-hours and gigawatt-hours as the units of energy measurement.

Example 3.4 What is the energy cost of a 150 W television set that is used for 3 hours per day when the energy cost is €0.20/kWh?

Solution

If the TV set is used for three hours per day, then it will be in use for $365 \times 3 = 1095$ hours per year (h/year). Therefore, the energy usage per year is $0.15 \times 1095 = 164.25$ kWh and the annual energy cost is $164.25 \times 0.2 =$ €32.85.

Example 3.5 The average bath requires about 85 l of water, while a low flow-rate shower head delivers 10 l/min. The average temperature of the bath and shower water is 42°C. If the water is heated by an electrical element from 13°C at a cost of €0.21/kWh, what is the cost difference between a bath and a 3.4 min shower each day per month?

Solution

The volume of water used per shower is

$$V_s = 3.4 \times 10 = 34\,\text{l}$$

[3]The specific heat capacity of water is 4.1813 J/(g K). Assuming water boils at 100°C, the energy required is $4.1813 \times (100 - 15) \times 1000$ J.

The monthly cost per shower is therefore

$$\begin{aligned}\text{Cost}_S &= V_s\ \Delta T\ \frac{h_{H_2O}}{1000 \times 60^2}\text{price/kWh} \times \text{days in month}\\ &= 34(42-13)\frac{4181.3}{3\,600\,000}0.21 \times 30\\ &= €7.22.\end{aligned}$$

Likewise, the cost of bathing for a month is

$$\begin{aligned}\text{Cost}_B &= V_b\ \Delta T\ \frac{h_{H_2O}}{1000 \times 60^2}\text{price/kWh} \times \text{days in month}\\ &= 80(42-13)\frac{4181.3}{3\,600\,000}0.21 \times 30\\ &= €16.98.\end{aligned}$$

The price difference is therefore €16.98 – €7.22 = €9.76 per person per month. This is a sizeable difference should you be paying the bill for a family of four.

The energy cost represents the cost of generating one unit (kWh) of electrical energy, including a good profit. However, consider the hypothetical situation where a customer connects a large capacitor to the electricity supply and nothing else. As we have seen in Section 3.6.1, the average power in this case is zero and this customer will therefore not have to pay anything. That is all well and good, but the current is flowing in the network and the electricity provider therefore incurs real costs to provide the infrastructure to the costumer and in this case gets nothing in return.

Although it is not likely that someone will in fact connect a large capacitor to the electrical supply, the problem of equipment with a low power factor is a real concern. When a large appliance or an industrial process operates with a power factor of 0.5, it implies that the utility must supply two times more current than that really necessary. The utility (company that is engaged in the generation, transmission and distribution of electricity for sale) therefore has to invest in upgrading the transmission and distribution system to be able to deliver all the current to the customers. Most utilities introduced another charge for large costumers to recuperate some of these costs. This charge, sometimes called the *network access charge* or *demand charge*, takes several forms. The simplest and the most common charge is a cost levied per peak demand during the billing period.

Example 3.6 A company runs a heating process continually, this heater consumes 4 kW at unity power factor (as a heater is resistive). Once per month, a cleaning process is used for 4 hours that consumes 45 kW at a power factor of 0.65 lagging. The energy charge is € 0.11/kWh and the network access charge is € 2.10/kVA (of peak demand per month). Assume a 30-day month. Calculate the electricity cost per month both with and without the access charge.

The energy cost for the two processes is

$$\text{Cost-heating} = 4 \cdot 24 \cdot 30 \cdot 0.11 = €316.80$$

$$\text{Cost-cleaning} = 45 \cdot 4 \cdot 0.11 = €19.80$$

with a total energy cost of €336.80 per month. The peak demand is

$$S_{max} = 71.89\,\text{kVA}$$

and the access charge is

$$\text{access charge} = 71.89 \cdot 2.1 = €150.98.$$

The total monthly bill is therefore €336.80 per month without any access charges and €487.58 per month with the access charge. Interestingly, although the cleaning process accounts for only 5.9% of the energy use, it represents more than 30.0% of the monthly cost.

In general, large electrical energy users pay a penalty for a low power factor. It is, however, possible to correct the power factor by including a passive element in parallel with the offending equipment, or even at the incoming connection point. In general, most equipment in use today is inductive in nature, resulting in an overall lagging power factor and therefore the absorption of reactive power. To verify this statement, we see from (3.63) that when the current is lagging (i.e. $\theta_v > \theta_i$) then $\sin(\theta_v - \theta_i) > 0$ and therefore $Q > 0$. Remember from Definition 3.2 that positive power flow implies the absorption of energy. Conversely, a capacitor will deliver reactive power to the network. Although it would be theoretically possible to improve the power factor to one, it is not often done because apart from the fact that a large amount of extra capacitance is needed, it can also cause unwanted resonances in the power system. The process can be explained by revisiting Example 3.6.

Example 3.7 How can the energy cost for the company in Example 3.6 be decreased by using power factor correction? The terminal voltage is 230 V.

The heating process already has a unity power factor, so obviously that cannot be improved on. Taking the cleaning process, the current can be calculated as

$$|\mathbf{I}| = \frac{P}{V\text{pf}} = 301\ \text{A}$$

$$\arg\{\mathbf{I}\} = \arccos(\text{pf}) = 49.5°$$

but because we know that the current is lagging, we can write $\mathbf{I} = 301\angle -49.5°$ A. The complex power is therefore

$$\mathbf{S}_{old} = 45 - j52.61\ \text{kVA}.$$

If we improve the power factor to 0.9 lagging, then the complex power would be

$$\mathbf{S}_{\mathrm{new}} = 45 - j21.8 \text{ kVA}.$$

We would therefore have to generate $\Delta\mathbf{S} = 52.61 - 21.8 = 30.81$ kVAR to improve the power factor to 0.9 lagging. This can be achieved by adding a capacitor in parallel with the cleaning load. This capacitor needs to supply 30.81 kVAR, therefore the current in the capacitor must be

$$\mathbf{I}_{\mathrm{c}} = \frac{\Delta\mathbf{S}^*}{V} = \frac{30.81\angle -90^\circ}{220} = 133.96\angle 90^\circ \text{ A}.$$

Using the impedance of a capacitor at 50 Hz, we get the size of the capacitor as

$$C = \frac{|\mathbf{I}_{\mathrm{c}}|}{2\pi f V} = 5.5 \text{ mF}.$$

A graphical representation of the process is shown in Fig. 3.12 using a phasor diagram. You can verify that this power factor correcting intervention would save the company €40.38 per month (use Example 3.4 as reference).

3.10 FOURIER SERIES

A periodic signal is a signal that repeats itself every cycle, or more concisely for a period of T seconds and a counter $n \in \mathbb{Z}$

$$f(t) = f(t \pm nT). \tag{3.69}$$

In 1807, Jean Baptiste Joseph Fourier suggested a new method of harmonic analyses developed as a solution to his famous heat equation in a metal plate. The most

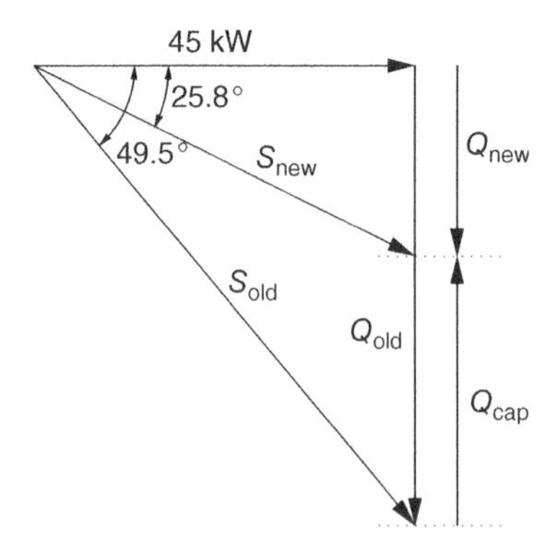

Figure 3.12 Phasor diagram of power factor correction.

general form of the Fourier series takes the form of the exponential Fourier series. Any periodic signal $f(t)$ with period T where $\omega = \frac{2\pi}{T}$ can be expressed as,

$$\boxed{f(t) = \sum_{n=-\infty}^{\infty} C_n e^{jn\omega t}.} \tag{3.70}$$

The coefficients of the series are found as

$$\boxed{C_n = \frac{1}{T} \int_{t_0}^{t_0+T} f(t) e^{-jn\omega t} \, dt.} \tag{3.71}$$

From this definition, it is clear that C_n is a complex number.

This definition of the Fourier series contains all the information necessary to investigate periodic signals and their interactions. However, working with positive and negative frequencies can be a little disconcerting. The task does become a little less daunting if we realise that $e^{jn\omega t}$ is simply a unit phasor rotating clockwise and $e^{-jn\omega t}$ is the same unit phasor but rotating anticlockwise. Although this definition of the Fourier series must be used if we investigate the multiplication of two periodic signals using frequency domain methods, for most other applications, the definition can be simplified to some extent. As C_n is a complex number, let us write it as

$$C_n = \alpha_n + j\beta_n \tag{3.72}$$

If we rewrite the summation in (3.70) to only use positive integers as counters we get

$$f(t) = C_0 + \sum_{n=1}^{\infty} \underbrace{(C_n e^{j\omega t} + C_{-n} e^{-jn\omega t})}_{=\delta_n}. \tag{3.73}$$

It is easy to prove that C_0 is simply the time average of $f(t)$ because from,

$$C_0 = \frac{1}{T} \int_{t_0}^{t_0+T} f(t) \, dt.$$

Let us define this average value as a_v therefore $C_0 = a_v$.

One of the fundamental theorems of Fourier analysis states that if we take the Fourier series of a real valued signal (i.e. a signal without any imaginary values, which for obvious reasons is what we are doing), then

$$C_n = C_{-n}^* \tag{3.74}$$

Using (3.72) and the fact that

$$\boxed{e^{j\theta} = \cos(\theta) + j\sin(\theta)} \tag{3.75}$$

it is possible to rewrite the summation term in (3.73) as

$$\begin{aligned}
\delta_n &= C_n e^{j\omega t} + C_{-n} e^{-jn\omega t} \\
&= (\alpha + j\beta)(\cos(n\omega t) + j\sin(n\omega t)) + (\alpha - j\beta)(\cos(n\omega t) - j\sin(n\omega t)) \\
&= \alpha\cos(n\omega t) + j\beta\cos(n\omega t) + j\alpha\sin(n\omega t) - \beta\sin(n\omega t) \\
&\qquad + \alpha\cos(n\omega t) - j\beta\cos(n\omega t) - j\alpha\sin(n\omega t) - \beta\sin(n\omega t) \\
&= 2\alpha\cos(n\omega t) - 2\beta\sin(n\omega t).
\end{aligned} \tag{3.76}$$

The terms α and β can be found in a similar way, from (3.10)

$$\begin{aligned}
\alpha + j\beta &= \frac{1}{T}\int_{t_0}^{t_0+T} f(t)(\cos(n\omega t) - j\sin(n\omega t))\,dt \\
&= \frac{1}{T}\left(\int_{t_0}^{t_0+T} f(t)\cos(n\omega t)\,dt - j\int_{t_0}^{t_0+T} f(t)\sin(n\omega t)\,dt\right)
\end{aligned} \tag{3.77}$$

therefore

$$\alpha = \frac{1}{T}\int_{t_0}^{t_0+T} f(t)\cos(n\omega t)\,dt \tag{3.78}$$

and

$$\beta = -\frac{1}{T}\int_{t_0}^{t_0+T} f(t)\sin(n\omega t)\,dt. \tag{3.79}$$

Finally, let us define two new variables

$$a_n = 2\alpha \tag{3.80}$$

$$b_n = -2\beta. \tag{3.81}$$

Now using these variables, it is possible to write the Fourier series as

$$\boxed{f(t) = a_v + \sum_{n=1}^{\infty}(a_n\cos(n\omega t) + b_n\sin(n\omega t)).} \tag{3.82}$$

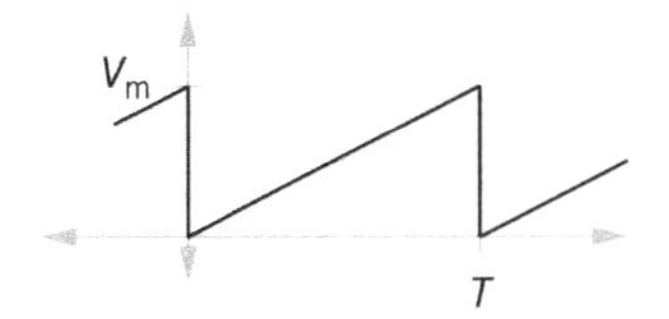

Figure 3.13 Periodic signal for Example 3.8.

where

$$a_v = \frac{1}{T}\int_{t_0}^{t_0+T} f(t)\,dt \tag{3.83}$$

$$a_n = \frac{2}{T}\int_{t_0}^{t_0+T} f(t)\cos(n\omega t)\,dt \tag{3.84}$$

$$b_n = \frac{2}{T}\int_{t_0}^{t_0+T} f(t)\sin(n\omega t)\,dt \tag{3.85}$$

for most applications, this simplified Fourier series is sufficient.

Example 3.8 Calculate the Fourier series coefficients for the signal in Fig. 3.13.

Solution

Choosing $t_0 = 0$, the expression for $f(t)$ in the integration time interval of $0 \le t \le T$ is

$$f(t) = \frac{V_m}{T}t. \tag{3.86}$$

The Fourier series coefficients C_n where $n \neq 0$ can be found as (remember that $\omega = \frac{2\pi}{T}$)

$$\begin{aligned}
C_n &= \frac{1}{T}\int_0^T \frac{V_m}{T} t e^{-jn\omega t}\,dt \\
&= \frac{V_m}{T^2}\int_0^T t e^{-jn\omega t}\,dt \\
&= \frac{V_m}{T^2}\left.\frac{e^{-jn\omega t}(jn\omega t + 1)}{n^2\omega^2}\right|_0^T \\
&= \frac{V_m}{T^2}\left(\frac{e^{-jn2\pi}(jn2\pi + 1)}{n^2\omega^2} - \frac{e^0(0+1)}{n^2\omega^2}\right)
\end{aligned}$$

$$= \frac{V_m}{T^2 n^2 \omega^2}(j2\pi n + 1 - 1)$$
$$= \frac{jV_m}{2n\pi}.$$

The average term or C_0 can be found as

$$C_0 = \frac{1}{T}\int_0^T \frac{V_m}{T} t dt = \frac{V_m}{2}.$$

The signal can therefore be rewritten as

$$f(t) = \sum_{n=-\infty}^{\infty} C_n e^{jn\omega t}.$$

where

$$C_n = \begin{cases} \frac{V_m}{2} & \text{for } n = 0 \\ \frac{jV_m}{2n\pi} & \text{otherwise.} \end{cases}$$

From inspection of the expression of C_n, the Fourier series can also be written in an alternate form as

$$f(t) = \frac{V_m}{2} - \frac{V_m}{\pi}\sum_{n=1}^{\infty}\frac{1}{n}\sin(n\omega t).$$

You can verify this by calculating a_v, a_n and b_n using (3.83)–(3.85).

3.11 HARMONICS IN POWER SYSTEMS

Although the ideal in all power systems is that both the voltage and current should be sinusoidal and free of harmonics, the modern power grid is full of harmonics. In a primitive power grid where all the loads are linear (i.e. a combination of resistors, capacitors and inductors), the current will be free of any harmonics as long as the voltage is harmonic free. However, non-ideal characteristics of power system equipment, such as generators and transformers, as well as the introduction of more sophisticated loads, such as computers and other electronic loads, introduce harmonics into the modern day grid.

Because any periodic voltage signal can be represented by the Fourier series as

$$v(t) = a_v + \sum_{n=1}^{\infty}(a_n \cos(n\omega t) + b_n \sin(n\omega t)) \tag{3.87}$$

we will begin from this point. For a start, let us ignore the possibility of a voltage signal with a dc offset. This leaves the possible distorted voltage waveform

$$v(t) = \sum_{n=1}^{\infty}(a_n \cos(n\omega t) + b_n \sin(n\omega t)). \tag{3.88}$$

However, the linear sum of a cosine and a sine signal is difficult to work with, so let us use the following mathematical trick shot. Let us define two new variables V_n and θ_n such that

$$a_n = V_n \cos\theta_n \tag{3.89}$$

$$b_n = V_n \sin\theta_n. \tag{3.90}$$

We can find the value of V_n by squaring both equations and adding them together

$$\begin{aligned} a_n^2 + b_n^2 &= V_n^2\cos^2\theta_n + V_n^2\sin^2\theta_n \\ &= V_n^2 \\ V_n &= \sqrt{a_n^2 + b_n^2} \end{aligned} \tag{3.91}$$

as $\cos^2\alpha + \sin^2\alpha = 1$. We can find θ_n by dividing the equations:

$$\begin{aligned} \frac{a_n}{b_n} &= \frac{V_n \cos\theta_v}{V_n \sin\theta_n} \\ \theta_n &= \arctan\frac{b_n}{a_n}. \end{aligned} \tag{3.92}$$

We can now use these new variables to rewrite (3.88) as

$$v(t) = \underbrace{V_1 \cos(\omega t + \theta_1)}_{\text{undistorted fundamental}} + \underbrace{\sum_{n=2}^{\infty} V_n \cos(n\omega t + \theta_n)}_{\text{distortion}}. \tag{3.93}$$

3.12 POWER AND NON-SINUSOIDAL WAVEFORMS

In our discussion of the Fourier series, we did not discuss the fundamental mathematical basis of the Fourier representation. Although this discussion falls outside of the scope of this book, one prerequisite of the discussion will be mentioned: the requirement that the describing functions are orthogonal. This requirement is simply the same as saying that in a Cartesian system (i.e. the xy plane), the x-axis must be orthogonal to the y-axis to allow us to describe any point in the plane using the x- and y-values. Although it is possible to uniquely describe any point in the plane with an x-axis and a y-axis that are not orthogonal, describing a locus in such a plane becomes tedious as any movement in the x-axis will imply movement in the y-axis. Therefore, to describe the movement as the sum of the x-axis movement and the y-axis movement, we have the requirement that the two axes must be orthogonal. We have the same requirement with the Fourier series. Two functions, $f(x)$ and $g(x)$ are orthogonal over an interval T if and only if

$$\int_T f(x)g(x)\,dx = 0.$$

For the Fourier series, the requirement that the functions are orthogonal can now be investigated. Using (3.93) as basis, the requirement boils down to the need that for $n \neq m$,

$$\int_{-\frac{T}{2}}^{\frac{T}{2}} V_n \cos(n\omega t + \theta_n) V_m \cos(m\omega t + \theta_m)\,dt = 0 \tag{3.94}$$

remembering that $\omega = \frac{2\pi}{T}$. Luckily, this is easy to prove for the case where the different frequencies are integer multiples of each other (which is always the case for the Fourier series). Use the trigonometric identity

$$\cos\alpha\cos\beta = \frac{\cos(\alpha - \beta) + \cos(\alpha + \beta)}{2}$$

to write

$$\cos(n\omega t + \theta_n)\cos(m\omega t + \theta_m) \\ = \frac{\cos(\omega(n+m)t + \theta_n + \theta_n) + \cos(\omega(n-m)t + \theta_n - \theta_n)}{2}. \tag{3.95}$$

It is easy to prove that for all $m, n \in \mathbb{Z}$

$$\int_{-\frac{T}{2}}^{\frac{T}{2}} \cos(\omega(n+m)t + \theta_n + \theta_n)\,dt = 0 \tag{3.96}$$

and

$$\int_{-\frac{T}{2}}^{\frac{T}{2}} \cos(\omega(n-m)t + \theta_n - \theta_n)\,dt = 0 \tag{3.97}$$

when $m \neq n$.

Now, although this characteristic of sinusoidal signals is one of the fundamentals of Fourier analysis, it has an important consequence in non-sinusoidal power

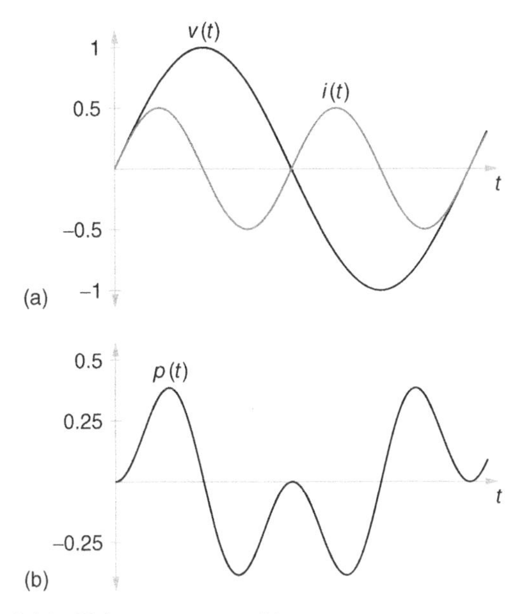

Figure 3.14 Voltage, current and instantaneous power waveforms.

systems. The average power delivered by different harmonics is zero. Take the example shown in Fig. 3.14 where

$$v(t) = 1\sin(\omega t)$$
$$i(t) = \frac{1}{2}\sin(2\omega t)$$

It is easy to show that the instantaneous power is

$$p(t) = \frac{\cos(\omega t) - \cos(3\omega t)}{4}.$$

From this equation and Fig. 3.14, we can see that the average power is zero.

Given this result, let us write a general equation for the power contained in non-sinusoidal systems. Using (3.93) as the basis, let the voltage and current be given as

$$v(t) = V_{dc} + \sum_{n=1}^{\infty} V_n \cos(n\omega t + \theta_n) \tag{3.98}$$

$$i(t) = I_{dc} + \sum_{n=1}^{\infty} I_n \cos(n\omega t + \varphi_n). \tag{3.99}$$

Now, as the average power can be found using

$$P = \frac{1}{T}\int_0^T v(t)i(t)\,dt$$

we can write the power in our non-sinusoidal system as

$$P = \frac{1}{T}\int_0^T \left(V_{\mathrm{dc}} + \sum_{n=1}^{\infty} V_n \cos(n\omega t + \theta_n)\right)\left(I_{\mathrm{dc}} + \sum_{n=1}^{\infty} I_n \cos(n\omega t + \varphi_n)\right) dt. \tag{3.100}$$

To evaluate this integral, we must multiply out the infinite series. However, this task becomes easier when we use the result of (3.94), which can be written as

$$\frac{1}{T}\int_0^T V_n I_m \cos(n\omega t + \theta_n)\cos(m\omega t + \varphi_m)\; dt = \begin{cases} 0 & \text{if } n \neq m \\ \dfrac{V_n I_n}{2}\cos(\theta_n - \varphi_n) & \text{if } n = m. \end{cases} \tag{3.101}$$

The average power can therefore be written as

$$\boxed{P = V_{\mathrm{dc}} I_{\mathrm{dc}} + \sum_{n=1}^{\infty} \frac{V_n I_n}{2}\cos(\theta_n - \varphi_n).} \tag{3.102}$$

We see therefore that average power is only transmitted when the Fourier series of the current and voltage contain terms of the same frequency.

Example 3.9 Consider a system where the voltage and current are

$$v(t) = 1\sin(\omega t + 45^\circ)\ \mathrm{V}$$

$$i(t) = \frac{1}{2}\sin(\omega t) + \frac{1}{5}\sin(4\omega t)\ \mathrm{A}$$

What is the instantaneous and average power?

Solution

Using the trigonometric identity that

$$\sin\alpha\sin\beta = \frac{\cos(\alpha - \beta) - \cos(\alpha + \beta)}{2}$$

the instantaneous power can be written as

$$p(t) = \frac{1}{4}\cos(45^\circ) - \frac{1}{4}\cos(2\omega t + 45^\circ) + \frac{1}{10}\cos(3\omega t - 45^\circ) - \frac{1}{10}\cos(5\omega t + 45^\circ)\ \text{W}.$$

We see from this equation that the average power is

$$P = \frac{1}{4}\cos(45^\circ) = 176.78\ \text{mW}.$$

It is also possible to get this same result using (3.102). The voltage, current, instantaneous power and average power waveforms are shown in Fig. 3.15.

3.13 EFFECTIVE OR RMS VALUE OF NON-SINUSOIDAL WAVEFORMS

Recall from Section 3.4 that the effective, or rms, value of a voltage waveform is defined as in (3.15), repeated for convenience as

$$V_{\text{rms}} = \sqrt{\frac{1}{T}\int_0^T v^2(t)\,dt}.$$

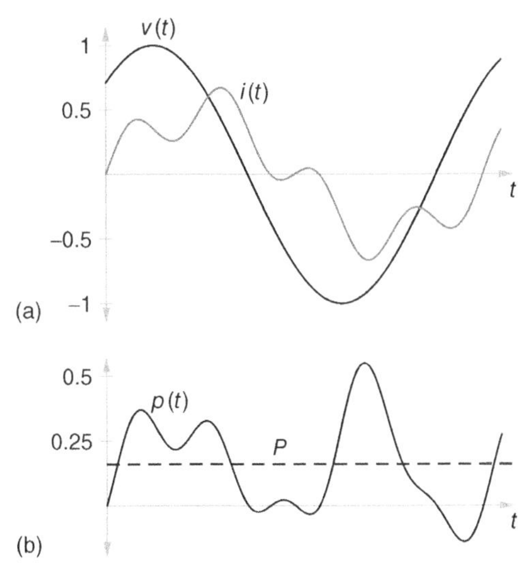

Figure 3.15 Example of power in harmonic systems. Voltage, current and instantaneous power waveforms.

The rms value of a non-sinusoidal waveform can be found by substituting (3.93) into (3.15)

$$V_{\text{rms}} = \sqrt{\frac{1}{T}\int_0^T \left\{ V_{\text{dc}}^2 + \left(\sum_{n=1}^{\infty} V_n \cos(n\omega t + \theta_n) \right)^2 \right\} dt}. \tag{3.103}$$

At first, this equation might seem intimidating; however, it can be simplified by using (3.101) and rewritten as

$$\boxed{V_{\text{rms}} = \sqrt{V_{\text{dc}}^2 + \sum_{n=1}^{\infty} \frac{V_n^2}{2}}.} \tag{3.104}$$

3.14 POWER FACTOR OF NON-SINUSOIDAL WAVEFORMS

In Section 3.7, we saw that it is advantageous to be able to express the average power in terms of the rms values of the current and the voltage. This statement is especially true as the rms values can be measured easily using standard equipment. In sinusoidal systems, the average power can be found as

$$P = (VI)\text{pf}, \tag{3.105}$$

where pf is the power factor.

Let us try to get the same definition for non-sinusoidal systems. *But first, let us assume that the voltage is not distorted.* Although it is possible to describe systems where both the current and voltage are distorted, it is mathematically rather complex. In all discussions, we will therefore assume that the voltage source is undistorted. That said, this assumption does not imply that all voltages will be undistorted but we will assume all voltage *sources* to be undistorted.

Let us take the following example, assuming that a distorted current is flowing through a resistor with resistance R. We can write

$$i(t) = I_{\text{dc}} + \sum_{n=1}^{\infty} I_n \cos(n\omega t + \varphi_n) \tag{3.106}$$

$$v(t) = I_{\text{dc}}R + \sum_{n=1}^{\infty} I_n R \cos(n\omega t + \theta_n). \tag{3.107}$$

As it is a resistive circuit, the current and voltage are in phase not only at the fundamental frequency but at all frequencies. Therefore, for all harmonics the term

$$\cos(\theta - \varphi) = 1. \tag{3.108}$$

Using (3.102), the average power can be calculated by using (3.104)

$$\begin{aligned} P &= V_{dc}I_{dc} + \sum_{n=1}^{\infty} \frac{V_n I_n}{2} \cos(\theta_n - \varphi_n) \\ &= I_{dc}^2 R + \sum_{n=1}^{\infty} \frac{I_n^2 R}{2} \\ &= I_{rms}^2 R. \end{aligned} \tag{3.109}$$

In this case, we see that the power factor of the circuit is unity and

$$P = VI. \tag{3.110}$$

This makes sense because we know that the power factor of a resistive circuit is always unity. This is also true if the current through (or equivalently the voltage across) the resistor is non-sinusoidal. This is because owing to the resistive nature, all harmonics in the current will also be present in the voltage across the resistive circuit.

But let us generalise the circuit a bit further to a circuit excited by an undistorted sinusoidal voltage source that draws a distorted current from the source. Therefore, using rms values,

$$v(t) = \sqrt{2}V \cos(\omega t + \theta) \tag{3.111}$$

$$i(t) = I_{dc} + \sum_{n=1}^{\infty} \sqrt{2} I_n \cos(n\omega t + \varphi_n). \tag{3.112}$$

We know that the current and voltage must have the same frequency for any real power to be dissipated. Therefore, only the harmonic components with the same frequency are of interest and the average power can be expressed as

$$P = VI_1 \cos(\theta - \varphi_1). \tag{3.113}$$

However, the rms current is, from (3.104),

$$I = \sqrt{I_{dc}^2 + \sum_{n=1}^{\infty} I_n^2} \tag{3.114}$$

and, of course, the rms value of the voltage is simply V. Therefore, if we use (3.105) as the basis, we can express the power factor as

$$\begin{aligned} \text{pf} &= \frac{P}{VI_1} \\ &= \frac{VI_1 \cos(\theta - \varphi_1)}{V\sqrt{I_{\text{dc}}^2 + \sum_{n=1}^{\infty} I_n^2}} \end{aligned} \quad (3.115)$$

Rewriting this yields

$$\text{pf} = \underbrace{\frac{I_1}{\sqrt{I_{\text{dc}}^2 + \sum_{n=1}^{\infty} I_n^2}}}_{\text{distortion factor}} \underbrace{\cos(\theta - \varphi_1)}_{\text{displacement factor}} . \quad (3.116)$$

With this result, we introduce two further definitions. In non-sinusoidal systems, the distortion factor is defined as

$$(\text{Distortion factor}) = \frac{I_1}{\sqrt{I_{\text{dc}}^2 + \sum_{n=1}^{\infty} I_n^2}} = \frac{\text{rms of fundamental component}}{\text{rms of total waveform}} \quad (3.117)$$

and the definition of the power factor in non-sinusoidal systems is expressed as

$$\text{pf} = (\text{displacement factor})(\text{distortion factor}). \quad (3.118)$$

It is true that we did not change the original definition of the power factor as the distortion factor in sinusoidal systems without harmonic distortion is equal to 1.

Sometimes it is beneficial to speak of the amount of distortion in a signal. Let us define a new term the *total harmonic distortion* or THD which is defined as the ratio of the rms value of the distortion present in the waveform to the rms value of the fundamental. Mathematically, this can be expressed, for a current, as

$$\text{THD} = \frac{\sqrt{\sum_{n=2}^{\infty} I_n^2}}{I_1}. \quad (3.119)$$

This means that in systems where the voltage source is undistorted, the distortion factor and the THD are closely related. From (3.117) and (3.119)

$$\boxed{(\text{Distortion factor}) = \frac{1}{\sqrt{1+\text{THD}^2}}.} \tag{3.120}$$

Example 3.10 Calculate the power factor, displacement factor, distortion factor and the THD for the circuit parameters in Example 3.9.

Solution

The rms value of the voltage is $V = \frac{1}{\sqrt{2}}$ and the rms value of the current is

$$I = \sqrt{\left(\frac{1}{\sqrt{2}} \cdot \frac{1}{2}\right)^2 + \left(\frac{1}{\sqrt{2}} \cdot \frac{1}{5}\right)^2} = 0.38 \text{ A}.$$

The displacement factor is $\cos(45° - 0°) = \frac{\sqrt{2}}{2}$. The distortion factor is

$$(\text{Distortion factor}) = \frac{\frac{1}{\sqrt{2}} \cdot \frac{1}{2}}{\sqrt{\frac{29}{200}}} = 0.9285$$

and the power factor is therefore

$$\text{pf} = \frac{\sqrt{2}}{2} \times 0.9285 = 0.6565.$$

The THD of the current is

$$\text{THD} = \frac{\frac{1}{5}}{\frac{1}{2}} = \frac{2}{5} = 0.4.$$

Naturally, the average power can also be calculated as

$$\begin{aligned} P &= (VI)\text{pf} \\ &= \frac{1}{\sqrt{2}} \cdot 0.38 \cdot 0.6565 \\ &= 176.78 \text{ mW} \end{aligned}$$

which is the same result as in Example 3.9.

3.15 HARMONICS IN POWER SYSTEMS

Most of the equipment connected to the electrical distribution system will either malfunction or suffer damage if the electrical grid is polluted with harmonics. This damage is mainly caused by increased losses due to the harmonic content. However, harmonics can cause damage through other means than just increased losses. A good example is that of equipment with a capacitor in the input port, or systems equipped with power factor-correcting capacitors. The impedance of a capacitor decreases with frequency. To see how this is influenced by the presence of harmonics, let us consider a PFC (power factor-correcting) capacitor connected to the grid. When the source is undistorted, the capacitor draws

$$I = V\omega C. \tag{3.121}$$

However, if there is an 8% third harmonic and a 4% ninth harmonic distortion, which does not sound like much, the current will increase to an rms value, using (3.104), of

$$\begin{aligned} I &= \sqrt{V^2\omega^2C^2 + (0.08V)^2 3^2\omega^2C^2 + (0.04V)^2 9^2\omega^2C^2} \\ &= V\omega C\sqrt{1 + 0.0579 + 0.1296} \\ &= 1.123\ V\omega C \end{aligned} \tag{3.122}$$

which is 12.3% larger than without the distortion. If the capacitor was selected to operate safely with the assumption that the grid voltage is free of harmonics, this increase in current might be enough to cause failure.

To limit the harmonic content in the grid, the international community has adopted several measures, mostly expressed as standards of good conduct that are enforced at some levels. This method of using legalisation takes two approaches. Firstly, it places limits on the voltage distortion of all sources. This is important as no system can be free of current harmonics if the voltage contains harmonics. Secondly, the user is limited in terms of the current harmonics that can be generated (or drawn from the grid, depending on your perspective). That said however, it is very difficult to monitor the amount of voltage distortion as the system must be completely free of current harmonics before an accurate measurement can be made. For the sake of simplicity, we will assume that the voltage is always free of harmonics.

It is however easier to enforce the current harmonic limits imposed on consumers. For example, in Europe, all electrical equipment that is for sale must bear the CE mark; this mark is affixed to prove that the equipment adheres to the applicable European Community directives. The letters CE stand for ‘Conformité Européene’ translated as European Conformity. For electrical equipment, it means that, among other requirements, it adheres to European standards EN61000-x that impose limits on the harmonic content. As an example, the EN61000-3-2 standard is applicable to equipment drawing less than 16 A per phase and EN61000-3-12 for equipment drawing more than 16 A but less than 75 A per phase. Therefore,

TABLE 3.1 Harmonic Limits Imposed by EN61000-3-2 on Home Audio Equipment

Harmonic Number	Maximum Permissible Harmonic Current (A)	Harmonic Number	Maximum Permissible Harmonic Current (A)
Odd Harmonics		Even Harmonics	
3	2.3	2	1.08
5	1.4	4	0.43
7	0.77	6	0.30
9	0.40	$8 \leq n \leq 40$	$\frac{8(0.23)}{n}$
11	0.33		
13	0.21		
$15 \leq n \leq 40$	$\frac{8(0.15)}{n}$		

the requirement that all equipment sold within the union carries the CE mark has the implication that the current harmonic content is controlled without in fact controlling the current drawn at every household. For example, the limits imposed on household audio equipment and classified in EN61000-3-2 as Class-A equipment are tabulated in Table 3.1. The standards do not allow a dc offset.

Example 3.11 In the past, a cheap method of converting ac to dc was to use only a diode bridge rectifier; this method is called a *peak detection rectifier*. However, although cheap and simple, this rectifier is notorious for generating harmonics. The harmonic distortion of this type of rectifier is shown in Fig. 3.16. Assume that the

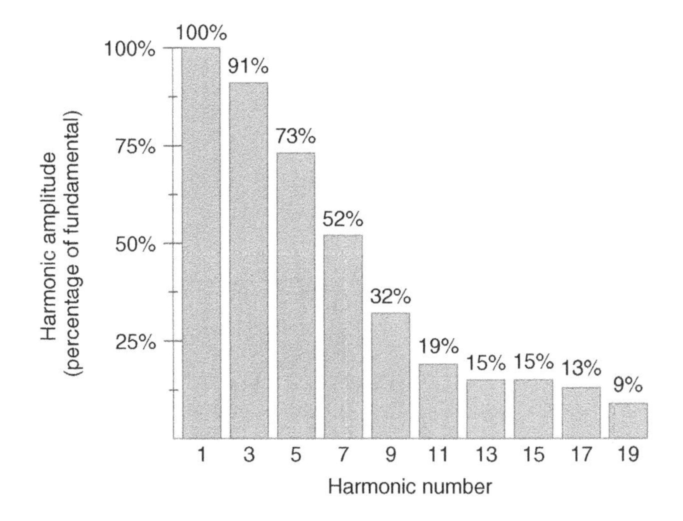

Figure 3.16 Typical harmonic spectrum of a peak detection rectifier.

displacement factor is zero. Calculate the THD and distortion factor of this load type. Also calculate what the maximum power rating would be of a device using this rectifier that will still conform to the limits of EN61000-3-2 as tabulated in Table 3.1.

Solution

Using (3.117), the distortion factor is found as

$$(\text{Distortion factor}) = \frac{1}{\sqrt{2.84}} = 59.3\%.$$

The THD can be found using either (3.119) or (3.120). The THD is calculated as

$$\text{THD} = \frac{\sqrt{1.84}}{1} = 136\%.$$

The maximum power that can be taken from the grid with a device using this type of rectifier can be found by comparing Table 3.1 and Fig. 3.16. If the device is taking 2.3 A of the third harmonic current, then the fifth harmonic current will be (2.3)(0.73)/0.91 = 1.85 A which is clearly above the limits specified in Table 3.1. In fact, when the harmonic limits are compared to the harmonics contained in the peak detection rectifier input current, we see that the fifteenth harmonic is the limiting factor. According to EN61000-3-2, the fifteenth harmonic current must be smaller than 80 mA. When the amplitude of the fifteenth harmonic current of the peak detection rectifier is set to this value, all the other harmonics are within the limits specified by EN61000-3-2. This is shown in Fig. 3.17.

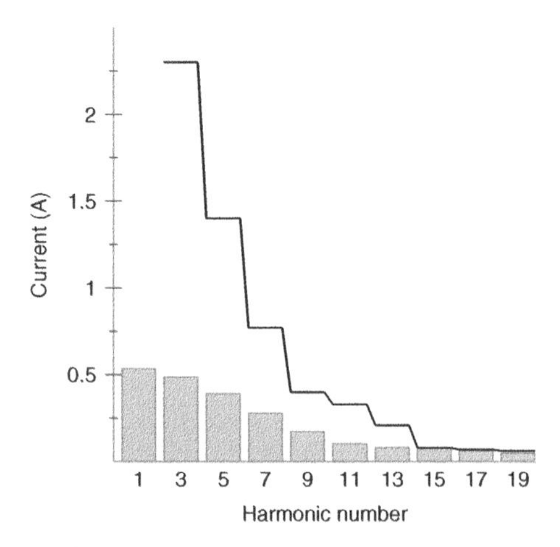

Figure 3.17 Normalised harmonic spectrum of the peak detection rectifier and the EN61000-3-2 limits.

The maximum fundamental component is therefore 0.08/0.15 = 0.53 A, much less than the possible 16 A of the supply. The maximum power that such a device can therefore draw from the supply, assuming a 230 V supply and no displacement factor, is 123 W. This is indeed bad news for anyone hoping to sell 600 W stereo audio amplifiers while using a peak detection rectifier.

3.16 THREE-PHASE SYSTEMS

When it comes to power delivery on a large scale, we see something interesting. You probably are aware that all power lines consist of three wires. A photograph of a steel lattice power line pylon that forms part of a power line close to Madrid, Spain is shown in Fig. 3.18. Now, if we need two wires to close an electrical circuit what is the third wire doing? This is an interesting problem.

Most electrical power is used, generated and transmitted using a three-phase system. A three-phase system consists of three periodic voltage sources that are phase shifted with respect to one another in such a manner that the phase shift between each phase and its neighbours is the same. Although the world did initially flirt with the idea of a two-phase system, the three-phase system introduced by Nicola Tesla in 1887 is the polyphase system in almost exclusive use today. The main advantages of three-phase systems are

1. Less conductor volume is needed to transmit the same amount of power over a distance, with a fixed transmission efficiency, when three phases are used rather than a single phase. The difference is typically in the region of 25%.
2. As the transmission lines are lighter, considerable savings are realised in the construction of the transmission lines because the supporting structures can be smaller and further apart.

Figure 3.18 A pylon carrying two three-phase power lines.

3. The power delivery with the three-phase system is completely smooth, while the power arrives in 'humps' when a single phase is used. This pulsating nature of power delivery in single-phase systems causes noise and accelerates equipment failure.
4. A three-phase electrical motor can be self-starting as one can develop a circulating magnetic flux with more than one phase.

Let us generate three voltages with an equal phase shift between them. As there are 2π radians in a full circle, to have an equal phase shift between all three voltages, we will need a $\frac{2}{3}\pi$-radian phase shift between each of the voltages. Therefore let

$$v_a(t) = V_m \cos(2\pi f t) \tag{3.123}$$

$$v_b(t) = V_m \cos\left(2\pi f t + \frac{2}{3}\pi\right) \tag{3.124}$$

$$v_c(t) = V_m \cos\left(2\pi f t - \frac{2}{3}\pi\right). \tag{3.125}$$

The three voltages are shown, for $V_m = 1$, in the time domain in Fig. 3.19. We can naturally also represent the three voltages in the frequency domain using the phasor. As we know the frequency, we can write the three voltages as

$$\mathbf{V_a} = V\angle 0° \tag{3.126}$$

$$\mathbf{V_b} = V\angle 120° \tag{3.127}$$

$$\mathbf{V_c} = V\angle -120°. \tag{3.128}$$

Here, we have chosen to use degrees to indicate the phase shift. Although it is also possible to use radians, it is not quite as easy to visualise as by using degrees. Note that according to our definition of the phasor, we are using the rms value and therefore $V = \frac{1}{\sqrt{2}}V_m$. The three phasors of a three-phase system are shown in Fig. 3.20.

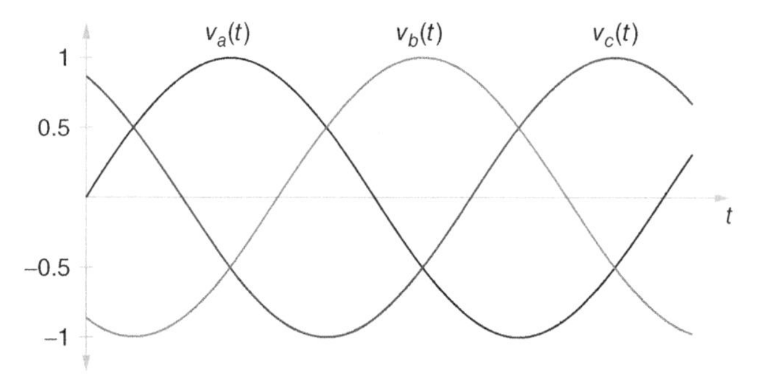

Figure 3.19 Three-phase voltages plotted in the time domain.

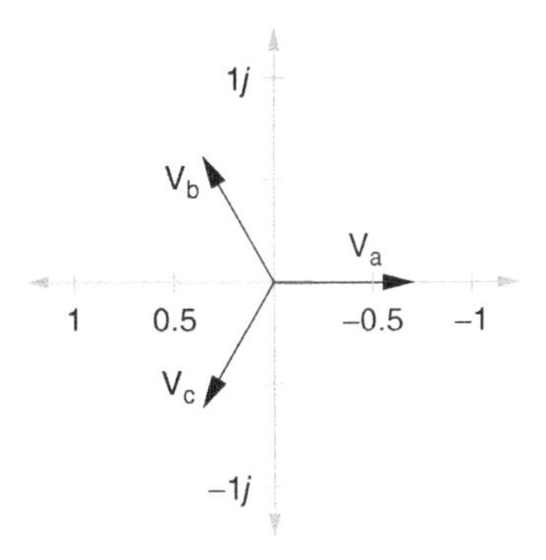

Figure 3.20 Three-phase voltages plotted as phasors in the frequency domain.

An interesting point to be noted is that for all time, the algebraic sum of the three voltages is equal to zero. It is not particularly easy to prove this statement in the time domain (although it can be done using trigonometric identities) but by using phasor notation, it is easy to show that

$$V\angle 0^\circ + V\angle 120^\circ + V\angle - 120^\circ = 0. \tag{3.129}$$

This result leads us to the simplest form of three-phase circuits, the balanced three-phase system.

A three-phase system is balanced when all of the following conditions are met:

1. The magnitude of the voltage in all three phases is the same.
2. The phase shift between all three phases is the same.
3. The magnitude of the current magnitude in all three phases is the same.
4. The loads connected to each of the phases are identical.

Although we have mentioned four conditions, we only need three of the four conditions as that will imply that the fourth condition is met. You can verify this statement for yourself.

3.17 HARMONICS IN BALANCED THREE-PHASE SYSTEMS

Let us only consider balanced three-phase systems. It is well known that when there are no harmonics present in the system that a three-phase four-wire system, with a wye-connected load as shown in Fig. 3.21 does not have any current flowing in the neutral connector. This can easily be proven as

$$I\angle 0^\circ + I\angle 120^\circ + I\angle - 120^\circ = 0. \tag{3.130}$$

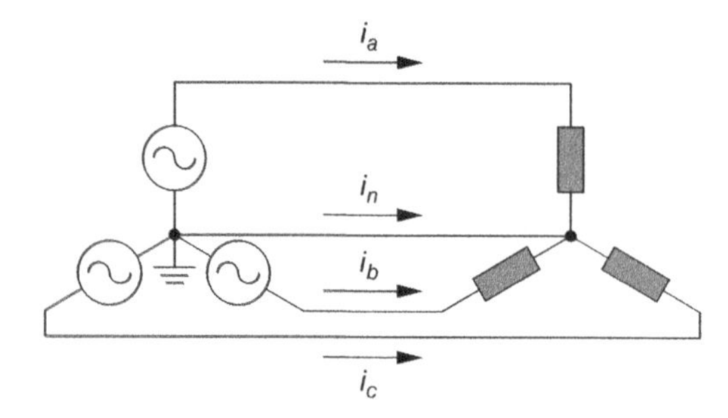

Figure 3.21 Three-phase four-wire system with wye-connected load.

However, this changes when current harmonics are present. Consider the fact that

$$I\cos(n\omega t) + I\cos\left(n\left(\omega t + \frac{2\pi}{3}\right)\right) + I\cos\left(n\left(\omega t - \frac{2\pi}{3}\right)\right)$$
$$= \begin{cases} 0 & \text{if } n \neq 3\mathbb{Z} \\ 3I\cos(n\omega t) & \text{if } n = 3\mathbb{Z}. \end{cases} \tag{3.131}$$

We see that the neutral wire carries all the triplen harmonic currents as the third harmonics add up in the neutral conductor. The rms current in the neutral conductor can be calculated, where I_n is the rms value of the line current at harmonic n, as

$$\boxed{I_n = 3\sqrt{\sum_{n=3,6,9}^{\infty} I_n^2}.} \tag{3.132}$$

Naturally, in a three-wire three-phase system, there is no neutral wire and no neutral current. Such a circuit is shown in Fig. 3.22. That said, in all three-wire three-phase circuits, the star point voltage is of concern. In unbalanced systems, the

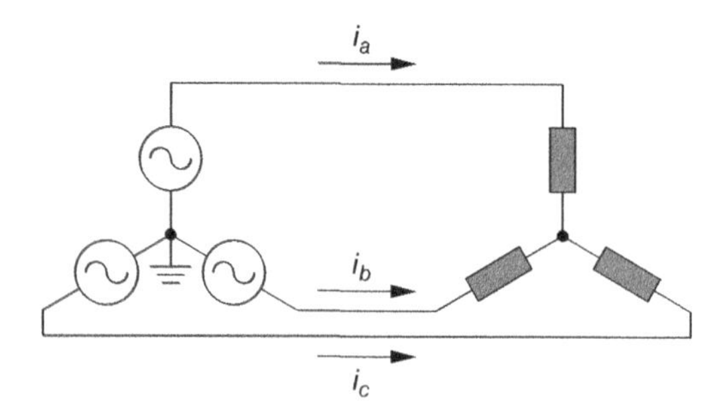

Figure 3.22 Three-phase three-wire system with wye-connected load.

star point voltage is no longer equal to the neutral point of the generator (ground in Fig. 3.22) but the unequal current flow through the three phases causes a voltage shift. In balanced systems with harmonic distortion, we see the same result. The fact that there is no neutral wire combined with the fact that the only path for triplen harmonic currents to flow is the neutral wire implies that there are no triplen harmonic currents present in the system. The only way in which the triplen harmonic currents are eliminated from the line currents is if a voltage is induced on the star point voltage to cancel the triplen harmonic currents. Therefore, although we do not see triplen harmonic currents in the system, we get another unwanted characteristic.

In a three-phase three-wire system with a delta-connected load, as shown in Fig. 3.23, there is also no neutral connection, so the line currents do not contain any triplen harmonics. However, the loads are connected line to line and the voltages across them are therefore undistorted. As the third harmonics are generated and cannot flow in the line currents, they circulate through the delta connection.

3.18 HIGHLIGHTS

- Using the passive sign convention, positive power is defined as power dissipated, while negative power implies power supplied to the system.
- The effective, or rms value, of a periodic signal $v(t)$ is calculated as

$$V_{\text{eff}} = \sqrt{\frac{1}{T}\int_{t_0}^{t_0+T} v^2(t)\ dt}.$$

In the case of single-phase ac circuits, it becomes

$$V_{\text{eff}} = \frac{V_m}{\sqrt{2}}.$$

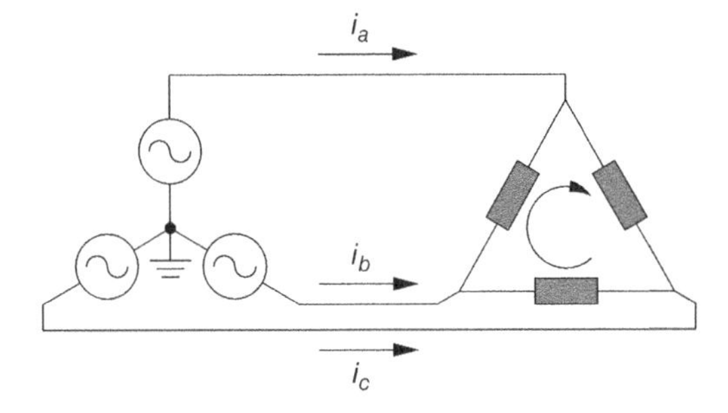

Figure 3.23 Three-phase three-wire system with delta-connected load.

- In ac circuits, we use the phasor notation to represent sinusoidal signals. We defined the phasor to use the rms value, therefore

$$V\angle\theta = \sqrt{2}V\cos(\omega t + \theta).$$

- In single-phase ac circuits, the power delivery to a load is not constant, but the average power can be calculated as

$$P = VI\cos(\theta_v - \theta_i),$$

where V and I are the rms values. This can be rewritten as

$$P = V, I\text{pf where pf} = \cos(\theta_v - \theta_i).$$

- The complex power is defined as

$$\mathbf{S} = \mathbf{VI}^*$$

and can be rewritten as

$$\mathbf{S} = P + jQ,$$

where P is the average power and Q is the reactive power.
- Non-sinusoidal periodic signals can be represented using the Fourier series and expressed by

$$f(t) = \sum_{n=-\infty}^{\infty} C_n e^{jn\omega t},$$

where

$$C_n = \frac{1}{T}\int_{t_0}^{t_0+T} f(t)e^{-jn\omega t}\, dt.$$

- The Fourier series can also be rewritten as

$$f(t) = a_v + \sum_{n=1}^{\infty}(a_n\cos(n\omega t) + b_n\sin(\omega t)),$$

where

$$a_v = \frac{1}{T}\int_{t_0}^{t_0+T} f(t)\, dt$$

$$a_n = \frac{2}{T}\int_{t_0}^{t_0+T} f(t)\cos(n\omega t)\, dt$$

$$b_n = \frac{2}{T}\int_{t_0}^{t_0+T} f(t)\sin(n\omega t)\, dt.$$

- The average power of sinusoidal voltage and sinusoidal current is zero if the frequencies of the two sinusoids are unequal, or, if the frequencies are equal but the phase shift between the signals is equal to 90°. Therefore, the average power can be written as

$$P = V_{dc} I_{dc} + \sum_{n=1}^{\infty} V_n I_n \cos(\theta_n - \varphi_n),$$

where V_n and I_n are the rms values of the frequency components.

- The rms value of a non-sinusoidal signal is

$$V_{rms} = \sqrt{V_{dc}^2 + \sum_{n=1}^{\infty} V_n^2}.$$

- Harmonics are undesired in an ac power system and regarded as a distortion:

$$v(t) = \underbrace{V_1 \cos(\omega t + \theta_1)}_{\text{undistorted fundamental}} + \underbrace{\sum_{n=2}^{\infty} V_n \cos(n\omega t + \theta_n)}_{\text{distortion}}.$$

- In circuits with sinusoidal voltages but non-sinusoidal currents, the power factor is defined as

$$\text{pf} = (\text{displacement factor})(\text{distortion factor}),$$

where the displacement factor is $\cos(\theta_v - \theta_i)$ and the distortion factor is

$$(\text{Distortion factor}) = \frac{I_1}{\sqrt{I_{dc}^2 + \sum_{n=1}^{\infty} I_n^2}} = \frac{\text{rms of fundamental component}}{\text{rms of total waveform}}.$$

- Another method of expressing the amount of distortion is by using the THD defined as

$$\text{THD} = \frac{\sqrt{\sum_{n=2}^{\infty} I_n^2}}{I_1}.$$

- The distortion factor and the THD are closely related.

$$\text{pf} = \frac{1}{\sqrt{1 + \text{THD}^2}}.$$

- Harmonics in three-phase systems are particularly unwanted because, in wye-connected systems, the third harmonics (i.e. harmonics at frequency multiples) add up and therefore the neutral conductor must carry a large current.

PROBLEMS

3.1 What is the meaning of rms voltage and current? Should we interpret it differently for dc, sinusoidal and non-sinusoidal ac?

3.2 Which frequency components contribute to real power when non-sinusoidal voltages and currents are applied to an electrical load? (A Fourier analysis of the voltage and current waveforms that are applied to a load results in dc, fundamental ac and harmonic ac components.)

3.3 Why must energy delivery equipment such as the power line be rated for apparent power?

3.4 Some power systems use 50 or 60 Hz ac, while other systems use dc. Can we apply the definitions of power factor and distortion factor to dc power?

3.5 Why must energy delivery equipment such as power lines and transformers be rated for the apparent power?

3.6 Calculate the resistor current, resistor voltage, source apparent power, source reactive power, source average power and the source power factor for each of the circuits shown in Fig. 3.24. Take the frequency of the source as 50 Hz.

1. Why is the power in (a), (b), (d) and (f) equal?
2. Why is the power in (c) and (e) equal?
3. Compare the reactive power in (b), (c), (d) and (e). What can you conclude?
4. What is happening in (f)?

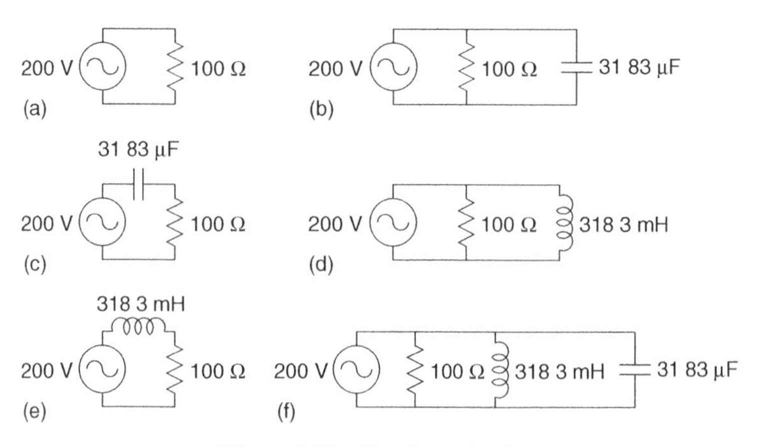

Figure 3.24 Simple ac circuits.

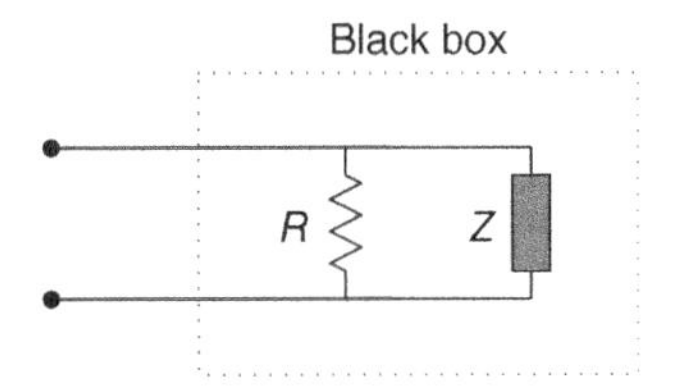

Figure 3.25 Black box.

3.7 You have a black box, as indicated in Fig. 3.25. When the box is connected to a 120 V, 60 Hz supply, it consumes 1 kVA at a power factor of 0.78 lagging. Assume that the elements inside the box can be approximated as in Fig. 3.25. Z can be either a capacitor or an inductor. Can you calculate R and Z? Is Z an inductor or a capacitor?

3.8 A 100 Ω resistor is connected to a 230 V 50 Hz single-phase source. What are the average power and peak instantaneous power? Three 100 Ω resistors are connected to the three phases of a 230 V 50 Hz voltage source. What is the average and the peak power to the resistors collectively?

3.9 The power of the heating element of a warm water geyser is 4 kW. If the specific heat capacity of water is 4.1813 J/(g K), then calculate the time it takes to heat 100 l of water from 20° to 60°C.

3.10 Consider the following voltage and current: $v(t) = 300\cos(\omega t)V$ and $i(t) = 8\sin(\omega t - 30) + 2\sin(5\omega t)$, $\omega = 1000$. Calculate

1. the rms current,
2. the average power,
3. the power factor.

3.11 Consider the circuit in Fig. 3.26. Calculate the apparent power delivered by the source as well as the active and reactive power in all the elements at a 50 Hz frequency. Do the apparent power vectors add up to give the same value as at the input? Explain.

3.12 An industrial process operates from a 220 V supply and consumes 3 kW at a power factor of 0.6 lagging. How much capacitance must be added to improve the power factor to 0.88 lagging?

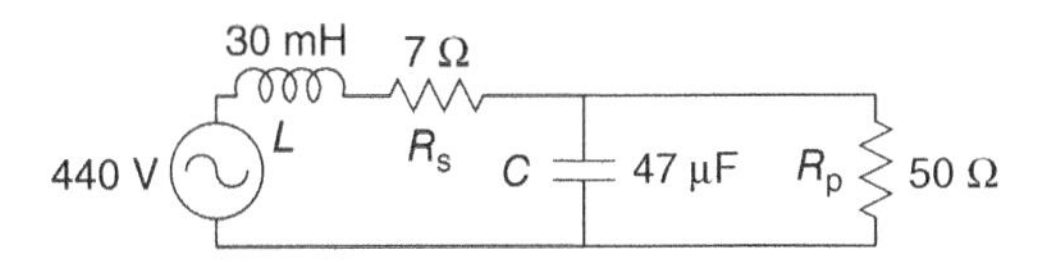

Figure 3.26 Parallel ac circuit.

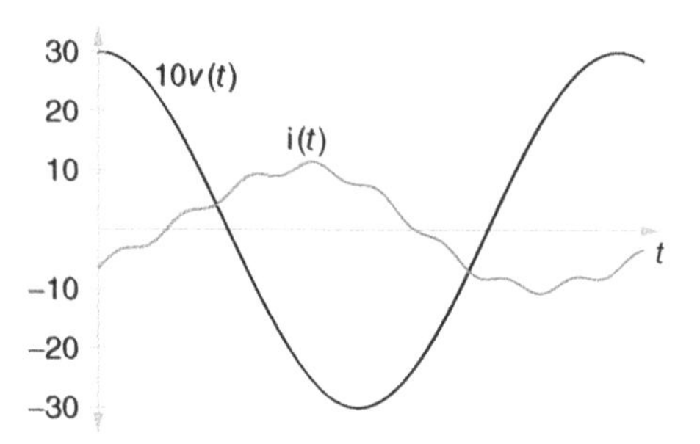

Figure 3.27 Measured current and voltage waveforms.

3.13 Calculate the capacitance needed to correct the power factor of a 3 kW process operating at 0.7 lagging to 0.9 leading. What is the rms current rating of the capacitor? Now assume the system has an 8% third harmonic component. What is the current rating of the capacitor in this case?

3.14 A $(50 + j50)\,\Omega$ impedance is connected to a 230 V 50 Hz single-phase source. What is the average and peak instantaneous power? Three $(50 + j50)\,\Omega$ impedances are connected to the three phases of a 230 V 50 Hz voltage source. What is the average and peak power to the impedances collectively?

3.15 The voltage and current waveforms as shown in Fig. 3.27 are measured on a load.
The voltage is

$$v(t) = 300\cos(\omega t)\ \text{V}$$

and the current

$$i(t) = 10\sin\left(\omega t - \frac{\pi}{4}\right) + 1\sin\left(3\omega t + \frac{\pi}{8}\right) + 3\sin\left(5\omega t + \frac{\pi}{4}\right).$$

Calculate

1. the distortion factor,
2. the displacement factor,
3. the power factor,
4. the average power.

FURTHER READING

Boylestad R.L., *Introductory Circuit Analysis*. 8th edition, Prentice-Hall International, Upper Saddle River, New Jersey, 1997.

Erickson R.W., Maksimoviè D., *Fundamentals of Power Electronics*. 2nd edition, Kluwer Academic Press, Norwell, 2001.

Irwin J.D., Nelms R.M., *Engineering Circuit Analysis*. 10th edition, Wiley, Hoboken, New Jersey, 2011.

Nilsson J.W., Riedel S.A., *Electric Circuits*. 6th edition, Prentice-Hall, Upper Saddle River, New Jersey, 2000.

CHAPTER 4

MAGNETICALLY COUPLED NETWORKS

4.1 INTRODUCTION

Magnetic fields play an important role in electrical systems. When a current flows a magnetic field is created; when a magnetic field exists, then there will be current or moving charges somewhere that have generated it. It is a chicken and egg situation—both exist simultaneously.

Magnetic fields are important in power conversion because they make it possible for one electrical circuit to exchange energy with another and, as shown in later chapters, make it possible to convert electrical energy into mechanical energy.

We have to start with the basics, and this is what is contained in this chapter. The laws that govern magnetic fields are applied to simple coils and transformers, and magnetic circuit analysis is introduced. In power conversion, we find inductors and transformers of all shapes and sizes. Those that operate at 50 or 60 Hz are usually heavy cubic blocks of copper and steel, whereas the high frequency ones are small and vary a lot in shape. The photograph in Fig. 4.1 shows two common shapes of magnetic cores: toroidal and rectangular.

4.2 BASIC CONCEPTS

In the next four sections, we introduce the concepts of magnetic field strength, magnetic flux, magnetic flux density and inductance. Although there are a number

The Principles of Electronic and Electromechanic Power Conversion: A Systems Approach, First Edition.
Braham Ferreira and Wim van der Merwe.

Figure 4.1 Square and round inductor shapes in a high frequency filter of a power electronics converter.

of ways to look at this we first introduce magnetic field strength using Ampére's law and magnetic flux using Faraday's law. We then discuss how these concepts are related in a magnetic circuit and introduce the concept of permeability. Finally, by combining all these concepts, we discuss the inductor.

4.2.1 Ampère's Circuital Law

In 1820 the Danish physicist Hans Christian Ørsted discovered that the needle of a compass held close to a wire deflected whenever a current flowed through the wire. This was an important discovery as it was the first proof that electricity and magnetism are linked to each other in some way. This discovery prompted the French physicist André-Marie Ampère to investigate the phenomenon in more detail. He found that parallel current-carrying wires will either attract or repel each other depending on whether the current is flowing in opposite (attracting) or in the same (repulsion) direction. This discovery laid the foundation for what we now call Ampère's circuital law, which describes the relationship between the movement of electrons and the magnetic field that it creates. In simple form, this is more than sufficient for our purposes. Ampère's circuital law states that for a very long and thin wire carrying a current $i(t)$, the magnetic field strength at a distance r from the wire will be

$$\boxed{H(t) = \frac{i(t)}{2\pi r}.} \tag{4.1}$$

Here, H is the *magnetic field strength*. This is a measure of the strength of the 'magnet' created by the current flow—or, expressed in another way, how much a compass magnet placed at this point would deflect when the current is switched on. If we investigate (4.1), we see that the term $2\pi r$ is simply the length of the circular path around the conductor at the radius, r, of interest. By substituting $\ell = 2\pi r$, we can rewrite (4.1) as

$$H(t) = \frac{i(t)}{\ell}. \tag{4.2}$$

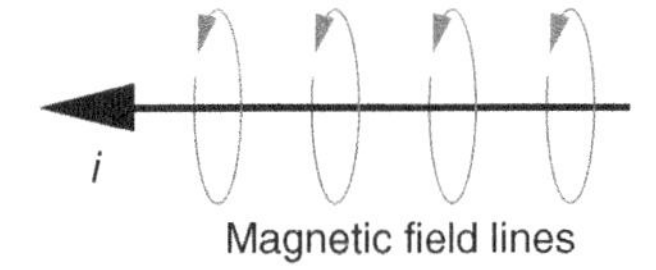

Figure 4.2 Ampère's law.

The magnetic field lines close to a current-carrying wire are shown in Fig. 4.2. It is important to note that the field lines are circular and in a plane perpendicular to the current direction. From (4.1), we also observe that the magnetic field depends on the current and on the inverse of the distance from the wire.

The direction of the magnetic field can be determined by using the right-hand rule. If the thumb of your right hand is pointing in the direction of the current, then if you make half a fist, your fingers would point in the direction of the magnetic field. This can be tested on Fig. 4.2.

Thus far, we have assumed that the path of interest is circular and that there is only one current-carrying conductor. However, Ampère's law does not prescribe any of these conditions; the path of interest can have any shape with the prerequisite that it is both closed and does not cross itself. Furthermore, the magnetic field strength is determined by all the current that is enclosed (or contained) in the area formed by the path of interest. Consider Fig. 4.3; here we see that we have three conductors, each carrying a current i, enclosed by the paths of interest. The magnetic field strength along the paths will therefore be (where we generalise for any number of conductors, we will use N-conductors)

$$\boxed{H(t) = \frac{Ni(t)}{\ell}.} \tag{4.3}$$

4.2.2 Faraday's Induction Law

In 1831, the English chemist and physicist Michael Faraday observed that when two coils are placed next to one another, a current is induced in the second coil whenever the current in the first is switched either on or off. Incidentally, across the Atlantic in Albany, New York, Joseph Henry made the same discovery at roughly

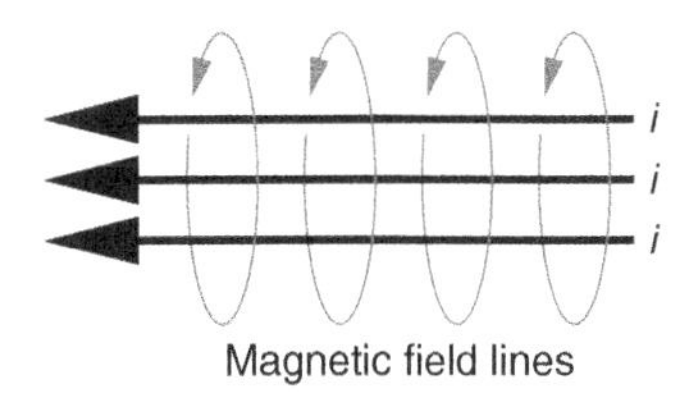

Figure 4.3 Ampère's law with more than one conductor.

the same time. However, Faraday was the first to publish his results, and the law describing this phenomenon bears his name today. That said, as we will see later, inductance is a logical consequence of the induction law, and Henry's contribution is recognised by naming the unit of inductance after him.

Let us first define two new concepts: magnetic flux and magnetic flux density. *Magnetic flux* is a term used to describe the amount of magnetic field in a given region. Now, it is true that we cannot see the magnetic field but can observe the results thereof. Think of the well-known experiment with a permanent magnet, a sheet of paper and a fist full of iron filings. If we place the magnet under the sheet of paper and spread the iron filings on top of the paper, we observe the magnetic flux lines through the iron filings. Magnetic flux is denoted by ϕ and is measured in Weber (Wb), named after Wilhelm Eduard Weber, a German physicist who first devised a system of absolute measurements for electrical currents. *Magnetic flux density* is the averaged magnetic flux in an area. Because it is an averaged value, we can only use the term with any meaning if the area where we average the flux is homogeneous. We often speak of magnetic flux density rather than of magnetic flux as it is more convenient and useful to normalise the magnetic flux measurement to the area where we measure. This is important because we could measure all the flux in the world if we were not careful. Magnetic flux density is denoted as B and is measured in Tesla (T), which equals 1 Wb/m^2, named after the brilliant Croatian-American inventor and engineer Nikola Tesla. Formally, for an area A, the flux and flux density are related by

$$\boxed{B = \frac{\phi}{A}.} \tag{4.4}$$

Faraday's induction law states that any change in the flux 'contained' by a wire loop will induce a voltage in the wire. This voltage will be induced in such a direction that the change in flux is opposed. It may sound complex, so let us take it step by step. In basic terms, the law states that

$$\boxed{e(t) = \frac{d\phi(t)}{dt}.} \tag{4.5}$$

However, we need to figure out some of the strict definitions. Firstly, the amount of flux 'contained' by a wire loop is simply the amount of flux that passes through the inside of the coil, or alternately the flux *linked* by the loop; see Fig. 4.4. Of course, if we make a coil with two turns, then the linked flux will be double. Therefore, we can express the linked flux λ, of a coil with N turns and an area of A as

$$\lambda = N\phi \tag{4.6}$$

$$\text{but } \phi = BA \tag{4.7}$$

$$\therefore \lambda = NBA. \tag{4.8}$$

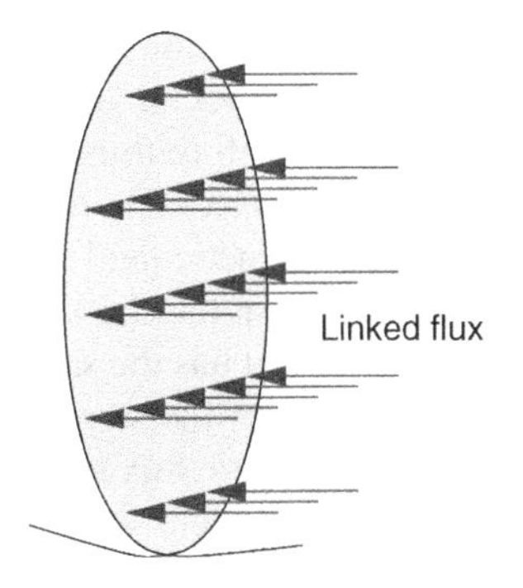

Figure 4.4 Flux linkage in a wire loop.

Therefore, for a coil with multiple turns, the induced voltage becomes

$$\boxed{e(t) = \frac{d\lambda(t)}{dt} = NA\,\frac{dB(t)}{dt}.} \tag{4.9}$$

Secondly, let us consider the statement that the voltage induced will be in a direction opposite to the change in flux. This statement speaks for itself. If the converse was true, we would have perpetual motion machines. Let us think of it in this way by doing a hypothetical experiment. When we push a mass along a surface, friction opposes the movement. If it did not, then we could give the mass a small push and then it would accelerate forever by virtue of this weird friction in our hypothetical experiment. Clearly this is not the way the world operates! The direction of the induced effect is described by Lenz's law, which states "*An induced current is always in such a direction as to oppose the motion or change causing it.*"

4.2.3 Relationship between Magnetic Flux and Magnetic Field Strength

One of the fascinating things we observe when playing with magnets is that they do not attract some metals, such as copper or aluminium. This is an interesting problem, especially when we speak about the strength of the magnet. Does the strength of the magnet change when it is brought close to copper? Is aluminium to a magnet what kryptonite is to Clark Kent? Of course not. The magnet always has the same strength; it is just that some metals such as copper and aluminium are not as readily magnetised as some other metals. In fact, wood, is more readily magnetised than copper while electrical steel (a special type of steel used in electrical equipment such as motors and transformers) is magnetised up to 4 000 times more easily than copper.

We see therefore that there is a difference between the 'strength' of the magnet and the magnetic result. We can think of the magnetic field strength as the 'strength' of the magnet. Therefore, using our discussion of the long thin wire as expressed in (4.1), the magnetic field strength at a specific distance associated with a certain current is always the same. We can therefore think of the magnetic field strength as a force that tries to create *magnetic flux*. The magnitude of the magnetic flux depends on two factors: the applied magnetic field strength and the material.

With these definitions, we can take another look at the problem of a magnet, a copper coin and a steel washer. A magnet has the same strength at all times; therefore, the magnetic field strength is always the same. When the magnet is brought close to the copper coin, not much magnetic flux is produced. This is because copper is not readily magnetised. As there is very little magnetic flux, the interaction between the magnet and the copper coin is low. When the magnet is brought close to the steel washer, much more magnetic flux is produced. We observe this flux in the interaction between the magnet and the washer.

A typical relationship between magnetic field strength and magnetic flux density is shown in Fig. 4.5. From an analysis point of view, we see that there are a couple of factors that complicate our discussion. Firstly, the relationship between H and B is not linear, and secondly, the forward and return paths are not the same; the relationship clearly exhibits a form of hysteresis. *The definition of hysteresis is a process whose output not only depends on the value of the input but also on the historical input.*

Although it is possible to analyse a system that exhibits a non-linear behaviour (inclusive of hysteresis) using advanced mathematical methods and specialised computer algorithms, we can make a few assumptions to simplify the system. As with all assumptions, it implies that we are simplifying reality and the resulting answer will not be 100% correct, but if we make our assumptions in such a way that we do not cheat reality, we can get an answer that is accurate enough. If a more accurate answer is needed, we can always use all the advanced tools at our disposal and calculate the answer to the required accuracy, albeit with the added time and cost expenditure.

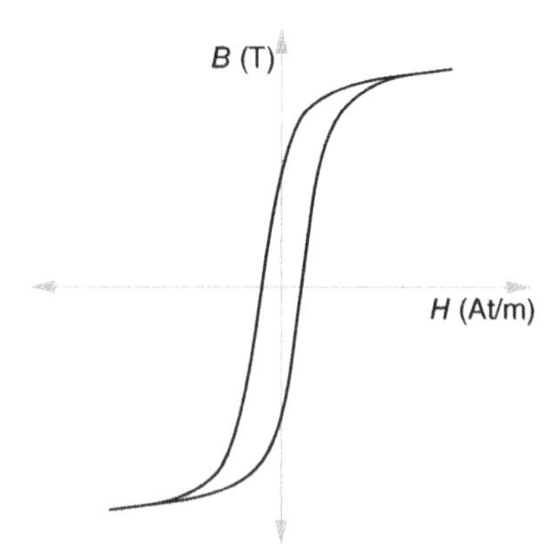

Figure 4.5 The relationship between magnetic field strength and magnetic flux density for a typical material.

Consider Fig. 4.5. Let us make the simplification that the B–H relationship does not exhibit hysteresis. Now we see that in the area around the zero point, the relationship between B and H is almost linear. However, when B (and by implication H) becomes large, the relationship is no longer linear. We call this area, where the linear relationship no longer holds, the saturation area. This area is analogous to many phenomena we observe in nature; let us discuss the following analogy: a student studying for an examination. If he does not study, he can expect to score a zero in the examination and the more he studies the higher he can expect his mark to be. We therefore can approximate the expected mark versus time-invested relationship as a linear relationship. But this result does not hold forever, does it? As one cannot score more than full marks in a test at a certain point, no matter how much more time he spends studying he cannot improve his mark; he has therefore entered the saturation region. Although we do not discuss the exact physics of the magnetisation curve here, the saturation region operates on a principle similar to this analogy.

Comparing Fig. 4.5 and the more detailed Fig. 4.6, we make the following assumptions:

1. We ignore the hysteresis nature of the relationship. We therefore assume that the forward and backward paths are the same.
2. We assume that we do not operate in the region of saturation, that is, $B < B_S$. Luckily, this is easy to verify because we usually know the value of B_S; we can therefore solve the circuit for B and then verify that $B < B_S$. If it is not, then we know that we cannot solve the circuit using our simplified model,

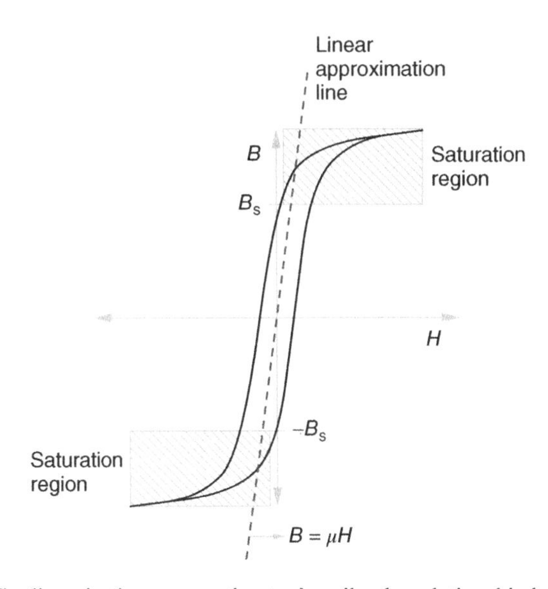

Figure 4.6 The linearisation assumption to describe the relationship between H and B.

and off to the advanced mathematics and computer algorithms we go. In the area of interest, we can now *approximate* the relationship as

$$\boxed{B = \mu H.} \tag{4.10}$$

The term μ in (4.10) is called the *permeability*. Permeability is a measure of the ability of a material to support the formation of a magnetic field. One difficulty when comparing the permeability of different materials is the fact that (in SI units) the permeability is normally a very small number and it is difficult to get a good feel of the variations in permeability from one material to the next. Luckily, this problem can be solved quite easily by normalising the permeability measure to the permeability of free space or, in other words, comparing how readily a material is magnetised as against how easily a vacuum is magnetised. Let us denote the permeability of a vacuum (free space) as μ_0. The permeability of free space is one of the fundamental constants of nature and is given by

$$\boxed{\mu_0 = 4\pi \times 10^{-7}.} \tag{4.11}$$

We can now describe the permeability of any material using only the relative permeability, μ_r, where

$$\boxed{\mu = \mu_r \mu_0.} \tag{4.12}$$

We can broadly categorise the magnetic characteristics of the materials likely to be encountered in electrical power processing systems into three broad categories:

Diamagnetic Materials with a permeability slightly less than that of free space.

Paramagnetic Materials with a permeability slightly higher than that of free space.

Ferromagnetic Materials that are very readily magnetised; they have permeabilities hundreds or even thousands of times higher than that of free space.

The relative permeabilities of some materials are listed in Table 4.1. We see that for all practical purposes (especially when compared with the large permeabilities of ferromagnetic materials), for diamagnetic and paramagnetic materials we can assume that $\mu_r \approx 1$.

4.2.3.1 *A Side Note on the Relationship between B and H* Even when we do not use the assumptions, we can still use the equation

$$B = \mu H \tag{4.13}$$

to describe the relationship between B and H. However, if we try to describe all the characteristics of a magnetic circuit, we see that μ is no longer a constant but it is a function of the applied magnetic field strength, the history of the applied magnetic

TABLE 4.1 The Relative Permeability of Some Materials Encountered in Electrical Power Processing Systems

Material	μ_r	Material	μ_r	Material	μ_r
Diamagnetic		Paramagnetic		Ferromagnetic	
Water	$1-\frac{8}{10^6}$	Air	$1+\frac{37}{10^8}$	Steel	100
Copper	$1-\frac{6}{10^6}$	Wood	$1+\frac{43}{10^8}$	Ferrite	>600
		Aluminium	$1+\frac{22}{10^6}$	Electrical steel	4 000
				Permalloy	8 000
				Mumetal	20 000

field strength, the frequency and the temperature. It is also true that permeability can be a complex number. It is clearly not a good idea to taunt this beast without a good reason.

That said, depending on the answers we require, substituting the complex function μ with a constant value will, in most cases, yield an answer with acceptable accuracy. We just need to ensure that we do not violate the main assumption that states that the magnetic circuit is not operating in the saturation region, or $B < B_s$.

4.2.4 Inductance

We can now put together much of what we have discussed this far. Consider the coil, wound around a toroidal core, shown in Fig. 4.7. We know that according to Ampère's law, a magnetic field will be induced by current flowing in the wire. Now, when we look at Fig. 4.2 we can see, using the right-hand rule, that the direction of the magnetic field is, as indicated, in a clockwise direction through the core.

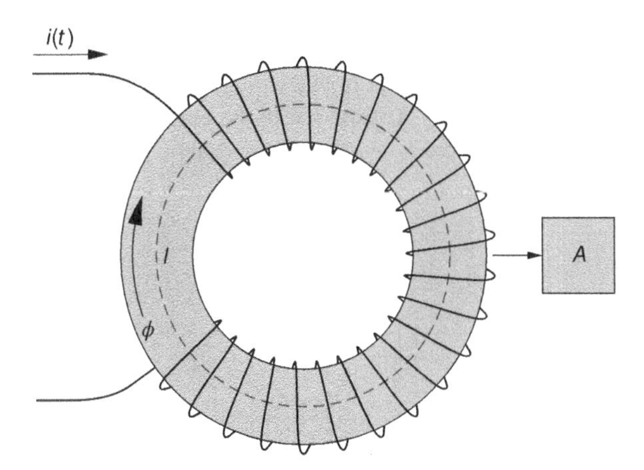

Figure 4.7 Flux in a toroid.

According to Ampère's law, the strength of this magnetic field will be a function of the amount of current and the length of the magnetic path.

Now, it is true that the flux that takes the inside lane will travel a shorter distance than that in the outside lane. However, we can get a good approximation by using the average length, as shown in Fig. 4.7. Let us denote this average length by ℓ; now, the following relationships hold

$$H(t) = \frac{Ni(t)}{\ell} \tag{4.14}$$

$$B(t) = \mu H(t) = \mu\frac{Ni(t)}{\ell}. \tag{4.15}$$

Using (4.10) and (4.8), we can write the flux linked by the coil as

$$\phi(t) = \mu N\frac{A}{\ell}i(t). \tag{4.16}$$

The voltage induced in the coil is therefore, by using Faraday's induction law (4.5),

$$\begin{aligned} v(t) &= N\frac{d\phi(t)}{dt} \\ &= \mu N^2\frac{A}{\ell}\frac{di(t)}{dt}. \end{aligned} \tag{4.17}$$

Here, we see the foundation of inductance. Recall that for an inductor

$$v(t) = L\frac{di(t)}{dt} \tag{4.18}$$

therefore,

$$\boxed{L = \mu N^2\frac{A}{\ell}.} \tag{4.19}$$

Let us consider the circuit shown in Fig. 4.8 to see how the inductor operates using this newly defined terminology. At time $t < 0$, the current i is zero. As the current is zero, we know from Ampère's law that the magnetic field strength is zero and,

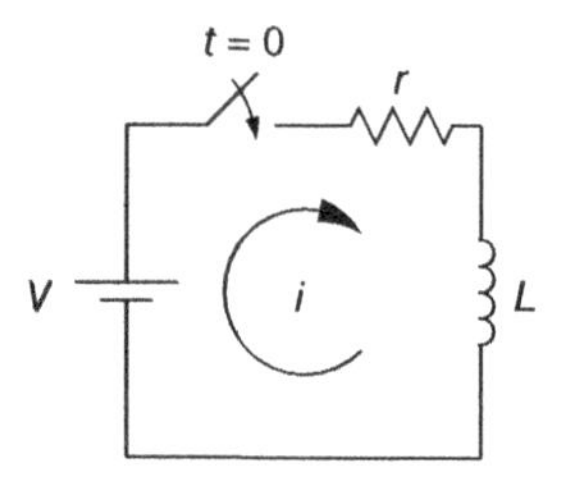

Figure 4.8 A simple LR circuit.

therefore, the magnetic flux is also zero. At $t = 0$, when the switch closes, the voltage source forces a current to flow. If there was no inductance, the current would immediately increase to $i = \frac{V}{r}$. However, when the inductor is in the circuit, we know from Faraday's induction law that the flux must slowly build up to its final value. If the current, and therefore the flux, are to increase immediately from zero to their final values, it would mean that the voltage induced in the coil will have to get to infinity for an infinitely short time, the stuff sci-fi spaceships are made of. However, in real life, we do not like to think in terms of infinitely short periods of time, and even less of infinitely large voltages. With this 'easy' solution out of the way, the only solution is for the flux in the inductor to slowly build up to its final value. From circuit theory, we know that

$$V = i(t)r + L\frac{di(t)}{dt}$$

with solution

$$i(t) = \frac{V}{r}\left(1 - e^{(-r/L)t}\right).$$

As an exercise can you explain what will happen if the switch opens after being closed for a few minutes? Not so easy, is it? We will discuss this question further in Chapter 6.

As a side note, if we investigate (4.19), we see that the inductance of a coil depends on the number of turns and the dimensions of the inductor and on how easy it is to magnetise the area around the coil. As electrical steel can be magnetised about 4 000 times more easily then it would mean that if we can keep the dimension the same, an inductor wound around an electrical steel core would require about 63 times fewer turns than one wound without a core (in other words an air-core inductor) or even a wooden core. This constitutes a considerable saving in wire.

4.2.5 Basic Magnetic Circuits

With the tools developed in the previous sections, we can now attack quite a few magnetic circuit problems with confidence. The interactions of the current, magnetic field and the voltage are visualised in Fig. 4.9. We use these steps and the theory developed to discuss a few circuits in the next few examples.

Example 4.1 Toroidal Inductor with an Air Core

Consider the inductor wound as a toroid using an air core, as shown in Fig. 4.10. We have developed an expression for the inductance of a toroidal inductor wound around a core; see Fig. 4.7. Develop an expression for the inductance of the air-core inductor, assuming that the inductor is wound as a perfect toroidal helix with N-turns.

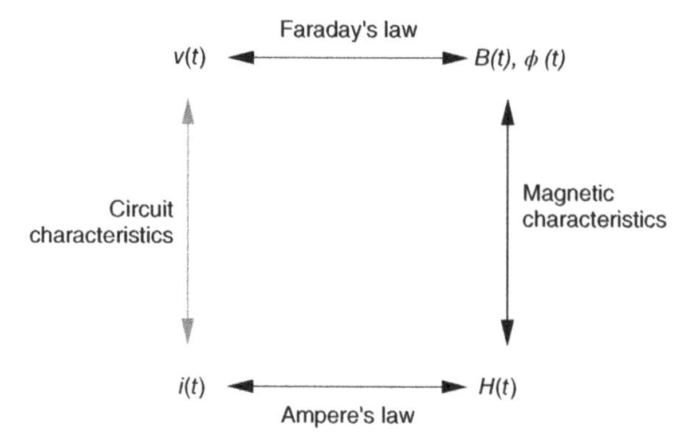

Figure 4.9 Summary of the steps in the determination of the electrical terminal characteristics of a magnetic element.

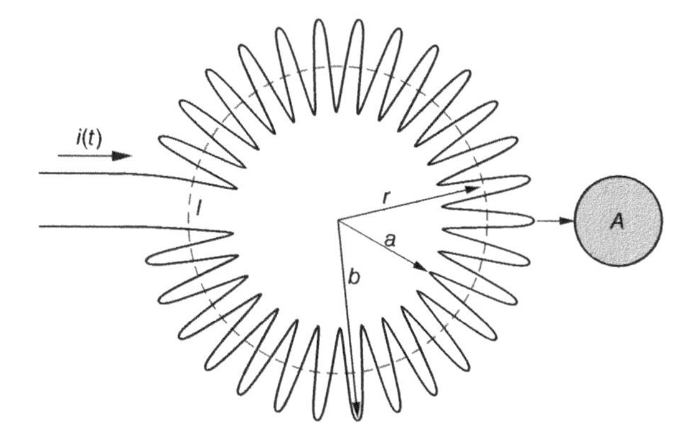

Figure 4.10 An air-core toroid.

Solution

From Fig. 4.10, we see that the inner radius of the toroid is a and the outer radius is b. From our knowledge of magnetic fields, we know that the current through the conductor will create a magnetic field, according to Ampère's law. The average length of the magnetic field is ℓ as indicated in Fig. 4.10. We can relate ℓ and the dimensions of the toroid by

$$\begin{aligned}\ell &= 2\pi r \\ &= 2\pi\left(a + \frac{1}{2}(b-a)\right).\end{aligned}$$

Now, according to Ampère's law, the magnetic field strength is

$$H(t) = \frac{Ni(t)}{\ell} = \frac{Ni(t)}{2\pi\left(a + \frac{1}{2}(b-a)\right)}.$$

The (averaged) flux density in the centre of the coil can be found as

$$B(t) = \mu_0 H(t) = \mu_0 \frac{Ni(t)}{2\pi\left(a + \frac{1}{2}(b-a)\right)}.$$

To get the flux contained in the coil, we need to multiply the flux density with the area in question. The area, A, as shown in Fig. 4.10 is

$$A = \pi\left(\frac{1}{2}(b-a)\right)^2 = \frac{\pi}{4}(b-a)^2$$

so the flux is

$$\phi(t) = B(t)A = \mu_0 \frac{Ni(t)(b-a)^2}{8\left(a + \frac{1}{2}(b-a)\right)}.$$

The induced voltage in the coil is therefore

$$e(t) = N\frac{d\phi(t)}{dt} = \mu_0 \frac{N^2(b-a)^2}{8\left(a + \frac{1}{2}(b-a)\right)} \frac{di(t)}{dt}$$

and as $e(t) = L\frac{di}{dt}$, we know that

$$L = \mu_0 \frac{N^2(b-a)^2}{8\left(a + \frac{1}{2}(b-a)\right)}.$$

In general, an air-core inductor has a very low value of inductance, mainly due to the fact that air is not very magnetisable. If we manufacture an air-core inductor with 100 turns, $a = 2\,\text{cm}$ and $b = 4\,\text{cm}$, the inductance will be

$$\begin{aligned} L &= \mu_0 \frac{N^2(b-a)^2}{8\left(a + \frac{1}{2}(b-a)\right)} \\ &= 4\pi \times 10^{-7} \frac{100^2(0.04 - 0.02)^2}{8(0.02 + 0.01)} \\ &= 21\ \mu\text{H}. \end{aligned}$$

Example 4.2 Inductor with a Square Core

Consider the inductor wound using a square core, as shown in Fig. 4.11. Develop an expression for inductance; the core has a relative permeability of μ_r.

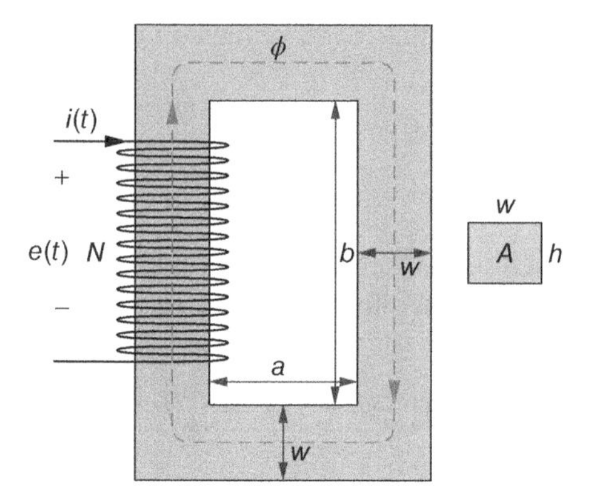

Figure 4.11 An inductor with a square core.

Solution

We can calculate the average path length, ℓ, from the dimensions of the core as

$$\begin{aligned}\ell &= 2(a + w) + 2(b + w)\\ &= 2(a + b + 2w).\end{aligned}$$

Now, according to Ampère's law, the magnetic field strength is

$$H(t) = \frac{Ni(t)}{\ell} = \frac{Ni(t)}{2(a + b + 2w)}.$$

The (averaged) flux density in the centre of the coil can be found as

$$B(t) = \mu_0 H(t) = \mu_r \mu_0 \frac{Ni(t)}{2(a + b + 2w)}.$$

To get the flux contained in the coil, we need to multiply the flux density by the area in question. The area, A, as shown in Fig. 4.10 is

$$A = h\, w$$

and the flux is therefore

$$\phi(t) = B(t)A = \mu_r \mu_0 \frac{Ni(t)wh}{2(a + b + 2w)}.$$

The induced voltage in the coil is therefore

$$e(t) = N\frac{d\phi(t)}{dt} = \mu_r \mu_0 \frac{N^2 wh}{2(a + b + 2w)} \frac{di(t)}{dt}$$

and as $e(t) = L\frac{di}{dt}$, we know that

$$L = \mu_r \mu_0 \frac{N^2 wh}{2(a+b+2w)}.$$

An inductor wound around a core can have a very high value of inductance because of the high permeability of the core material. If we manufacture an inductor with 100 turns and we let $a = 2$ cm, and $b = 4$ cm, $w = 1$ cm, $h = 1$ cm and $\mu_r = 4000$, the inductance will be

$$\begin{aligned} L &= \mu_r \mu_0 \frac{N^2 wh}{2(a+b+2w)} \\ &= 16\,000\pi \times 10^{-7} \frac{100^2\ 0.01^2}{2(0.02+0.04+2(0.01))} \\ &= 31.4 \text{ mH}. \end{aligned}$$

There is, however, a price to pay for this high value of inductance. If the core saturates at $B_s = 0.4$ T, then the maximum current can be calculated as

$$\begin{aligned} I_{max} &= \frac{2(a+b+2w)B_s}{\mu_r \mu_0 N} \\ &= \frac{0.8(0.02+0.04+2(0.01))}{16\,000\pi \times 10^{-7} \times 100} \\ &= 0.127\text{A}. \end{aligned}$$

If the current is higher than 12.7 A, then the flux density will be more than B_s and the system will be operating in the saturation region. As we have to assume that the system is not in saturation to make the approximation that μ is a constant, the answers we get will not be valid should $i > I_{max}$.

4.2.6 Magnetic Circuit with an Air Gap

What will happen if the system is not homogeneous? To answer this question, let us investigate the circuit of Fig. 4.7 which has been modified slightly to have an air gap. This modified magnetic circuit is shown in Fig. 4.12.

If we define the average magnetic length of the magnetic circuit as

$$\ell = 2\pi r \tag{4.20}$$

then, we can write the magnetic path length through the core as

$$\ell_c = 2\pi r - g \tag{4.21}$$

and the path length through the air gap is

$$\ell_g = g. \tag{4.22}$$

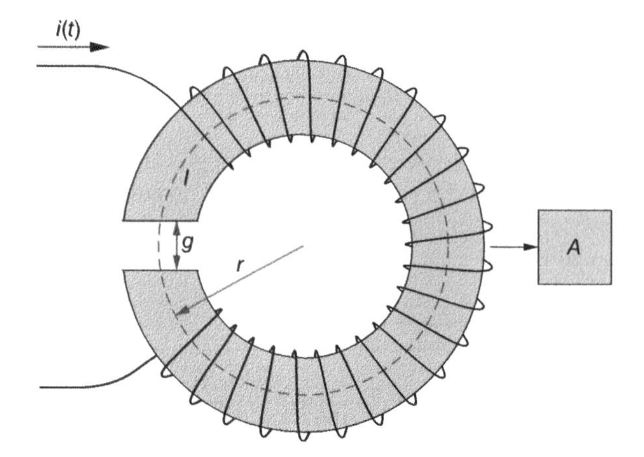

Figure 4.12 A magnetic circuit with an air gap.

Naturally,

$$\ell = \ell_c + \ell_g. \tag{4.23}$$

Now, let us define the magnetic field strength in the core as $H_c(t)$ and in the air gap as $H_g(t)$. We can rewrite Ampère's induction law as

$$\boxed{Ni(t) = H_c(t)\ell_c + H_g(t)\ell_g.} \tag{4.24}$$

We can visualise magnetic flux in the same way as we visualise the behaviour of electrical current. Current is only present in a closed circuit and we know that the current in a wire, without any nodes, is equal everywhere in the wire. The same is true with magnetic flux: it always originates and terminates at the same point (another way of saying that it must exist as a closed line). Therefore, the flux in the air gap and the flux in the core are the same. If we assume the area of the air gap to be equal to the area of the core, then

$$B_g = B_c \tag{4.25}$$

where B_g is the magnetic flux density in the air gap and B_c is the magnetic flux density in the core. Now, let B be the flux density anywhere in the magnetic circuit; then, the magnetic field strength in the core is

$$H_c(t) = \frac{B(t)}{\mu_0 \mu_r} \tag{4.26}$$

and in the air gap

$$H_g(t) = \frac{B(t)}{\mu_0}. \tag{4.27}$$

Substituting into (4.24) yields

$$Ni(t) = \frac{B(t)}{\mu_0 \mu_r}\ell_c + \frac{B(t)}{\mu_0}\ell_g$$

and therefore,

$$\boxed{B(t) = \frac{Ni(t)}{\frac{\ell_c}{\mu_r \mu_0} + \frac{\ell_g}{\mu_0}}.} \tag{4.28}$$

That said, most magnetic circuits with air gaps that we meet in electrical power processing systems will exhibit the characteristic that

$$\frac{\ell_c}{\mu_r} \ll \ell_g \tag{4.29}$$

and therefore the flux density can be approximated as

$$\boxed{B(t) \approx \mu_0 \frac{Ni(t)}{\ell_g}.} \tag{4.30}$$

When this is true, we say that the air gap dominates the behaviour of the magnetic circuit. In most systems, most notably electrical machines, this is a desirable characteristic. It is also interesting to note that when we use this assumption, the inductance of the system can be approximated as

$$\boxed{L \approx \mu_0 \frac{N^2 A}{\ell_g}.} \tag{4.31}$$

We will leave the derivation of this result to the reader (refer to Section 4.2.4).

4.3 MUTUAL INDUCTANCE

In the rest of this chapter, we investigate the behaviour of two inductors placed close enough to each other so that the magnetic field created by one couples with the other. We develop the equations describing the system from fundamental principles. This endeavour will enable us to describe and understand the operation of a device that makes our existence, as we know it, possible: the transformer. Without the humble transformer, our electricity network would have been impossible to realise and we would still be cooking our food with open fires at home or possibly with gas hobs. Transformers are all around us and come in many different shapes and sizes; as an example of a very large transformer, consider the 800 kV transformer manufactured by ABB in the small Swedish town of Ludvika (Fig. 4.13). These large, single-phase transformers are used in HVDC (high voltage direct current)

Figure 4.13 An 800 kV HVDC single-phase transformer manufactured by ABB (with a Volvo in the foreground).

transmission networks that enable the transmission of very large amounts of energy over huge distances. This specific transformer can handle around 2.4 GW of power! Instead of high voltage and high power, one also finds high frequency transformers. As frequency is increased, transformers become physically smaller. In Fig. 4.14, a 20 kHz transformer is shown that receives its power from a power electronics converter, shown next to the transformer. In Chapter 6, we discuss the operation of power electronics converters.

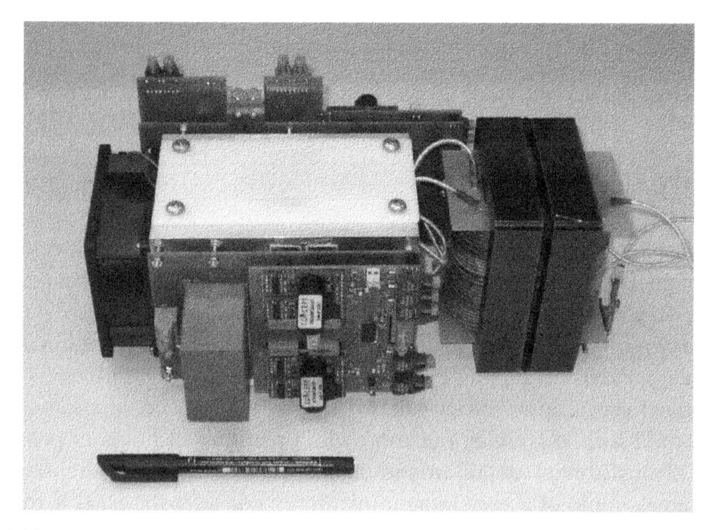

Figure 4.14 The 800 V transformer on the right is supplied with 20 kHz ac by the power electronic converter on the left.

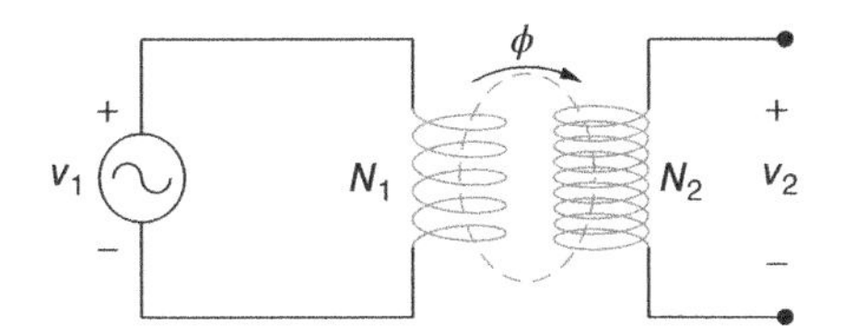

Figure 4.15 Two mutually coupled windings.

4.3.1 Simple Air-Core Transformer

What will happen if we place two coils next to one other and we are able to make sure that all the flux generated by coil 1 links to coil 2? Consider the circuit in Fig. 4.15. We can write the following relationship

$$v_1(t) = N_1 \frac{d\phi(t)}{dt}$$
$$\therefore \frac{d\phi(t)}{dt} = \frac{v_1(t)}{N_1}. \tag{4.32}$$

Using the same argument for coil 2,

$$v_2(t) = N_2 \frac{d\phi(t)}{dt} \tag{4.33}$$

and substituting (4.32) yields

$$\boxed{v_2(t) = \frac{N_2}{N_1} v_1(t).} \tag{4.34}$$

This is a very useful result, as we are now able to change the voltage level using two inductors. However, there is a catch in the sense that according to Faraday's law voltages are induced as the flux changes. We are, therefore, not able to transform dc voltages from one level to another using this method.

Let us now consider the circuit in Fig. 4.16. From (4.34), we know that $v_2(t) = \frac{N_2}{N_1} v_1(t)$. It is now easy to find the current through the resistor, but how much current

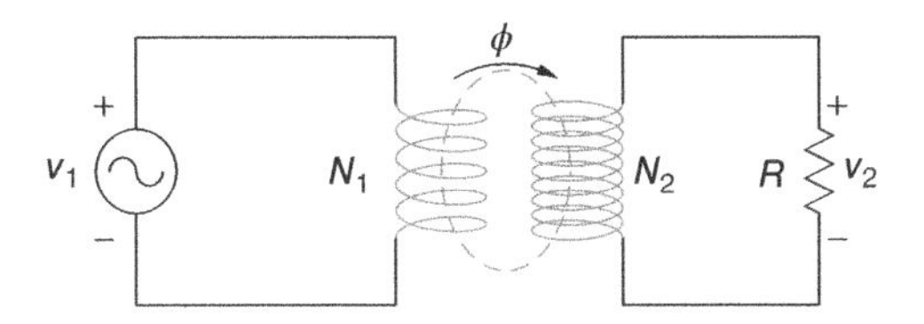

Figure 4.16 Two mutually coupled windings with a resistor connected to the secondary.

is flowing through the source? It is tempting to simply use the approach that there are no losses in the system and therefore $P_{\text{in}} = P_{\text{out}}$. As the load power is

$$P_{\text{out}} = \frac{v_2^2}{R} = \left(\frac{N_2}{N_1}v_1\right)^2 \cdot \frac{1}{R} \tag{4.35}$$

we can calculate the source current as

$$\begin{aligned} P_{\text{in}} &= v_1 i_1 = P_{\text{out}} \\ \therefore,\ i_1 &= \frac{v_1}{\frac{N_1^2}{N_2^2}R}. \end{aligned} \tag{4.36}$$

This is a very interesting result because for all practical purposes, it means that a load of $\frac{N_1^2}{N_2^2}R$ is connected to the source. This is discussed later.

4.3.2 Leakage Flux and the Transformer Core

The air-core transformer of Fig.4.16 is an oversimplification of reality. From Ampère's law, we know that a current flowing in a conductor will create magnetic flux all around the conductor. In fact, when we consider the coil and the magnetic flux around it we know that the flux should surround the coil in all directions. This is shown in Fig. 4.17.

One way of overcoming the problem of the flux surrounding the whole volume around the coil is by using a core that is readily magnetised. Consider the transformer shown in Fig. 4.18. If the core has a very high value of μ, implying that it is very readily magnetised, then nearly all the flux will be contained in the core. If we assume the ideal case where all the flux is contained in the core, then the discussion for the idealised transformer, as in the previous section, holds. However, what will

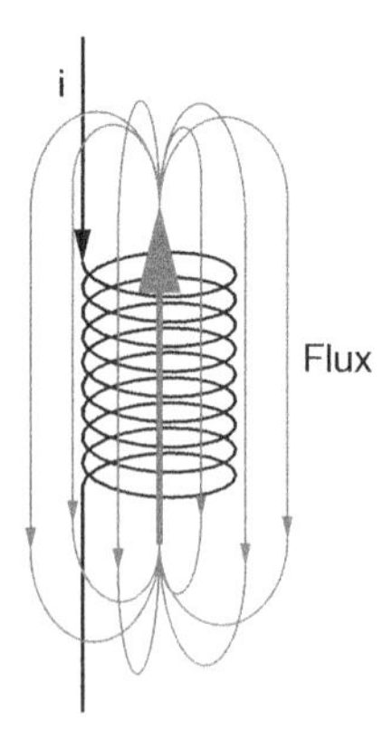

Figure 4.17 The flux generated by a winding is distributed all around the winding.

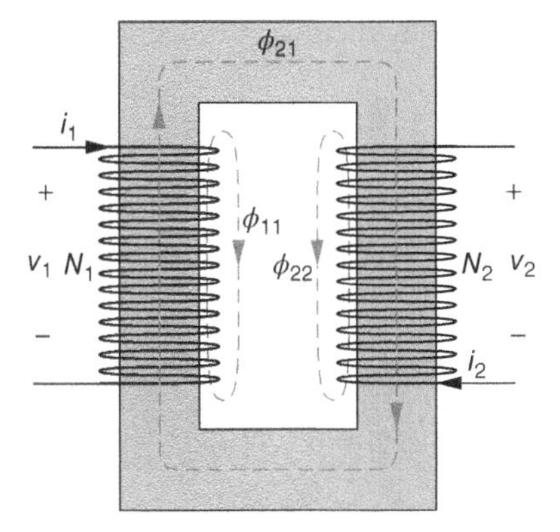

Figure 4.18 Basic transformer with core.

happen when μ is very high but not all the flux is contained in the core? In other words, with reference to Fig. 4.18, when $\phi_{11} \neq 0$ and $\phi_{22} \neq 0$.

To answer this question, we have to define a new term. Consider the flux paths ϕ_{11} and ϕ_{12} in Fig. 4.18; look particularly at the total length that the flux must travel to complete a loop. We see that the two lengths are different. To simplify our discussion, which thus far incorporates the length of the flux path, let us define a term *permeance* that we denote as $\mathscr{P}$. Officially, permeance is defined as

$$\mathscr{P} = \mu \frac{A}{\ell} \tag{4.37}$$

so that

$$\phi = \mathscr{P} N i. \tag{4.38}$$

The permeance therefore describes the magnetic properties of the space occupied by the flux. When the space is occupied by materials that are readily magnetised, then the permeance will be high.

Notation Note 4.1

Because we denote power by P, we will denote permeance by $\mathscr{P}$ to avoid confusion.

4.3.2.1 The Special Case Where $i_2 = 0$ Consider Fig. 4.18; we define ϕ_{11} as the flux generated by current i_1 that links with coil 1. Similarly, ϕ_{22} is the flux generated by current i_2 that links with coil 2. Lastly, ϕ_{12} is the flux due to current i_2 that links with coil 1.

Let us first consider the case where $i_2 = 0$. Define ϕ_1 as all the flux generated by current i_1 and therefore

$$\phi_1 = \phi_{11} + \phi_{21}. \tag{4.39}$$

If we define $\mathscr{P}_1$ as the permeance of the space occupied by ϕ_1 and, similarly, $\mathscr{P}_{11}$ as the permeance of the space occupied by ϕ_{11} and $\mathscr{P}_{21}$ the permeance connected with ϕ_{21}, then

$$\phi_1 = \mathscr{P}_1 N_1 i_1 \tag{4.40}$$

$$\phi_{11} = \mathscr{P}_{11} N_1 i_1 \tag{4.41}$$

$$\phi_{21} = \mathscr{P}_{21} N_1 i_1 \tag{4.42}$$

As

$$\phi_1 = \phi_{11} + \phi_{21} \tag{4.43}$$

we know that

$$\mathscr{P}_1 = \mathscr{P}_{11} + \mathscr{P}_{21}. \tag{4.44}$$

Let us use Faraday's induction law to derive an expression for v_1:

$$\begin{aligned} v_1 &= \frac{d\lambda_1}{dt} = N_1 \frac{d\phi_1}{dt} = N_1 \frac{d}{dt}\left(\phi_{11} + \phi_{21}\right) \\ &= N_1^2 \left(\mathscr{P}_{11} + \mathscr{P}_{21}\right) \frac{di_1}{dt} = N_1^2 \mathscr{P}_1 \frac{di_1}{dt} = L_1 \frac{di_1}{dt} \end{aligned} \tag{4.45}$$

We see that we can express v_1 by using any of these expressions. The final relationship was constructed by using (4.19) and (4.37) to form $L = N^2\mathscr{P}$. We can also use Faraday's induction law to derive expressions for v_2 as

$$v_2 = \frac{d\lambda_2}{dt} = N_2 \frac{d\phi_{21}}{dt} = N_2 \mathscr{P}_{21} N_1 \frac{di_1}{dt}. \tag{4.46}$$

The term $N_1^2 \mathscr{P}_1 \frac{di_1}{dt}$ in (4.46) refers to the self-inductance of coil 1. However, the term $N_2 \mathscr{P}_{21} N_1 \frac{di_1}{dt}$ in (4.46) refers to the mutual inductance between coils 1 and 2. We call this *mutual inductance* as a current in coil 1 induces a voltage in coil 2. Let us define the mutual inductance from the current in coil 1 to the voltage in coil 2 as

$$\boxed{M_{21} = N_2 N_1 \mathscr{P}_{21}.} \tag{4.47}$$

Using this definition, we can write from (4.46)

$$\boxed{v_2 = M_{21} \frac{di_1}{dt}.} \tag{4.48}$$

4.3.2.2 *The Special Case Where* $i_1 = 0$

We use the same reasoning to develop an expression for the mutual induction from the current in coil 2 to the voltage in coil 1 with $i_1 = 0$. The flux ϕ_{12} is not shown explicitly in Fig. 4.18, but it is the same as the flux ϕ_{21} but in an opposite direction. The total flux linking coil 2 is

$$\phi_2 = \phi_{22} + \phi_{12} \tag{4.49}$$

which is related to the current i_2 as

$$\phi_2 = \mathscr{P}_2 N_2 i_2 \tag{4.50}$$

$$\phi_{22} = \mathscr{P}_{22} N_2 i_2 \tag{4.51}$$

$$\phi_{12} = \mathscr{P}_{12} N_2 i_2. \tag{4.52}$$

However, it is important to note that in Fig. 4.18, the definition of v_2 is chosen such that it is now opposite to the definition of the current direction. The voltage induction in coil 2 can be written, with these definitions,[1] as

$$v_2 = -N_2 \frac{d\phi_2}{dt}. \tag{4.53}$$

Therefore, using Faraday's induction law, voltages v_1 and v_2 can be expressed as

$$v_2 = \frac{d\lambda_2}{dt} = -N_2 \left(\mathscr{P}_{22} + \mathscr{P}_{12}\right) \frac{di_2}{dt} = -N_2^2 \mathscr{P}_2 \frac{di_2}{dt} = -L_2 \frac{di_2}{dt} \tag{4.54}$$

$$v_1 = \frac{d\lambda_1}{dt} = -N_1 \frac{d\phi_{12}}{dt} = -N_1 N_2 \mathscr{P}_{12} \frac{di_2}{dt}. \tag{4.55}$$

The mutual inductance from the current in coil 2 to the voltage in coil 1 is

$$\boxed{M_{12} = N_1 N_2 \mathscr{P}_{12}.} \tag{4.56}$$

However, because the flux takes the same path from coil 1 to coil 2 as in the opposite direction, for the systems we discuss

$$\mathscr{P}_{12} = \mathscr{P}_{21} \tag{4.57}$$

and therefore,

$$\boxed{M_{12} = M_{21} = M.} \tag{4.58}$$

[1]There are many different definitions for the current and voltage directions and many textbooks will choose a different secondary voltage definition such that the signs in the second row of the matrix in (4.61) reverses. Although this yields the more recognisable form of the equation, we have chosen to be consistent with the standard definitions of an ideal transformer and live with the ensuing inconvenient form of (4.61).

4.3.2.3 The General Case Where $i_1, i_2 \neq 0$ We can combine our discussions for the two special cases to write the following equations for v_1 and v_2

$$v_1 = \frac{\lambda_1}{dt} = N_1 \left(\frac{d\phi_1}{dt} - \frac{d\phi_{12}}{dt} \right) = L_1 \frac{di_1}{dt} - M \frac{di_2}{dt} \tag{4.59}$$

$$v_2 = \frac{\lambda_2}{dt} = -N_2 \left(\frac{d\phi_2}{dt} - \frac{d\phi_{21}}{dt} \right) = -L_2 \frac{di_2}{dt} + M \frac{di_1}{dt} \tag{4.60}$$

or in matrix form

$$\boxed{\begin{bmatrix} v_1 \\ v_2 \end{bmatrix} = \begin{bmatrix} L_1 & -M \\ M & -L_2 \end{bmatrix} \frac{d}{dt} \begin{bmatrix} i_1 \\ i_2 \end{bmatrix}} \tag{4.61}$$

This expression is, maybe, not quite as clear, especially when we consider the initial discussion regarding the flux linked by the different coils. Let us revisit the expressions for the self-inductances:

$$L_1 = N_1^2 \mathscr{P}_1 \tag{4.62}$$

$$L_2 = N_2^2 \mathscr{P}_2 \tag{4.63}$$

multiplying these two relationships yields

$$L_1 L_2 = N_1^2 N_2^2 \mathscr{P}_1 \mathscr{P}_2. \tag{4.64}$$

We can rewrite this expression, keeping in mind that for the linear systems we are interested in, $\mathscr{P}_{12} = \mathscr{P}_{21}$,

$$\begin{aligned} L_1 L_2 &= N_1^2 N_2^2 (\mathscr{P}_{11} + \mathscr{P}_{21})(\mathscr{P}_{22} + \mathscr{P}_{12}) \\ &= N_1^2 N_2^2 \mathscr{P}_{12}^2 \left(1 + \frac{\mathscr{P}_{11}}{\mathscr{P}_{12}}\right)\left(1 + \frac{\mathscr{P}_{22}}{\mathscr{P}_{12}}\right) \\ &= M^2 \frac{1}{k^2}, \end{aligned} \tag{4.65}$$

where

$$\frac{1}{k^2} = \left(1 + \frac{\mathscr{P}_{11}}{\mathscr{P}_{12}}\right)\left(1 + \frac{\mathscr{P}_{22}}{\mathscr{P}_{12}}\right). \tag{4.66}$$

Maybe it seems as if we have complicated the situation to some extent. However, consider the fact that now

$$M^2 = k^2 L_1 L_2 \tag{4.67}$$

or equivalently

$$\boxed{M = k\sqrt{L_1 L_2}.} \tag{4.68}$$

Now it makes a little more sense; we have defined a constant k, which is the *coefficient of coupling* between the two coils. If $k = 0$, then there is no common flux between the two coils and the mutual inductance is zero. On the other hand, if $k = 1$ there is perfect coupling between the two coils and $\phi_{11} = \phi_{22} = 0$ and represents the ideal case. The allowable range for the coefficient of coupling is $0 \leq k \leq 1$. In practical systems, it is not possible to get a perfect coupling, that is, $k = 1$. However, with a well-designed system, where the intention is to maximise the coupling, it is possible to achieve a coupling coefficient in excess of 0.99.

4.3.2.4 Dot Markings Consider Fig. 4.18. As we know that the flux generated by the current i_1 opposes the flux generated by current i_2 (see Section 4.2.4), we can determine that the directions of the two currents as shown in Fig. 4.18 are correct. The procedure to verify this can be summarised as follows:

1. Pick any side and assign a direction to this current.
2. Use your right hand and curl your fingers in the direction of the current flow.
3. Make a note of the direction of the flux, as indicated by your right thumb.
4. Consider the other coil and place your right thumb in the direction opposing the flux direction found in Step 3.
5. Curl the fingers of your right hand, they will now indicate the direction of current flow.

However, it is not convenient to draw two coupled inductors using such an elaborate sketch as in Fig. 4.18. The accepted method for drawing two coupled inductors in a circuit diagram is shown in Fig. 4.19. Although this diagram shows the two inductors and how they are coupled, the information regarding the current directions is missing. To solve this problem, a *dot convention* was adopted by the engineering fraternity.

Definition 4.1

When the reference direction of a current enters the dotted terminal of a coil, the reference polarity of the voltage that it induces on the other coil is positive at the dotted terminal. We can also state it differently. When the current enters a coil at the dotted terminal, then the current flow due to the induced voltage on the other coil will leave the coil at its dotted terminal. The standard definitions of the voltages and currents used in conjunction with the dot convention are shown in Fig. 4.19.

Using this definition of the dot convention, it is trivial to find the dot markings of a transformer. Let us discuss the method of determining the dot markings of the coupled inductors in Fig. 4.20 so that we can represent them using the simplified circuit of Fig. 4.19. The steps are as follows:

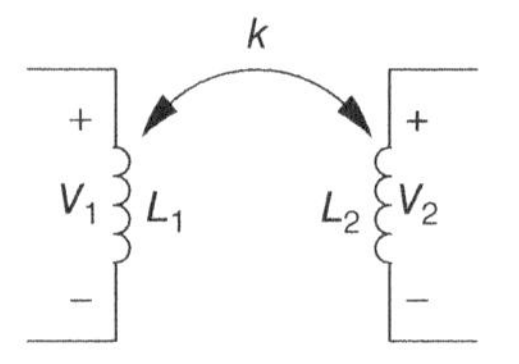

Figure 4.19 Circuit diagram of coupled inductors.

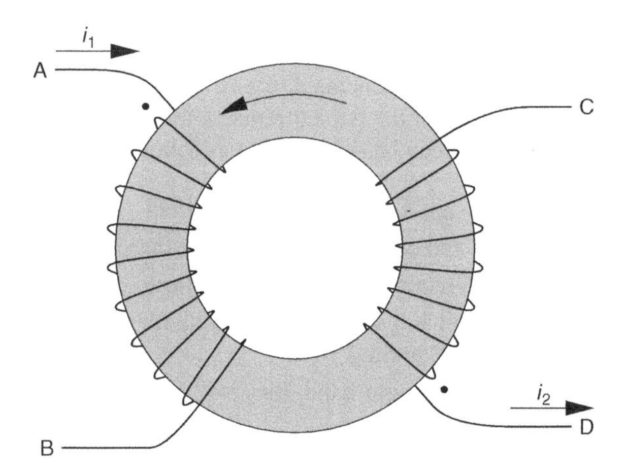

Figure 4.20 Determination of the dot markings.

1. Pick any side and draw a current in any direction. In this case we have chosen the left-hand coil, then choose a current flowing from terminal A into the coil.
2. Place a dot marking at the point where the current enters the coil.
3. Use the right-hand rule to determine the flux direction and draw an arrow. The flux direction corresponding to the current i_1 in coil 1 is shown in Fig. 4.20.
4. Use the right-hand rule to find the current direction in coil 2 so that the flux produced by this current opposes the flux produced by the current in coil 1 (this is why we indicated the flux produced by i_1 in the figure).
5. Draw the direction of current found on the sketch. This current is shown in Fig. 4.20 as i_2.
6. Place the last dot where the current exits coil 2.

Example 4.3 Consider the circuit in Fig. 4.21; calculate i_1 and i_2. The values of the inductors are $L_1 = 400$ mH and $L_2 = 100$ mH.

The mutual inductance is

$$
\begin{aligned}
M &= k\sqrt{L_1 L_2} \\
&= 0.9\sqrt{0.4 \times 0.1} = 180\,\text{mH}.
\end{aligned}
$$

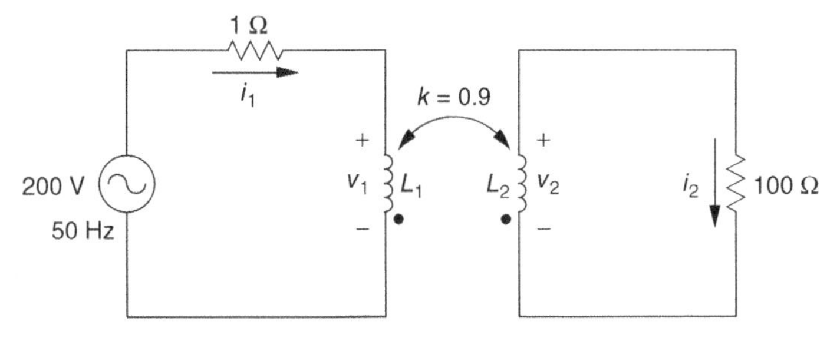

Figure 4.21 Example 1: Coupled inductors.

The impedance of the different inductors at the circuit frequency is

$$\mathbf{Z}_{\mathrm{L}_1} = j125.67\,\Omega$$
$$\mathbf{Z}_{\mathrm{L}_2} = j31.42\,\Omega$$
$$\mathbf{Z}_{\mathrm{M}} = j56.55\,\Omega.$$

We can now write the following mesh current equations

$$0 = 200 - 1\mathbf{I}_1 - \mathbf{V}_1$$
$$0 = -\mathbf{V}_2 + 100\mathbf{I}_2.$$

From (4.60), we can write

$$\mathbf{V}_2 = -\mathbf{Z}_{\mathrm{L}_2}\mathbf{I}_2 + \mathbf{Z}_{\mathrm{M}}\mathbf{I}_1$$

from which it follows that

$$0 = \mathbf{Z}_{\mathrm{L}_2}\mathbf{I}_2 - \mathbf{Z}_{\mathrm{M}}\mathbf{I}_1 + 100\mathbf{I}_2$$

or equivalently

$$\mathbf{I}_1 = \frac{100 + j31.42}{j56.55}\mathbf{I}_2.$$

Similarly, from (4.59), we can write

$$\mathbf{V}_1 = \mathbf{Z}_{\mathrm{L}1}\mathbf{I}_1 - \mathbf{Z}_{\mathrm{M}}\mathbf{I}_2$$

which we can use to rewrite the first mesh current equation as

$$0 = 200 - 1\mathbf{I}_1 - \mathbf{Z}_{\mathrm{L}_1}\mathbf{I}_1 + \mathbf{Z}_{\mathrm{M}}\mathbf{I}_2.$$

Let us substitute $\mathbf{I}_1$ with the expression for $\mathbf{I}_1$ in terms of $\mathbf{I}_2$ to yield

$$0 = 200 - (1 + j125.67)\mathbf{I}_1 + j56.55\mathbf{I}_2$$
$$= 200 - (1 + j125.67)\frac{100 + j31.42}{j56.55}\mathbf{I}_2 + j56.55\mathbf{I}_2$$

$$\therefore \mathbf{I}_2 = 0.896\angle - 2.96^\circ\text{ A}$$

and $\mathbf{I}_1 = 1.66\angle - 75.5^\circ$ A.

4.4 IDEAL TRANSFORMER

In many instances, it is easier to work with the concept of an ideal transformer. Let us define the transfer ratio, or equivalently, the turns ratio of this ideal transformer as

$$\boxed{a = \frac{N_1}{N_2}.} \tag{4.69}$$

With this definition, the following relationships hold for the ideal transformer:

$$\frac{v_1(t)}{v_2(t)} = \frac{N_1}{N_2} = a \tag{4.70}$$

$$\frac{i_1(t)}{i_2(t)} = \frac{N_2}{N_1} = \frac{1}{a}. \tag{4.71}$$

These equations are, indeed, the same as those developed in Section 4.3.1 for the ideal case where all the flux from the first coil couples with the second coil. This is, the case where $k = 1$.

We draw this ideal transformer in a circuit diagram as shown in Fig. 4.22.

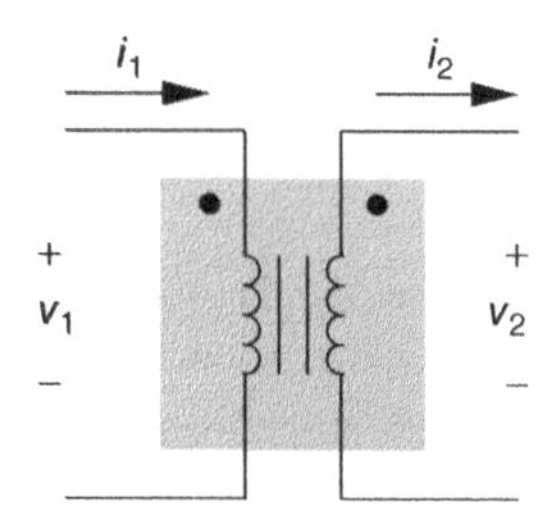

Figure 4.22 An ideal transformer.

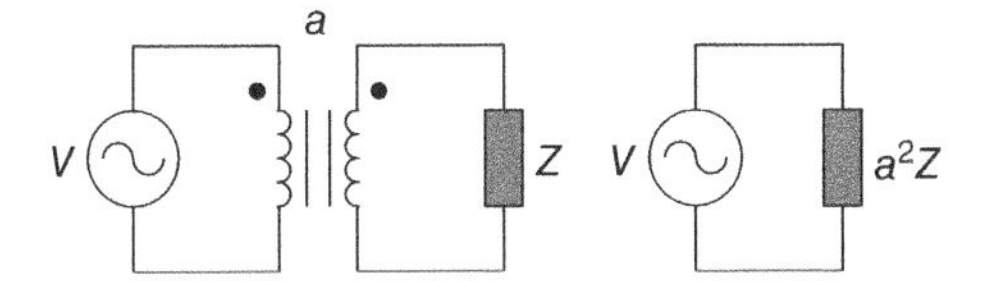

Figure 4.23 Referring an impedance from the secondary to the primary.

4.4.1 Referral of an Impedance

It is possible to refer an impedance from one side of an ideal transformer to another. This process implies that we replace an impedance on one side of an ideal transformer with an *equivalent* impedance on the other side of the transformer. Let us investigate this with a concrete example.

Consider the circuit in Fig. 4.23. Let us first consider the original circuit, with the ideal transformer, on the left. The voltage across the impedance is

$$V_{\mathrm{Z}} = aV \tag{4.72}$$

The current through the impedance on the secondary I_{Z} and the current through the source I_{S} can be calculated as

$$I_{\mathrm{Z}} = \frac{V_{\mathrm{Z}}}{Z} = \frac{1}{a}\frac{V}{Z} \tag{4.73}$$

$$I_{\mathrm{S}} = \frac{1}{a}I_{\mathrm{Z}} = V\frac{1}{a^2 Z}. \tag{4.74}$$

Therefore, we see that from the source's point of view, there is no difference whether an impedance with a magnitude a^2Z is directly connected to it, as shown on the right-hand side of Fig. 4.23 or whether an impedance with a magnitude Z is connected through a transformer with a turns ratio a. If we denote the equivalent impedance on the primary as Z', then we can refer to the impedance as

$$\boxed{Z' = a^2 Z.} \tag{4.75}$$

Notation Note 4.2

When an impedance is moved from one side of a transformer to another, we speak of a referred impedance. In mathematical shorthand, we use Z' to denote an impedance Z that was referred to the other side of the transformer. However, this notation does not explicitly state whether the impedance was referred from the primary to the secondary or vice versa. Therefore, we have to add a one-off explanatory note whenever we first introduce this notation in a solution.

4.4.2 Leakage and Magnetising Inductances

Replacing coupled inductors with the ideal transformer circuit is very handy as it makes the calculation of currents and voltages very easy. However, in real-life circuits, the coupling is never perfect and we cannot just replace the coupled inductors with the ideal transformer model. That said, we can add some external components to *approximate* the behaviour of the non-ideal coupling.

Let us consider Fig. 4.18 and Section 4.3.2 again. The inductance of coil 1 without the influence of coil 2 is

$$L_1 = N_1^2 \mathscr{P}_1. \tag{4.76}$$

However, not all the flux couples from coil 1 to coil 2, hence we have two permeance paths which we know are related such that

$$\mathscr{P}_1 = \mathscr{P}_{11} + \mathscr{P}_{21} \tag{4.77}$$

and we can therefore rewrite the self-inductance of coil 1 as

$$L_1 = N_1^2 \left(\mathscr{P}_{11} + \mathscr{P}_{21} \right) = L_{11} + L_{21}. \tag{4.78}$$

We see therefore that we have two inductors, one representing the flux that is 'leaked' into the surrounding air and the other representing the flux that is coupling with the second coil. We call the inductance associated with the leaked flux the *leakage inductance* (L_{L_1}) and the inductance associated with the coupled flux the *magnetising inductance* (L_m).

However, as we do not have knowledge about the relative magnitudes of $\mathscr{P}_{11}$ and $\mathscr{P}_{21}$, it is difficult to determine the precise ratio of L_{L_1} to L_m. However, it is possible to determine the relative size of each because we know the coupling coefficient of the coils. We have defined the mutual inductance as

$$M = k\sqrt{L_1 L_2} = N_1 N_2 \mathscr{P}_{21}. \tag{4.79}$$

Because the magnetising inductance is defined as $L_\mathrm{m} = N_1^2 \mathscr{P}_{21}$, we can determine the size of the magnetising inductance (on the primary side) as

$$\boxed{L_\mathrm{mp} = N_1^2 \mathscr{P}_{21} = \frac{N_1}{N_2} N_1 N_2 \mathscr{P}_{21} = \frac{N_1}{N_2} M = \frac{N_1}{N_2} k\sqrt{L_1 L_2}.} \tag{4.80}$$

The magnetising inductance does however pose an interesting question: how many magnetising inductors are there? Do we have one on the primary and another on the secondary? Let us use the same argument on the secondary side and find out.

We know that

$$\mathscr{P}_2 = \mathscr{P}_{22} + \mathscr{P}_{21}. \tag{4.81}$$

Therefore, using the same argument we can identify the secondary-side leakage inductance and magnetising inductance as

$$L_2 = N_2^2 \mathscr{P}_2 = N_2^2 \left(\mathscr{P}_{22} + \mathscr{P}_{21}\right) = L_{L_2} + L_{\text{ms}} \tag{4.82}$$

where

$$\boxed{L_{\text{ms}} = N_2^2 \mathscr{P}_{21} = \frac{N_2}{N_1} N_1 N_2 \mathscr{P}_{21} = \frac{N_2}{N_1} M.} \tag{4.83}$$

Now, this in an interesting result; note that

$$\boxed{L_{\text{mp}} = \frac{N_1^2}{N_2^2} L_{\text{ms}} = L'_{\text{ms}}.} \tag{4.84}$$

We see that the magnetising inductance that we have calculated on the secondary side is simply the referred magnetising inductance from the primary side. We see therefore that there is only one magnetising inductance. Actually, this makes a lot of sense because in Fig. 4.18 we see that there is only one flux associated with both coils, the linked flux. The magnetising inductance represents the energy stored in this flux and as there is only one flux path there is only one magnetising inductance.

Having seen that there is only one magnetising inductance, we can develop the equivalent circuit completely. Consider the two circuits in Fig. 4.24. We can approximate the circuit by making the assumption that the coefficient of coupling is high enough so that we can assume that the turns ratio and the transformation ratio of the ideal transformer are the same. This is a reasonable assumption for most coupling coefficients that we will encounter. Using this assumption and the fact that

$$L = N^2 \mathscr{P}$$

we can calculate the approximate turns ratio of the ideal transformer as

$$a \approx \frac{N_1}{N_2} = \sqrt{\frac{L_1}{L_2}}. \tag{4.85}$$

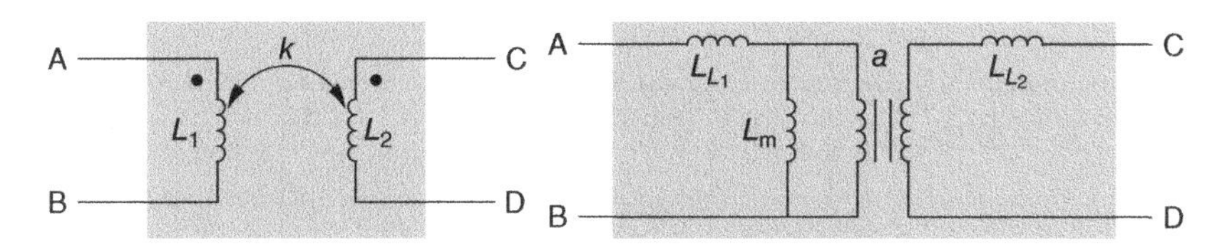

Figure 4.24 Approximating coupled inductors with an equivalent transformer model.

We calculate the mutual inductance as

$$M = k\sqrt{L_1 L_2}$$

and the magnetising inductance (which we will place on the primary side for convenience sake) is

$$L_{\rm m} = \frac{N_1}{N_2}M = \frac{N_1}{N_2}k\sqrt{L_1 L_2} = \sqrt{\frac{L_1}{L_2}}k\sqrt{L_1 L_2} = kL_1. \tag{4.86}$$

The primary-leakage inductance is therefore

$$\boxed{L_{L_1} = L_1 - L_{\rm m} = (1-k)L_1} \tag{4.87}$$

and similarly, the secondary-side leakage inductance becomes

$$\boxed{L_{L_2} = (1-k)L_2.} \tag{4.88}$$

Example 4.4 Calculate the currents i_1 and i_2 shown in Fig. 4.21 (as calculated in Example 4.3) using the leakage-magnetising inductance method.

We can redraw the circuit of Fig 4.21 using idealised components as shown in Fig. 4.25.

The turns ratio can be calculated from the inductances L_1 and L_2. As we know that

$$L = \mathscr{P}N^2$$

we can find

$$\frac{N_2}{N_1} = \sqrt{\frac{L_2}{L_1}} = 0.5 = 2:1.$$

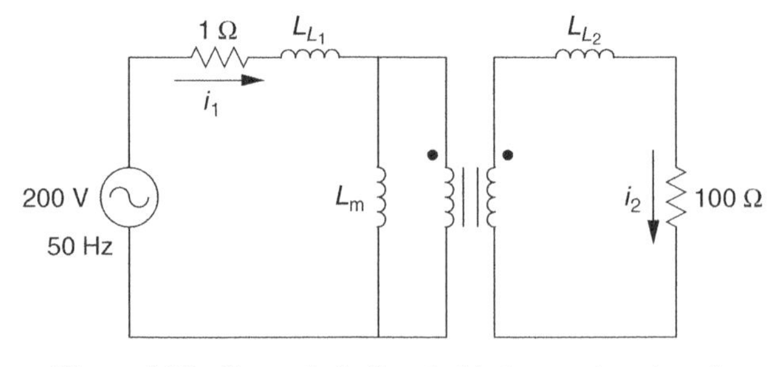

Figure 4.25 Example 2: Coupled inductors (continued).

The mutual inductance is

$$M = k\sqrt{L_1 L_2} = 180\,\text{mH}.$$

Let us refer the magnetising inductance to the primary side; then,

$$L_\text{m} = \frac{N_1}{N_2} M = 2 \times 180 = 360\,\text{mH}.$$

The primary-side leakage inductance is therefore

$$L_{L_1} = 400 - 360 = 40\,\text{mH}.$$

We can calculate the secondary-side leakage inductance using the same method

$$L_{L_2} = L_2 - \frac{N_2}{N_1} M = 10\,\text{mH}.$$

The load on the secondary side can be written in phasor form as

$$\mathbf{Z_2} = 100 + j2\pi 50 L_{L2} = 100\angle 1.8^\circ\ \Omega.$$

If we refer this value to the primary side we have

$$\mathbf{Z}_{2'} = \frac{N_1^2}{N_2^2}\mathbf{Z}_2 = 400\angle 1.8^\circ\ \Omega.$$

The parallel combination of the magnetising inductance and the referred secondary-side impedance is

$$\mathbf{Z_p} = \left(\frac{1}{j2\pi 50 L_\text{m}} + \frac{1}{400\angle 1.8}\right)^{-1} = 107.95\angle 74.35^\circ\ \Omega.$$

We can calculate the primary current as

$$\mathbf{I_1} = \frac{200}{1 + 2\pi 50 L_{L_1} + \mathbf{Z_p}} = 1.66\angle -75.51^\circ\ \text{A}.$$

The voltage across the primary of the transformer is therefore

$$\mathbf{V_1} = 200 - (1 + j2\pi 50 L_{L_1})\mathbf{I_1} = 179.4\angle -1.16^\circ\ \text{V}$$

and the secondary current

$$\mathbf{I_2} = \frac{\frac{N_2}{N_1}\mathbf{V_1}}{\mathbf{Z_2}} = 0.897\angle -2.96^\circ\ \text{A}.$$

For practice, redo the problem with both methods but change the coupling coefficient to $k = 0.99$.

4.5 HIGHLIGHTS

- Magnetic field strength is the magnetic force that establishes a magnetic field. The strength of the magnetic field strength can be calculated from Ampère's law as

$$H(t) = \frac{Ni(t)}{\ell}.$$

The magnetic field strength is measured in ampère-turns per metre (At/m).

- Magnetic flux, ϕ, is measured in Weber (Wb).
- Magnetic flux density is the amount that permeates through a surface area, or

$$B = \frac{\phi}{A}$$

and is measured in Tesla (T).

- According to Faraday's law, a voltage is induced in a coil because of a change in the magnetic flux contained in the coil:

$$e(t) = N\frac{d\phi(t)}{dt}.$$

- The relationship between H and B is highly non-linear but in making assumptions, mainly that the magnetic circuit is not operating in the saturation region, we can make a linear approximation that

$$B = \mu_0 \mu_r H,$$

where μ_r is the relative permeability and $\mu_0 = 4\pi \times 10^{-7}$ is the permeability of free space.

- The self-inductance of a coil can be found as

$$L = \mu N^2 \frac{A}{\ell}.$$

- If the magnetic system is not homogeneous (e.g. the magnetic circuit includes an air gap), then the flux density can be calculated as

$$B(t) = \frac{Ni(t)}{\frac{\ell_1}{\mu_1} + \frac{\ell_2}{\mu_2}}.$$

- With the assumption, that is normally true for most air gap magnetic systems we are interested in, that $\frac{\ell_c}{\mu_r} \ll \ell_g$ it is true that

$$B(t) \approx \mu_0 \frac{Ni(t)}{\ell_g}$$

and

$$L \approx \mu_0 \frac{N^2 A}{\ell_g}.$$

- We can model two coupled inductors using

$$\begin{bmatrix} v_1 \\ v_2 \end{bmatrix} = \begin{bmatrix} L_1 & -M \\ M & -L_2 \end{bmatrix} \frac{d}{dt} \begin{bmatrix} i_1 \\ i_2 \end{bmatrix},$$

  where

$$M = k\sqrt{L_1 L_2}.$$

- We use the dot-marking scheme when drawing a transformer symbol to help us distinguish the winding direction of a transformer.
- For ideal transformers, $k = 1$, we define the turns ratio as

$$a = \frac{N_1}{N_2}$$

  and get the following equations

$$\frac{v_1(t)}{v_2(t)} = a$$

$$\frac{i_2(t)}{i_1(t)} = a$$

$$Z' = a^2 Z,$$

  where Z is the impedance connected to the secondary side and Z' is the equivalent impedance referred to the primary side.
- We can model two coupled inductors, with non-ideal coupling, using an ideal transformer model, a magnetising inductance and two leakage inductances. The values of the magnetising inductance and leakage inductance are (magnetising inductance modelled on the primary side)

$$L_m = \frac{N_1}{N_2} M = \frac{N_1}{N_2} k\sqrt{L_1 L_2}$$

$$L_{L_1} = (1 - k)L_1$$

$$L_{L_2} = (1 - k)L_2$$

PROBLEMS

4.1 What consequences do the non-linear (saturating) magnetisation curves have on the design and performance of magnetic components?

4.2 What effect does the presence (or absence) of an air gap in the magnetic core have on the inductance of a magnetic component? What would happen if we inserted an air gap in a core that has two coils?

4.3 Propose methods to decrease/increase the coupling factor between two coils.

4.4 You have an inductor with 50 turns and an inductance of 1 mH. How many turns do you have to add to make the inductance 1.69 mH?

4.5 You have an inductor and are able to measure the inductance accurately as 540 μH. You take 15 turns off the inductor winding and measure the inductance again to find it has dropped to 437.4 μH. How many turns did the original inductor have?

4.6 Design an air-core toroid to generate a magnetic flux of 0.1 T that varies by less than 2% over a cross-sectional area of 10 cm. You have a current source of 10 A at your disposal.

4.7 Calculate the source current I_s and the secondary current I_2 for the circuit in Fig. 4.26. Use $L_1 = 400$ mH and $L_2 = 100$ mH.

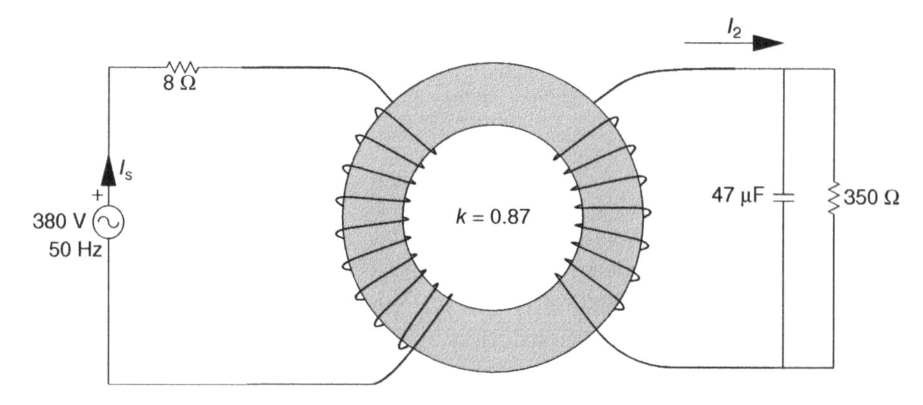

Figure 4.26 Transformer in an electrical circuit.

4.8 The dimensions of a square core inductor are $a = 1$ cm, $b = 4$ cm, $w = 1$ cm and $h = 1$ cm. The magnetic core saturates at 1.5 T, and the relative permeability of the material is 2 000. If the coil has 100 turns, what is the inductance value and at what current value will the core saturate?

4.9 The battery of a radio-controlled car is charged by means of an air-cored transformer. The primary is on the floor and the secondary is mounted on the vehicle. The equivalent circuit is measured, resulting in the following values: $L_m = 19\ \mu H$, $L_{L_1} = 89\ \mu H$ and $L_{L_2} = 13\ \mu H$. A 300 V 100 kHz source is connected between the primary terminals and a 100 Ω resistor is connected to the secondary terminals.

1. Calculate the voltage on the load.
2. Do you think such a transformer is suitable for charging the battery?
3. Could you suggest a method to improve the power transfer?

4.10 An electrical machine is largely made of high permeability steel to enhance the magnetic field produced by the currents. Assume that the stator and rotor of the machine are made from material that has an infinitely large permeability so that the air gap governs the inductance. Then calculate the flux (in Weber) per pole in a two-pole dc machine, set up by a 1.2 A current when the following dimensions apply: rotor length = 10 cm, rotor diameter = 5 cm, air gap = 1 mm, the number of turns on the field winding is 1 000. Assume that each pole covers 180° of the rotor.

FURTHER READING

Boylestad R.L., *Introducory Circuit Analysis*. 8th edition, Prentice-Hall International, Upper Saddle River, New Jersey, 1997.

Erickson R.W., Maksimoviĉ D., *Fundamentals of Power Electronics*. 2nd edition, Kluwer Academic Press, Norwell, 2001.

Mohan N., Undeland T.M., Robbins W.P., *Power Electronics: Converters, Applications, and Design*. 2nd edition, John Wiley and Sons, New York, 1995.

Nilsson J.W., Riedel S.A., *Electric Circuits*. 6th edition, Prentice-Hall, Upper Saddle River, New Jersey, 2000.

CHAPTER 5

DYNAMICS OF ROTATIONAL SYSTEMS

5.1 INTRODUCTION

If one critically observes electrical systems, it becomes clear that either the input or the output of the system is likely to be in the form of mechanical power. Think, for instance, about the steam turbines of power stations that generate mechanical power, which is then converted into electrical power by generators. The same is true of wind turbines. On the load side, there are washing machines and air conditioners in homes that use electrical motors. In factories, many more mechanical loads can be found, for example, pumps, cranes, hoists and ventilation fans.

Transportation systems have, for a long time, been the near exclusive domain of mechanical and chemical energy systems. However, as the world becomes more aware of the need for energy efficiency, we see electrical systems more and more often in transportation. As we saw in Chapter 1, this has been happening for a while in trains and rail transport. That said, recently we have seen this trend accelerating; consider all the examples visible in daily life—hybrid electrical and fully electrical automobiles as well as electrically assisted bicycles. This trend is also continuing in areas that are not quite so visible in daily life. For example, it is now extremely rare to find a large ship that uses a diesel engine directly coupled to the propulsion screw (propeller). For efficiency purposes and also because of the fact that a great many systems on a ship require electrical energy, the diesel engine is, instead, connected to a generator that converts the available mechanical

The Principles of Electronic and Electromechanic Power Conversion: A Systems Approach, First Edition.
Braham Ferreira and Wim van der Merwe.

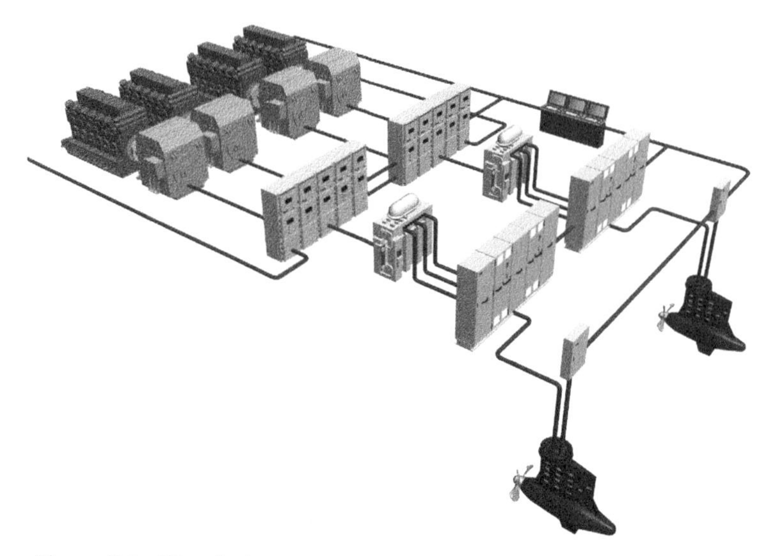

Figure 5.1 Electrical power generation and propulsion system on a ship.

energy to electrical energy. An electrical machine is then used to convert the electrical energy back to mechanical energy at a convenient location. This system is shown in Fig. 5.1 where we see four diesel engines connected to generators. The resulting electrical energy is then fed by wire to locations where it is needed throughout the ship. The ship propulsion is done by the two Azipod modules; Azipod is a product of the Swiss-Swedish company ABB, which allows for thrust in any direction as the module can swivel around as desired by the ship's navigation system or by the Captain for that matter. Two Azipod modules connected under the hull of a ship are shown in Fig. 5.2.

Owing to this symbiosis between the two systems, any study of the use of electrical energy conversion should include at least a cursory glance at rotational mechanical systems. Although it is not required to understand all the finer details, we need to know enough to understand that, for example, when designing a system, we often have the choice of controlling speed with a mechanical or an electronical/electromechanical gearbox. More importantly, the mechanical inertias and the properties of the mechanical loads usually determine the dynamics of the conversion system.

The purpose of this chapter is to briefly cover important mechanics knowledge that is relevant for conversion systems.

5.2 PRELIMINARIES

Rotational system dynamics is a part of classical mechanics and is in many respects similar to the dynamics of rigid body systems moving in free space. In fact, in a rotational system, there is only one direction of freedom—rotation about a fixed axis.

Figure 5.2 ABBs novel Azipod ship propulsion system.

Therefore, we can simplify this statement to specify that a rotational system is similar to rigid body dynamics with linear motion. Let us, for a start, review some of the equations describing linear motion.

Consider the system shown in Fig. 5.3. The mass in this case is restricted to move only in one plane, almost like a train on tracks. The *position* of the mass is the distance from a certain fixed point and is given by x.

If we define forward as moving to the right then the mass can either move forwards or backwards. All other movements are prohibited. The *velocity* of the mass is the change in position with time or

$$v = \frac{dx}{dt}. \tag{5.1}$$

Remember that velocity is speed but with a direction associated with it. In this case, because the movement can only be forwards or backwards, the speed will either be positive or negative. The *acceleration* of the mass is the change in speed with time

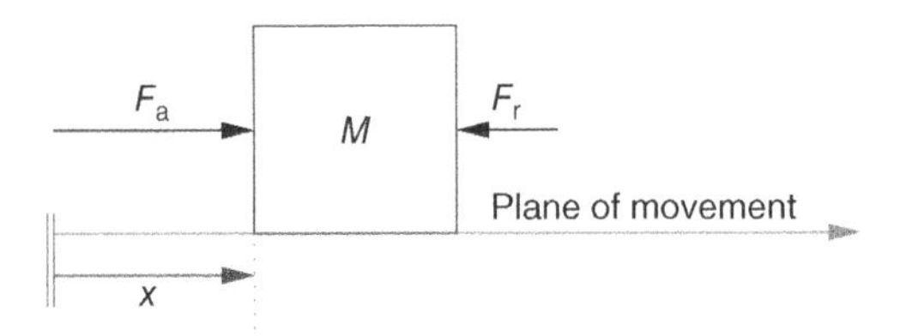

Figure 5.3 Linear motion of a mass.

or

$$a = \frac{dv}{dt} = \frac{d^2x}{dt^2}. \tag{5.2}$$

If we consider first the case of only one force, F_a, applied to the mass, we assume $F_r = 0$, then the acceleration of the mass is found using Newton's second law as

$$a = \frac{F_a}{M}. \tag{5.3}$$

Naturally, this implies that the mass will reach warp speed as it will accelerate until the end of time. However, in real life, things are never as exciting. Any moving mass will experience some force that opposes movement whether it be friction, air resistance or simply the extra effort required to move the mass uphill. What about a mass moving in space where there is no friction, you may ask. We will address that in a moment. The acceleration of the mass is simply due to the result of all the forces that act on the mass

$$\boxed{a = \frac{\sum F}{M}.} \tag{5.4}$$

Now, the force will perform a certain amount of work if it can displace the mass from one position to another. Remember that simply applying force is not enough. If this was true, then many a lazy construction worker leaning against a wall could successfully argue that he is doing work even though the wall is not moving. Of course when the wall moves we have another problem altogether. Formally, work completed (done) is defined as

$$W = F\Delta x. \tag{5.5}$$

From the work–energy theorem, we know that if we ignore any changes in potential energy (i.e. we assume the plane of movement to be horizontal), the change in energy simply reflects the work done. As the kinetic energy of a mass is given by

$$\boxed{E_k = \frac{1}{2}Mv^2} \tag{5.6}$$

we can also write the work done in a period of time as

$$W = \Delta E_k = \frac{1}{2}M(v_2^2 - v_1^2), \tag{5.7}$$

where v_1 is the velocity at the start of the period and v_2 the velocity at the end.

Power is the amount of work done in a second

$$P = \frac{dW}{dt}. \tag{5.8}$$

If the force remains constant, then

$$\boxed{P = F\frac{dx}{dt} = Fv.} \tag{5.9}$$

If we revisit our problem of the mass accelerating until the end of time when no friction is present, we can see that as the speed increases, the amount of power delivered by the constant source increases. Therefore, accelerating a 1 kg mass at 1 m/s requires 1 N of force when starting from zero speed, and when reaching light speed it requires nearly 300 MW of power. This is the equivalent of four GE90 jet engines as installed on a Boeing 777 (a 777 operates with only two of these bad boys). Because one engine weighs 8300 kg without any fuel, this is clearly not a feasible proposition (even ignoring the obvious fact that jet engines do not work in space).

5.3 ROTATIONAL DYNAMICS

The dynamics of a system rotating about a fixed axis is very similar to the dynamics of the rigid body described above.

5.3.1 Torque

Because a rotating system cannot move in a linear plane, the acting force of interest in a rotating system is the force that initiates rotation. This force is called *torque* and is measured in newton-metres (NM). Therefore, just as a linear force can be thought of as a pushing force or a pulling force, we can visualise torque as a twisting force.

Definition 5.1

The torque in a system rotating about a fixed axis is defined as the rotating force acting on the system. It is the force applied at a right angle to a radial line from the axis of rotation multiplied by its distance from the axis of rotation.

Consider the definition of torque in Fig. 5.4. Here, the torque is

$$\boxed{T = F\ell.} \tag{5.10}$$

It is clear from this figure that the force F tries to rotate the lever about the axis. The resulting torque is shown.

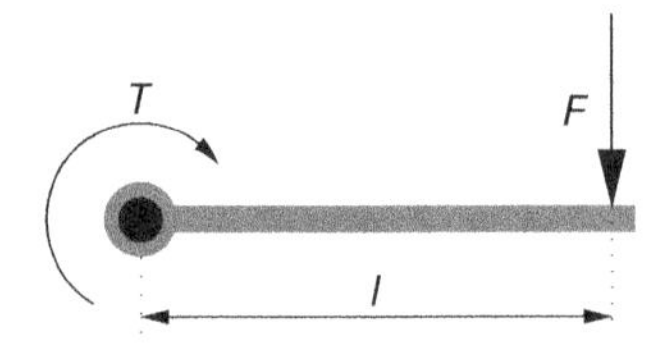

Figure 5.4 Definition of torque.

5.3.2 Angular Displacement, Speed and Acceleration

As with the movement of bodies along a linear axis, we need to define a suitable method of describing displacement from some reference point. Let us define angular displacement, θ, as the rotational displacement from a fixed reference point measured in radians. This definition is shown in Fig. 5.5.

Angular speed is denoted by ω and is measured in radians per second (rad/s). As in linear motion dynamics, it is defined as

$$\boxed{\omega = \frac{d\theta}{dt}.} \tag{5.11}$$

In some instances, it makes more sense to rewrite angular speed using a more human readable unit, rpm or rotations per minute. Angular speed expressed in rotations per minute is normally denoted by the symbol n. As there are 2π radians in a rotation, the conversion is easily found as

$$\boxed{n(\text{rpm}) = \omega(\text{rad/s}) \times \frac{60}{2\pi}.} \tag{5.12}$$

Angular acceleration is found in a similar manner; it is written as α with the unit (rad/s^2) and is defined as

$$\boxed{\alpha = \frac{d\omega}{dt} = \frac{d^2\theta}{dt^2}.} \tag{5.13}$$

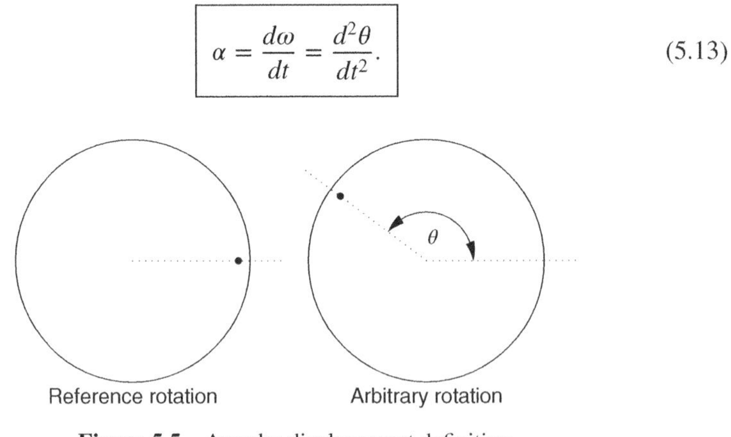

Figure 5.5 Angular displacement definition.

5.3.3 Equations of Rotational Motion

The dynamics of a rotating system follows the same form as the linear motion of a rigid body. We saw that in a linear motion system, the applied force and the system acceleration are related to each other by means of the mass. In a rotational system, we have a concept similar to the mass of an object called the *inertia*, which we denote by J. The inertia of an object is a measure of how much force it would take to rotationally accelerate the object. Similarly to linear motion, the applied torque and the rotational acceleration are related as

$$\boxed{\alpha = \frac{T}{J}.} \tag{5.14}$$

The angular kinetic energy is

$$\boxed{E_{\mathrm{k}} = \frac{1}{2} J\omega^2} \tag{5.15}$$

and the power applied to the system is

$$\boxed{P = T\omega.} \tag{5.16}$$

However, there is a difference between a linear system and a rotational system. A linear system operates the same irrespective of the direction of acceleration. To explain in what way the rotational system differs, consider an athletics javelin thrower and his javelin. Because the javelin is long we know that it is easier to spin the javelin about its long axis than trying to spin the javelin about its short axis. Although the weight of the javelin remains the same irrespective of the direction of rotation, the moment of inertia must change with the change in rotation axis.

5.3.4 Moment of Inertia

The moment of inertia of an object about a given axis describes how difficult it is to change its angular motion about that axis. Therefore, it encompasses not just how much mass the object has overall but also how far each bit of mass is from the axis. The further out the object's mass is, the more rotational inertia the object has and the more force required to change its rotational rate. This is clearly demonstrated by the figure skater who pulls her arms towards her body while spinning to increase the rotational speed of her spin. This phenomenon is explained later.

Inertia is denoted by J and is measured in kilogram square metre. As inertia describes how difficult it is to rotate a body about a fixed axis, it fulfils the same role in rotational systems as mass does in the description of rigid body dynamics.

Formally, the moment of inertia of a continuous solid body rotating about an axis can be found by the integral

$$J = \int_V \rho(\mathbf{r}) d(\mathbf{r})^2 dV(\mathbf{r}), \tag{5.17}$$

where $\mathbf{r}$ is the radius vector from the axis of rotation, $\rho(\mathbf{r})$ is the mass density at point $\mathbf{r}$ and V is the volume of the body. However, this formal definition is rarely used because most objects we are interested in will usually have a uniform (or nearly uniform) mass distribution.

Rather than calculating the moment of inertia from first principles for every rotating body, we can use the reference shape in Fig. 5.6 to good effect. This moment of inertia calculation is valid for any shape resembling Fig. 5.6 which is made from a single material, that is, having a uniform mass distribution. Using this reference shape, we can calculate the moment of inertia of a disc by choosing $r_1 = 0$; therefore, $J_{\text{disk}} = \frac{1}{2}Mr_2^2$. Similarly, the moment of inertia of a wheel can be approximated by assuming that all the mass is contained at a radius r from the axis. Therefore, using $r_1 = r_2 = r$, we get $J_{\text{wheel}} = Mr^2$.

5.3.5 Rotating System

Consider the rotating system shown in Fig. 5.7. We have, in this instance, an electrical motor that applies a torque T_{m} to a load that is connected to the machine by

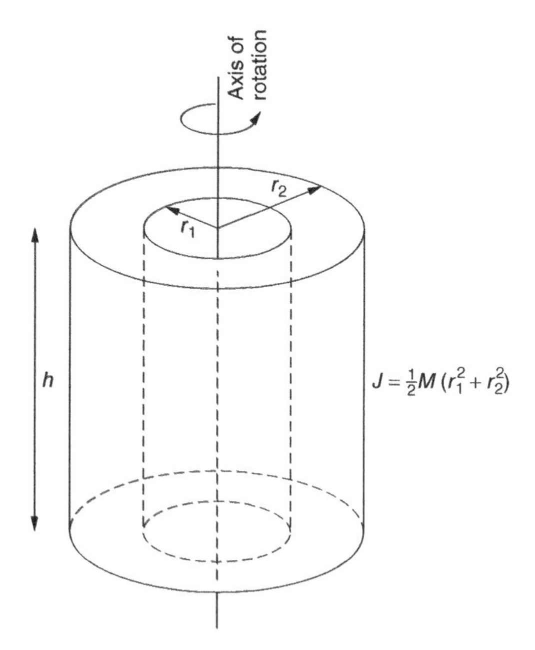

Figure 5.6 Moment of inertia of a uniform cylinder.

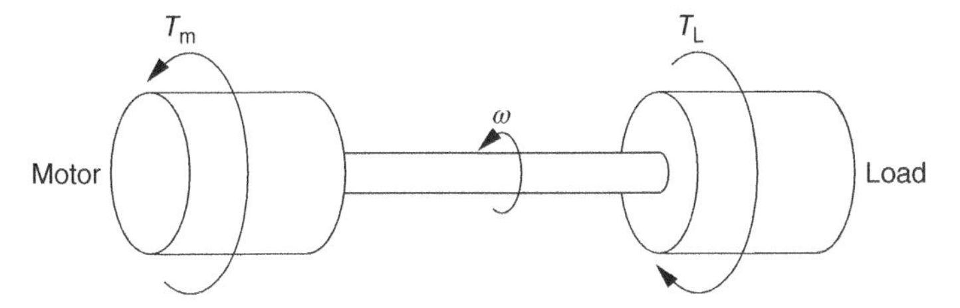

Figure 5.7 Torque in a rotational system.

a shaft. Let the inertia of the motor be J_m, and the inertia of the shaft and load are J_s and J_L, respectively. Firstly, recall from Chapter 3 that we defined the delivery of power as negative power. Positive power is therefore defined as the power that is dissipated. As this definition is difficult to reconcile with our definitions of the torque and angular speed direction as indicated in Fig. 5.7, we will specify the power direction in our discussions of rotational systems.

Now, we can find the angular acceleration of the system in Fig. 5.7 as

$$\alpha = \frac{T_m - T_L}{J_m + J_s + J_L}, \tag{5.18}$$

which we can represent as

$$\boxed{\alpha = \frac{\sum T}{\sum J}.} \tag{5.19}$$

The power delivered by the motor is

$$P_m = T_m \omega. \tag{5.20}$$

If the system is still accelerating, we know that $T_L < T_m$ and therefore

$$P_L = T_L \omega < P_m, \tag{5.21}$$

while the acceleration will continue until such time that $T_m = T_L$.

Now, let us revisit the example of the figure skater who increases her angular speed by pulling her arms in while spinning. Looking at the skater with our physicist hats on we know that the moment she pulls her arms towards her body she decreases her moment of inertia about an axis running through her spine. As she is hopefully spinning along the same axis, she is effectively changing the moment of inertia of a rotating system. As we know that the energy contained in a system cannot change without the influence of external forces, the angular kinetic energy before she pulled her arms in must be the same as that with her arms across her chest. Let the moment of inertia with arms out wide be J_1 and with arms pulled in be J_2. Then according to the conservation of energy

$$\frac{1}{2}J_1\omega_1 = \frac{1}{2}J_2\omega_2$$
$$\frac{\omega_2}{\omega_1} = \frac{J_1}{J_2}.$$

Example 5.1 An electrical motor is connected to a load using a shaft in an arrangement similar to Fig. 5.7. The inertia of the shaft and load combined is $30\,\text{kg m}^2$ and the load delivers 40 kW at a speed of 1 200 rpm. The load torque is proportional to the load speed. How much angular kinetic energy is stored in the system? Assuming no rotational friction how long will it take for the system to slow down to 1 rpm if all the power is removed from the motor? The inertia of the motor is $5\,\text{kg m}^2$.

The angular speed in the correct unit of rad/s is

$$\omega = 1200\frac{2\pi}{60} = 125.66\,\text{rad/s}.$$

The angular kinetic energy is therefore

$$E = \frac{1}{2}J\omega^2 = 276.35\text{ kJ}.$$

As the load torque is proportional to the load speed, the load torque is

$$T_\text{L} = \kappa\omega$$

and the constant κ can be calculated from the reference point at 1 200 rpm. If the load takes 40 kW at 1 200 rpm, the load torque is

$$T_\text{L} = \frac{P}{\omega} = \frac{40\,000}{125.66} = 318.3\text{ Nm}. \tag{5.22}$$

So the constant κ can be calculated as

$$\kappa = \frac{T_\text{L}}{\omega} = \frac{318.3}{125.66} = 2.53. \tag{5.23}$$

If all power is removed from the motor, the angular acceleration of the system is

$$\alpha = \frac{T_\text{m} - T_\text{L}}{J_\text{m} + J_\text{L}}$$
$$= \frac{-2.53\,\omega}{35}$$
$$\frac{d\omega}{dt} = -\frac{2.53}{35}\omega$$

a first-order differential equation with solution

$$\omega(t) = 125.66 \cdot e^{-2.53/35t}\ \text{rad/s}. \tag{5.24}$$

To slow down to 1 rpm (0.104 rad/s) will therefore take

$$\begin{aligned}
0.104 &= 125.66 \cdot e^{-\frac{2.53}{35}t} \\
e^{-\frac{2.53}{35}t} &= \frac{1}{1200} \\
\therefore -\frac{2.53}{35}t &= \ln\left(\frac{1}{1200}\right) \\
t &= -\frac{35}{2.53}\ln\left(\frac{1}{1200}\right) = 98\ \text{s}.
\end{aligned}$$

5.4 COUPLING MECHANISMS

Good common sense tells us that wherever possible it is better to have the load and the motor directly connected. However, more often than not we need to change the direction of motion. For a start, let us look at the translation from angular to linear motion and vice versa. Consider the vehicle shown in Fig. 5.8. The wheel at the back has a radius r. If we do not allow the drive wheel to slip, then the force applied to the vehicle F must be $F = rT$ to keep the vehicle at standstill. Compare this with the definition of torque in (5.10).

For every rotation of the wheel, the vehicle will move $x = 2\pi r$ m. The relationship between the angular and linear speed is therefore

$$\boxed{v = \omega r.} \tag{5.25}$$

As the front wheel is smaller than the back wheel, it makes sense that it must rotate faster; the angular speed of the front wheel with radius r_s is then

$$\omega_s = \frac{v}{r_s} = \omega\frac{r}{r_s}. \tag{5.26}$$

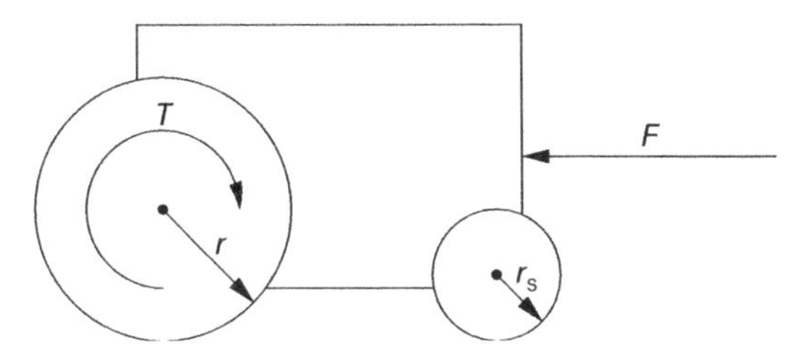

Figure 5.8 Translation from angular to linear motion.

The speed of the back wheel is different from the speed of the front wheel by a ratio of their respective radii.

In many instances, the angular speed of the load and the source can be very different. A good example is the difference in speed between the blades and spindle of a wind generator and an electrical machine used to generate electrical power from wind power. For most electrical machines, it is necessary to keep the speed in the range of 1 000–2 000 rpm to keep the physical size of the machine small and the efficiency high. However, because of mechanical stresses and noise considerations, it is imperative that the tip speed of the wind generator blade is less than the speed of sound. It does not sound like a big problem, does it? Let us look at the problem more closely. The amount of power that can be extracted from the wind is, among other things, a function of the blade length squared. A wind generator with a blade length of 30 m can extract four times more power from the wind as a generator with 15 m blades. Therefore, wind generators tend to be large; the Vestas V80 generator has a rotor diameter of 80 m—that is the wingspan of an Airbus A380! If we require the blade tip speed to be lower than 80% the speed of sound (340 m/s) then

$$\omega \leq \frac{0.8\, v_{\text{sound}}}{r} = 6.8\,\text{rad/s} = 65\ \text{rpm}. \tag{5.27}$$

Clearly, to make the electrical machine small and efficient, the angular speed must be increased by a factor of at least twenty five. In most instances, the difference in angular speed is even worse as the blade tip speed is normally limited to about 60 m/s.

5.4.1 Belt and Pulley

Luckily, it is relatively easy to change the rotational speed by using two wheels with different diameters; remember the two wheels of Fig. 5.8? One of the simplest ways to change the rotational speed is the belt and pulley mechanism shown in Fig. 5.9. This is similar to the gear systems in most bicycles. Let us connect the motor to the pulley with radius r_1 and the load to the pulley with radius r_2. If the angular speed of the motor is ω_1, then the linear speed of the belt at any point is $r_1\omega_1$. Remember that because the belt is continuous, the speed is the same at all points; if not, the belt would 'bundle up' in one area and stretch to an infinite length in another. Therefore, the tangential linear speed at any point along the outer radius of pulley 2 is also $r_1\omega_1$ but it can also be written as $\omega_2 r_2$. Therefore,

$$\boxed{\frac{\omega_1}{\omega_2} = \frac{r_2}{r_1}.} \tag{5.28}$$

If we want to decrease the speed, then $r_1 < r_2$, and to increase the speed $r_1 > r_2$.

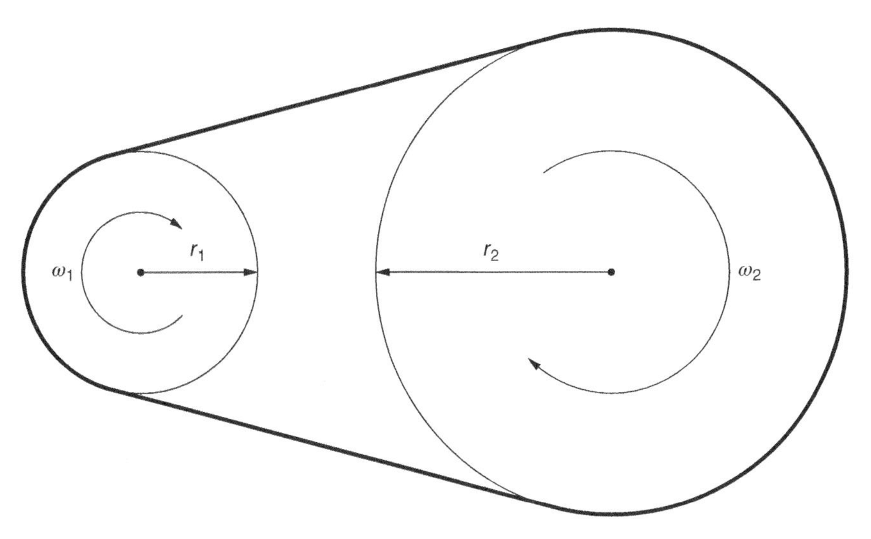

Figure 5.9 A pulley and belt system.

If we assume the belt and pulley system to be 100% efficient, then $P_{in} = P_{out}$. Therefore,

$$\boxed{T_1\omega_1 = T_2\omega_2 \Rightarrow \frac{T_1}{T_2} = \frac{\omega_2}{\omega_1} = \frac{r_1}{r_2}.} \tag{5.29}$$

This change in the torque available at the load is very important for large loads. Let the inertia of the drive pulley and motor be J_1 and the inertia of the load and load pulley be J_2. If the torque of the load is T_L, then the torque equation of the load side is

$$T_2 = T_L + J_2\frac{d\omega_2}{dt}. \tag{5.30}$$

Now, putting the motor back into the system the torque equation from the motor side is, where T_m is the torque developed by the motor,

$$\begin{aligned} T_m &= J_1\frac{d\omega_1}{dt} + T_2\frac{r_1}{r2} \\ &= J_1\frac{d\omega_1}{dt} + \frac{r_1}{r_2}\left(T_L + J_2\frac{d\omega_2}{dt}\right) \\ &= J_1\frac{d\omega_1}{dt} + \frac{r_1}{r_2}\left(T_L + J_2\frac{r_1}{r_2}\frac{d\omega_1}{dt}\right) \end{aligned} \tag{5.31}$$

and therefore

$$T_m = \underbrace{\left(J_1 + \left(\frac{r_1}{r_2}\right)^2 J_2\right)}_{J_{eq}} \frac{d\omega_1}{dt} + \frac{r_1}{r_2} T_L. \tag{5.32}$$

We see that the 'equivalent' inertia seen by the motor is $J_1 + \left(\frac{r_1}{r_2}\right)^2 J_2$. Therefore, if we need to connect a very large load with a high inertia to a motor, we can decrease the load on the motor significantly by using a gearbox. However, the maximum attainable speed at the load side will be limited by the maximum speed of the motor. Luckily, we rarely want to spin large loads at high speeds.

5.4.2 Gears

Many variants of the gearbox exist. In Fig. 5.10, a gearbox consisting of meshing gears is shown; note that in this case, the load side and drive side rotate in opposite directions. For a gear arrangement, the transfer ratio becomes

$$\frac{\omega_1}{\omega_2} = \frac{N_2}{N_1}, \tag{5.33}$$

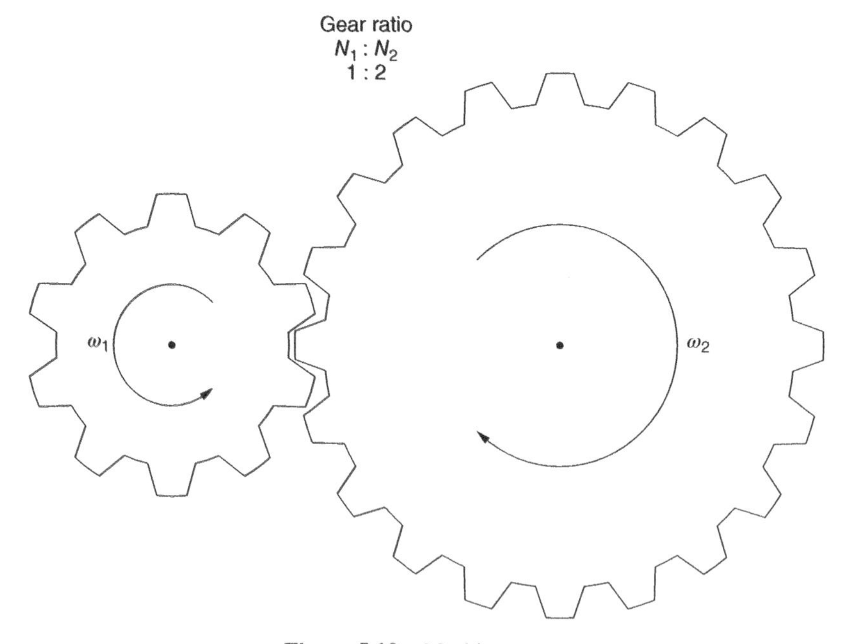

Figure 5.10 Meshing gears.

where N_1 and N_2 are the number of teeth on the two wheels. For the specific gear ratios shown in Fig. 5.10, it is true that $2\omega_1 = \omega_2$.

Example 5.2 A load with an inertia of 10 kg m^2 must be accelerated from rest to 100 rpm within 5 s. A motor with an inertia of 1 kg m^2 is able to deliver 2 Nm of torque from 0 rpm to its maximum speed of 800 rpm. Is it possible to find a gearbox to make this work? Assume there is no load torque.

Let us denote the gear ratio as a, that is, the ratio of the speed of the drive to the speed of load or $a = \frac{\omega_\mathrm{m}}{\omega_\mathrm{L}}$. We can rewrite the torque equation from the motor side as

$$T_m = \left(J_\mathrm{m} + a^2 J_\mathrm{L}\right) \frac{d\omega_1}{dt}.$$

As the torque is constant, the load will be accelerated linearly from standstill to 100 rpm, which means the acceleration of load must be

$$\alpha_2 = \frac{100 \times 2\pi}{60 \times 5} = 2.1 \text{ rad s}^2$$

and the acceleration of the motor is the load acceleration multiplied by the gear ratio, or

$$\alpha_1 = 2.1a \text{ (rad) s}^2.$$

Therefore, substituting all the known values into the torque equation of the motor yields

$$2 = (1 + 10a^2)2.1a$$
$$0 = 21a^3 + 2.1a - 2.$$

This is clearly not an easy equation to solve, not impossible however. We know that because the largest power is three, this polynomial will have at least one real root. If we look at the derivative of the function

$$f(a) = 21a^3 + 2.1a - 2$$
$$f'(a) = 63a^2 + 2.1$$

we can find the inflection points as

$$f'(a) = 0 \Rightarrow a = \pm\sqrt{\frac{-2.1}{63}}.$$

As $f(0) = -2$, we know that we have only one real root. Using the method of Newton–Rapson that states

$$x_1 = x_0 - \frac{f(x_0)}{f'(x_0)}$$

and choosing the first answer to be $a_0 = 0.5$, we find the required gearbox ratio as

$$a_1 = 0.5 - \frac{1.675}{17.85} = 0.406$$
$$a_2 = 0.406 - \frac{0.26}{12.5} = 0.385$$
$$a_3 = 0.385 - \frac{0.0109}{11.455} = 0.3844.$$

It is clear that the answer is accurate enough. One final check is necessary; will the maximum speed of the motor be fast enough?

$$a \times 800 \text{ rpm} = 307.5 \text{ rpm} > 100 \text{ rpm}.$$

Yes, it will work. As the maximum reduction in speed that will give the required 100 rpm output speed is 0.125, we can make the system work by using a reduction gearbox with a ratio anywhere between 0.125 and 0.3844.

Example 5.3 A 300-Nm, 350-rpm load is connected to a motor with a 1 : 10 gearbox that is 98% efficient. How much torque and power does the motor develop?

The load power is

$$P_\text{L} = T_\text{L}\omega_\text{L} = 300 \times 2\pi\frac{350}{60} = 11 \text{ kW}.$$

The power developed by the motor must be

$$P_\text{m} = P_\text{L} + P_\text{loss}$$
$$= P_\text{L} + (1 - \eta)P_\text{m}$$
$$\therefore P_\text{m} = \frac{P_\text{L}}{\eta} = 11\,220 \text{ W},$$

The torque developed by the motor is

$$T_\text{m} = \frac{P_\text{m}}{\omega_\text{m}} = \frac{11\,220}{350 \times 10}\frac{60}{2\pi} = 30.6 \text{ Nm}.$$

5.5 HIGHLIGHTS

- In linear systems, the following equations hold:

$$F = m\,a$$
$$E = \frac{1}{2}mv^2$$
$$P = F\,v.$$

- In a rotational system, the inertia of an object is a measure of how readily an object can be made to rotate about an axis. For most objects, the moment of inertia is dependent on the choice of the rotation axis. The moment of inertia is denoted by J.
- The force that tries to rotate a system is called *torque*. Torque is defined as the perpendicular force, F, acting on a lever with length ℓ as

$$T = F\,\ell.$$

- Rotational displacement is measured in radians and denoted by θ. The rotational speed, ω, and the rotational rotation, α, are therefore found as

$$\alpha = \frac{d\omega}{dt} = \frac{d^2\theta}{dt^2}.$$

- In a rotational system, the following equation holds:

$$\begin{aligned} T &= J\,\alpha \\ E &= \frac{1}{2}J\omega^2 \\ P &= T\,\omega. \end{aligned}$$

- A gear system made from two discs with radii r_1 and r_2 has the following characteristics

$$\begin{aligned} \frac{\omega_1}{\omega_2} &= \frac{r_2}{r_1} \\ \frac{T_1}{T_2} &= \frac{r_1}{r_2}. \end{aligned}$$

- The transfer ratio of a gearbox is defined as the rotational speed ratio or the ratio of the number of teeth

$$a = \frac{\omega_2}{\omega_1} = \frac{N_1}{N_2}.$$

- An object with inertia J connected to the side labelled '2' will equivalently appear as an object with inertia

$$J' = a^2 J$$

  on the side labelled '1'.

PROBLEMS

5.1 Name the loss mechanisms that could affect the efficiency of gear and pulley systems.

5.2 Will a figure skater be able to increase her rotational speed more by adding weight to her belt or by holding on to two weights in her hands? Assume that in both cases she can achieve the same rotational speed with her hands stretched out wide.

5.3 Compare the turns ratio of a transformer with the gear ratio and compare the similarities and differences.

5.4 A pulley system increases the speed of a flywheel $2\frac{1}{2}$ times. By how much should the flywheel below be made shorter so that the stored kinetic energy of inertia stays the same?

5.5 The speed of an automobile is 100 km/h. The engine turns at 3 000 rpm and the wheel diameter is 50 cm. What is the gear ratio of the gearbox?

5.6 The speed of a steam tractor is 5 km/h. A 50 cm pulley is connected on the shaft of the driving wheels that have a diameter of 110 cm. This 50 cm pulley is driven by a belt that is powered by a 10 cm diameter pulley on the steam engine. At what speed does the steam engine turn?

5.7 Estimate how much kinetic energy is stored in a 72 cm diameter bicycle wheel if the bicycle moves at 20 km/h? The mass of the wheel is 2 kg.

5.8 The ratio of a two-gear system is 25 : 60. The input torque is 10 Nm and the output torque is 22 Nm. What is the efficiency of the gearbox?

5.9 The diameters of the pulleys in a belt and a pulley system are 12 cm and 4 cm, respectively. A drive system applies 10 Nm at 10 rad/s to the large, driving pulley. We know that 2% of the torque is lost due to friction and that 1% of the speed is lost due to the slip between the pulleys and the belt. How much power is delivered to the load?

FURTHER READING

Hibbeler R.C., *Engineering Mechanics Dynamics*. 12th edition, Prentice Hall, Upper Saddle River NJ, 2010.

Mohan N., *Electric Drives: An Integrated Approach*. 1st edition, University of Minnesota Printing Services, Minneapolis, 2000.

CHAPTER 6

POWER ELECTRONIC CONVERTERS

6.1 INTRODUCTION

Not all forms of electrical energy are created equal; in fact, some are more equal than others (to paraphrase George Orwell). This is actually a fact with which we are very familiar, we know that you cannot charge a mobile phone by connecting it directly to a wall socket. Furthermore, it is not possible to power a household television set while camping using only a 12 V car battery. We have many different forms of electrical energy and the variations mainly occur by a change of voltage, a change in frequency and with restrictions of the maximum power flow.

Power electronic converters come in many different shapes and sizes; for the most part, a converter is designed for a specific purpose and cannot be easily used for another. Many different descriptors are used to describe converters; the following are some of the more common ones:

1. Describing the operational power flow characteristics. A converter can either be unidirectional, where the power can only flow from the input port to the output, or bidirectional, where the power can flow from any side to the other.
2. Describing the input and output characteristics in terms of the voltage shape. Using this descriptor, four different topologies exists: dc–dc converter, ac–dc converter often also called a *rectifier* if designed with unidirectional power

The Principles of Electronic and Electromechanic Power Conversion: A Systems Approach, First Edition.
Braham Ferreira and Wim van der Merwe.

flow, dc–ac often called an *inverter* when operating with unidirectional power flow, and ac–ac often called a cycloconverter.

3. Describing the application where the converter is used. Some examples of this would be a drive, a converter specially designed to operate with an electrical machine, or a grid-tied converter, which is a converter specially designed to connect to the distribution grid.
4. Describing the structure of the converter. Some of the examples are a modular converter, a converter consisting of many (normally similar) converters operating as one, or a back-to-back converter that is in essence two converters connected in series so that the output of the first is the input of the second.

As power electronic converters form the backbone of the modern electrical energy conversion systems, we will spend some time studying the basic operation of these devices. Although the field of power electronic topologies is extremely wide, it is possible to get a sufficient understanding of the technology by studying one class, namely, the dc–dc converter.

6.2 LINEAR VOLTAGE REGULATOR

The golden rule of power electronics, that can be abbreviated by the three e's, is efficiency, efficiency and, yes, efficiency. In an effort to improve the efficiency of power electronic converters, no resistors or semiconductors operating in the linear region are used in the main power flow path. To illustrate this point and introduce the concept of the switched circuit, let us look at the following examples.

Example 6.1 A sensitive electronic device is used in an experiment, and it is important that the input voltage remains exactly at 15 V. The device can be modeled as a 2 Ω resistor. Calculate the resistor value, R_s, needed if the available source is a stable 40 V dc source, and the efficiency of the system if the circuit of Fig. 6.1 is used to reduce the voltage.

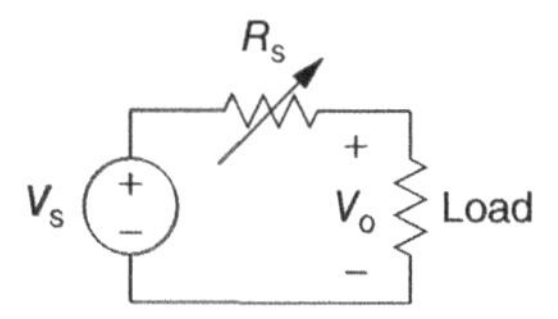

Figure 6.1 Reducing the voltage with a rheostat.

Solution

The output voltage of the circuit can be expressed as

$$V_o = \frac{R_l V_s}{R_l + R_s}$$

where $R_l = 2\,\Omega$ is the electronic device and the resistance required in series is

$$R_s = \frac{R_l(V_s - V_o)}{V_o} = 3\frac{1}{3}\,\Omega$$

and the efficiency of the system is

$$\eta = \frac{P_{in}}{P_{out}} = \frac{V_o I}{V_s I} = \frac{V_o}{V_s} = 37.5\%$$

Example 6.2 Let us continue with Example 6.1. If the input voltage is not a constant 40 V dc but rather a function of time and if $v_s(t) = V_s + \Delta V_s(t)$ then the output voltage of the circuit of Fig. 6.1 used in Example 6.1 will be $v_o(t) = \frac{R_l}{R_s+R_l}(V_s + \Delta V_s(t))$. Since we require the output voltage to be exactly 15 V, this is a problem. A possible solution is to use a transistor operating in its linear region in the power flow path, as shown in the circuit of Fig. 6.2. With $V_z = 15.7$ V, the transistor gain $\beta = 120$ and the base-emitter voltage of the transistor $V_{be} = 0.7$ V; calculate the efficiency of the circuit at $V_s = 40$ V. Assume that the input voltage can vary between 18 and 45 V and that the zener diode must have at least 5 mA of reverse current to bias the voltage accurately.

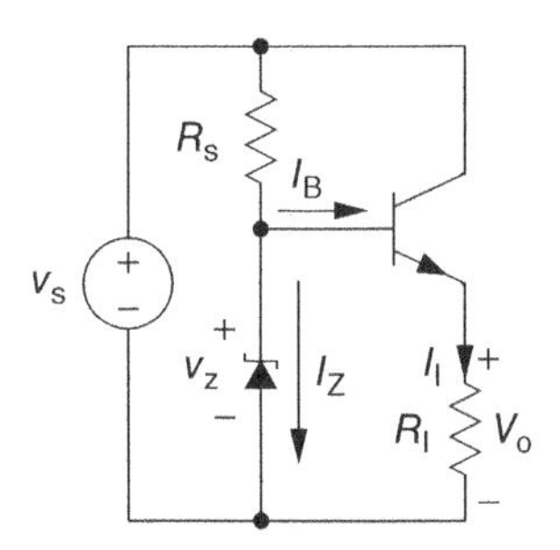

Figure 6.2 Linear regulator: emitter follower circuit.

Solution

The emitter current is

$$I_e = \frac{V_o}{R_l} = 7.5\text{ A}$$

and therefore the base current is

$$I_b = \frac{I_e}{\beta + 1} = 62\text{ mA}$$

As the minimum reverse current of the zener diode is 5 mA, the current through the bias resistor R_s should be at least

$$I_{R_s} > I_B + I_z = 62 + 5 = 67\,\text{mA}$$

At an input voltage of 18 V, this implies that the value of the set resistance is

$$R_s = \frac{18 - 15.7}{67 \times 10^{-3}} = 34.35\,\Omega$$

When $V_s = 40$ V, the source current is

$$I_s = \frac{V_s - V_z}{R_s} + I_l \frac{\beta}{\beta + 1} = 8.15\,\text{A}$$

and the efficiency is

$$\eta = \frac{P_{out}}{P_{in}} = \frac{15 \times 7.5}{40 \times 8.15} = 34.5\%$$

In the early days of power electronics, a linear regulator was often used to build power supplies. Figure 6.3 shows a laboratory power supply. It is rated for 500 W and is very heavy. Owing to the large 50 Hz transformer and the large heatsink that is needed to cool the bipolar transistor, its weight is about 20 kg. A modern day equivalent can be as much as ten times lighter notwithstanding not to speak about the improvement in efficiency.

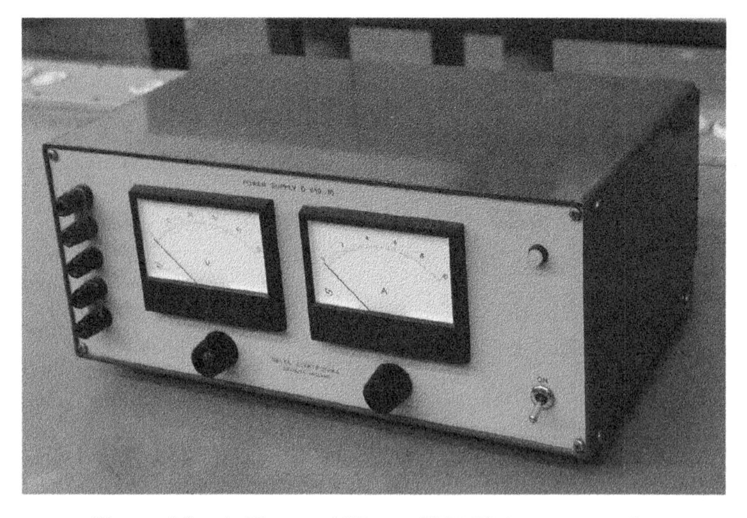

Figure 6.3 A 40 year-old linear 50 V, 10 A power supply.

6.3 SWITCHED APPROACH

Let us assume that the electronic device in Example 6.1 is not sensitive to very high frequency voltages. Now we can use the circuit of Fig. 6.4 to control the output voltage to any required value. Let us assume that the switch can be easily switched on or off and that it is ideal in every way. What we mean exactly by an ideal switch is defined in Section 6.4.1.

Now consider the circuit of Fig. 6.4. If the requirement is that the output voltage must be equal to the input voltage, the problem is solved relatively easily by just switching the switch on and leaving it in the on-state permanently. Likewise, if we require the output voltage to be zero then leaving the switch in the off-state should do the trick.

However, leaving these trivial problems aside, we can also control the output voltage to take *any* value between zero and the value of V_s. Let us assume that we switch on for one second and then off again for a second and we repeat this process endlessly. Now the average output voltage will be equal to $\frac{1}{2}V_s$, but because it is on for a second and then off for a second, we would probably be able to see this variation of the input voltage in the behaviour of the load. This is probably not such a good solution … but what will happen if we switch the switch on for 1 ms and off for 1 ms? Will we still be able to see the difference between the off-state and the on-state? Maybe this answer depends on the type of load. But then if we decrease the on and off times to 1 μs, it would be nearly impossible to see the difference. If you are still not convinced, let us decrease the on and off times to 1 ns. If you still think that you would be able to observe the difference between the on-state and the off-state, keep in mind that light only travels 30 cm in 1 ns. That is a really short period of time!

We can visualise the output voltage, v_o, of the circuit as shown in Fig. 6.5. Here, we have used T_s to denote the switching period. In Fig. 6.5, the circuit is operating in such a manner that the switch is off for one-third of the time and on for the remainder. However, because it is not feasible to speak of the length of time that a switch is on and off, let us introduce a definition. We will define the duty cycle, denoted as D, as the portion of the time that the switch is on. If t_{on} is the length of time that the switch is on and t_{off} is the length of time that the switch is off then

$$D = \frac{t_{on}}{t_{on} + t_{off}} = \frac{t_{on}}{T_s} \tag{6.1}$$

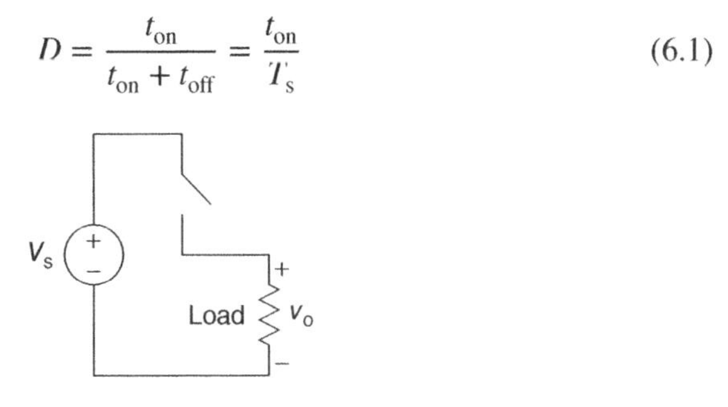

Figure 6.4 Reducing the voltage with a switched approach.

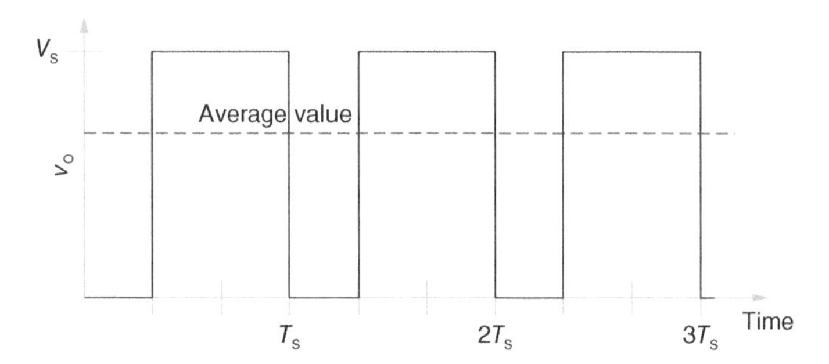

Figure 6.5 Switched approach waveform.

because by definition

$$T_s = t_{on} + t_{off}. \tag{6.2}$$

Notation Note 6.1

The periodic signals observed in power electronic systems are not, as a rule, sinusoidal. For the voltage waveforms $v_z(t)$ (the same assumptions also apply to current waveforms), the peak value is denoted by $\hat{v}_z$ and the average value by $\overline{v}_z$. There are cases where the signal is relatively constant and $v_z(t) \approx V_z = \overline{v}_z$. In this case, we can refer to the signal as $v_z(t) = V_z + \tilde{v}_z$ where $\tilde{v}_z$ refers to the small variations in the signal.

The question now becomes: how do we generate the control signal for the switch? From the requirements thus far, we know that we want the switch to switch with a high switching frequency (so that the switching behaviour is not noticeable), the duty cycle must be continuously variable from 0 to 1 and the process must be repeatable and easy to implement. It is actually not that difficult to create the control signal of Fig. 6.5. Let us create a sawtooth signal, which we we will call the carrier. We make the frequency of this carrier signal equal to our desired switching frequency, f_s and we can express it mathematically as

$$v_c(t) = \hat{V}_c\left(\frac{1}{2} + \frac{1}{\pi}\arctan\left(\tan\left(\pi f_s\, t\right)\right)\right), \tag{6.3}$$

where $\hat{V}_c$ is the peak value of the carrier signal. Let us now compare this signal with a reference value, which we will call V_r. A circuit showing this comparison is shown in Fig. 6.6. We can now express the output, which becomes our control signal, as

$$s(t) = \begin{cases} 1 & \text{if } V_r > v_c(t) \\ 0 & \text{otherwise.} \end{cases} \tag{6.4}$$

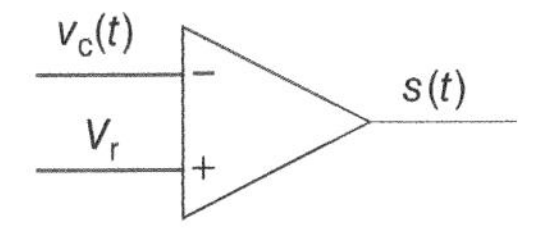

Figure 6.6 Control signal generation circuit.

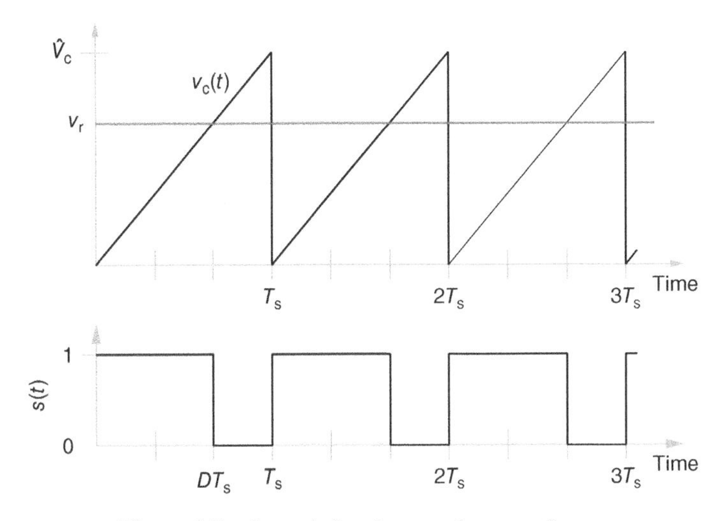

Figure 6.7 Control signal generation waveforms.

The resulting waveforms are shown in Fig. 6.7. We can see that by changing the value of V_r we can adjust the duty cycle of the control signal. In fact,

$$D = \frac{V_r}{\hat{V}_c}. \tag{6.5}$$

As we are effectively controlling our circuit by varying the duty cycle we will call this control process pulse-width modulation, abbreviated as PWM. It is also possible to use a triangular carrier rather than the sawtooth carrier. In this case, the carrier can be expressed as

$$v_c(t) = \hat{V}_c \left(\frac{1}{2} + \frac{1}{\pi} \arcsin\left(\sin\left(2\pi f_s\, t\right)\right) \right). \tag{6.6}$$

We can see from Figs. 6.5 and 6.7 that the control signal, and therefore, by implication, the output voltage across the load, is now a periodic signal with a fixed frequency. The analysis of this circuit is discussed later using time-domain methods, but let us first look at the circuit in the frequency domain using the Fourier series as discussed in Section 3.10. We can write the output voltage, $v_o(t)$, of Fig. 6.4

using the Fourier series, see (3.70), as

$$s(t) = \sum_{n=-\infty}^{\infty} C_n \, e^{j2\pi n f_s t}. \tag{6.7}$$

We can calculate the Fourier series coefficients, C_n, using (3.71). To make the calculation easier let us translate the waveforms of Fig. 6.7 slightly so that the on-cycle starts at $t = 0$. Then, when $n \neq 0$,

$$\begin{aligned} C_n &= \frac{1}{T_s} \int_0^{T_s} s(t)\, e^{-j2\pi n f_s t} \, dt \\ &= \frac{1}{T_s} \int_0^{DT_s} e^{-j2\pi n f_s t} \, dt \\ &= \frac{1}{T_s} \frac{-1}{2\pi n f_s} e^{-j2\pi n f_s t} \Big|_{t=0}^{t=DT_s} \\ &= \frac{j}{2\pi n} (e^{-j2\pi n D} - 1). \end{aligned} \tag{6.8}$$

Likewise when $n = 0$,

$$\begin{aligned} C_0 &= \frac{1}{T_s} \int_0^{T_s} s(t)\, e^{-j2\pi 0 f_s t} \, dt \\ &= \frac{1}{T_s} \int_0^{DT_s} 1 \, dt \\ &= D. \end{aligned} \tag{6.9}$$

Therefore,

$$C_n = \begin{cases} \dfrac{j}{2\pi n} \left(e^{-j2\pi n D} - 1 \right) & \text{when } n \neq 0 \\ D & \text{when } n = 0. \end{cases} \tag{6.10}$$

This seems like a very complicated way of analysing the circuit. However, once we have the Fourier coefficients of $s(t)$, we can write the Fourier series of $v_o(t)$ as

$$v_o(t) = V_s \sum_{n=-\infty}^{\infty} C_n \, e^{j2\pi n f_s t} \tag{6.11}$$

because we know that

$$v_o(t) = V_s \, s(t). \tag{6.12}$$

Our case is helped here, no doubt, by the fact that V_s is a constant and not a function of time.

Investigating the dc value of $v_o(t)$ now simply requires us to look at the Fourier series of $v_o(t)$, at the special case where $n = 0$, therefore,

$$\bar{v}_o = DV_s. \tag{6.13}$$

There is, however, one major problem with our circuit. We can see this problem clearly when we look at the frequency spectrum of the voltage $v_o(t)$ in Fig. 6.8. Here we have chosen $D = 0.75$, and for the sake of simplicity, we have included only the terms $n \geq 0$ in this figure. We have previously made the assumption that the load is not sensitive to high frequency components, and we have made the case that if we increase the switching frequency to a high enough value then this is indeed the case for nearly all types of loads. However, in real life, it becomes very difficult to switch a reasonable amount of power at frequencies higher than 250 kHz and we can no longer ignore the high frequency components. So is this method now invalid?

Most definitely not! Although we cannot do anything about the high frequency components of the switched waveforms, we can shield the load from them. Consider the circuit of Fig. 6.9. Here we have added an LC low pass filter between the switched waveform, $v_{pwm}(t)$ and the load. We have also included a diode and will discuss exactly why this diode is necessary a little later. If we can now design our LC filter to remove all frequencies higher than say $0.1f_s$ then the output voltage

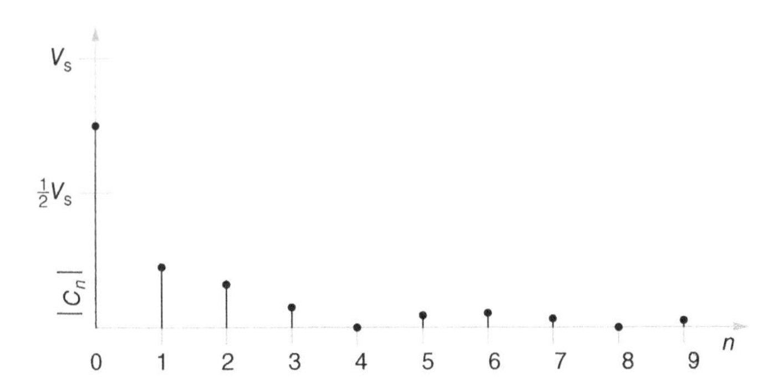

Figure 6.8 The spectrum of the voltage $v_o(t)$ of Fig. 6.4 when $d = 0.75$.

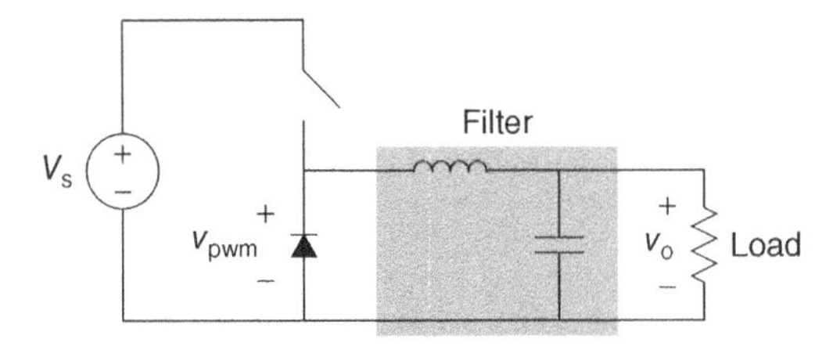

Figure 6.9 Simple buck converter.

$v_o(t)$ will, for all practical purposes, be a dc value, although the signal $v_{pwm}(t)$ is harmonically rich.

6.4 BASIC ASSUMPTIONS

To analyse power electronic circuits, we need to make a number of assumptions with regards to the components that we will use. To simplify our analysis, we will always firstly replace all components with idealised versions of themselves.

6.4.1 Switching Components

The switching components used in power electronic converters can be roughly divided into two sections: controlled switches such as transistors and uncontrolled switches called *diodes*. For all switches, we will assume the followings.

1. When in the off-state, the switch can handle any voltage across it and will not fail.
2. When in the on-state, the switch can handle any current through it without failing.
3. The switch can switch instantaneously from the off-state to the on-state and vice versa.
4. There are no losses associated with the switching cycle.
5. The switch only requires a signal to control the state of the switch. We will assume that if the control signal is equal to '1', the switch will be closed (i.e. in the on-state) and when the signal is equal to '0', the switch is in the off-state. We will call this signal $s(t)$, and by definition, $s(t) \in \{0, 1\}$ or, in other words, the value of s can be either 1 or 0 for any and all points in time.

We will represent the idealised components in our circuit diagrams using the components shown in Fig. 6.10.

6.4.2 Linear Components

The capacitor and the inductor are used extensively in the power electronic circuits. Although all capacitors and inductors exhibit some losses, which are normally represented by an ESR (equivalent series resistance), we will assume that both the elements are lossless. Consider the capacitor shown in Fig. 6.11. We can write the following equations to describe the behaviour of the capacitor.

$$i_c = C\frac{dv_c}{dt} \leftrightarrow v_c = \frac{1}{C}\int i_c dt \tag{6.14}$$

$$E = \frac{1}{2}Cv_c^2 \tag{6.15}$$

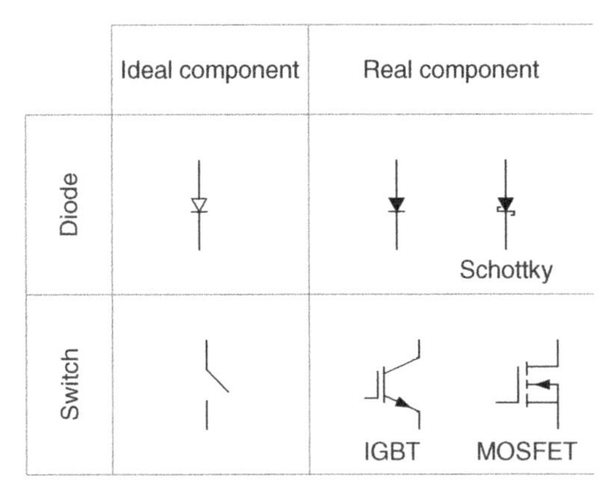

Figure 6.10 Representation of components.

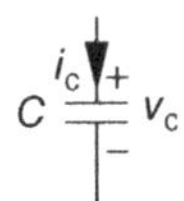

Figure 6.11 A capacitor with definitions.

It is important to note the following:

1. The voltage across a capacitor cannot change instantaneously. This can be seen by the fact that should the voltage change abruptly the time derivative of the voltage and by implication the current through the capacitor will be infinitely large. If we allow such a large current to flow (to force this abrupt change in the capacitor voltage), it will break everything in its path.
2. In a power electronic circuit where the currents and voltages are periodic functions, the capacitor voltage is assumed to be in steady state when, where $q \in \mathbb{Z}$,

$$v_c(qT_s) = v_c((q+1)T_s). \tag{6.16}$$

This implies that the average current through the capacitor must be equal to zero, or

$$v_c(qT_s) = v_c((q+1)T_s) \leftrightarrow \frac{1}{C}\int_{qT_s}^{(q+1)T_s} i_c\,dt = 0. \tag{6.17}$$

3. Power electronic circuits are normally designed in such a manner that the voltage across a capacitor remains nearly constant over a switching period.

Figure 6.12 An inductor with definitions.

Likewise, consider the inductor shown in Fig. 6.12. We can write the following equations to describe the behaviour of the inductor.

$$v_{\mathrm{L}} = L\frac{di_{\mathrm{L}}}{dt} \leftrightarrow i_{\mathrm{L}} = \frac{1}{L}\int v_{\mathrm{L}}\,dt. \tag{6.18}$$

$$E = \frac{1}{2}Li_{\mathrm{L}}^2 \tag{6.19}$$

It is important to note the following:

1. The current through an inductor cannot change instantaneously. This can be seen by using the same argument as with the capacitor voltage. As the time derivative of the current would be infinitely large, the resulting voltage across the inductor would also be infinitely large. Unfortunately, or maybe fortunately, the universe does not allow infinitely large voltages because the sparks will literally fly.
2. For periodic signals, the inductor current is assumed to be in steady state when, where $q \in \mathbb{Z}$,
$$i_{\mathrm{L}}(qT_{\mathrm{S}}) = i_{\mathrm{L}}((q+1)T_{\mathrm{S}}). \tag{6.20}$$
This implies that the average voltage across the inductor must be equal to zero, or
$$i_{\mathrm{L}}(qT_{\mathrm{S}}) = i_{\mathrm{L}}((q+1)T_{\mathrm{S}}) \leftrightarrow \frac{1}{L}\int_{qT_{\mathrm{s}}}^{(q+1)T_{\mathrm{s}}} v_{\mathrm{L}}\,dt = 0. \tag{6.21}$$
3. Power electronic circuits are normally designed in such a manner that the current through an inductor remains nearly constant over a switching period.
4. As the current through an inductor cannot change instantaneously, *each inductor must be associated with a current path for every state of the circuit operation.* This alternate current path is often called a *freewheeling path.*

6.5 BUCK CONVERTER

The buck converter is a basic dc–dc converter that is used to reduce a dc voltage to a lower dc voltage. It might not be clear that why the converter is called the *buck converter* but the word buck can be used as a verb that can be defined as ‘to resist’.

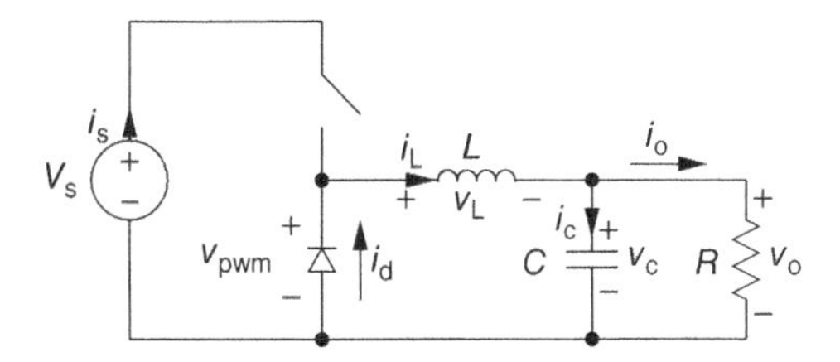

Figure 6.13 A buck converter.

We will see in our discussion of the converter that the inductor resists the current flow.

Consider the basic circuit of the buck converter shown in Fig. 6.13. The structure of the converter is the same as the basic switching circuit discussed in Section 6.3. To minimise the harmonic components in the voltage supplied to the load, a low pass filter is added. However, because we have an inductor in the direct power flow path a freewheeling diode is added to the circuit to ensure that the current through the inductor does not change abruptly.

Two states can be identified in the converter. In State I, the switch is closed, corresponding to the time periods where $s(t) = 1$. State II is identified by the points in time where the switch is open, or equivalently, where $s(t) = 0$. The sub-circuits associated with each of the states are shown in Fig. 6.14.

Using the variables as defined in Fig. 6.13, we can now write equations to describe the circuit operation. Let us first write these equations for each state separately.

6.5.1 State I

Using Fig. 6.14a as our basis, we can write

$$L\frac{d\,i_{\mathrm{L}}}{dt} = V_{\mathrm{s}} - v_{\mathrm{c}} \tag{6.22}$$

$$C\frac{d\,v_{\mathrm{c}}}{dt} = i_{\mathrm{L}} - \frac{v_{\mathrm{c}}}{R}. \tag{6.23}$$

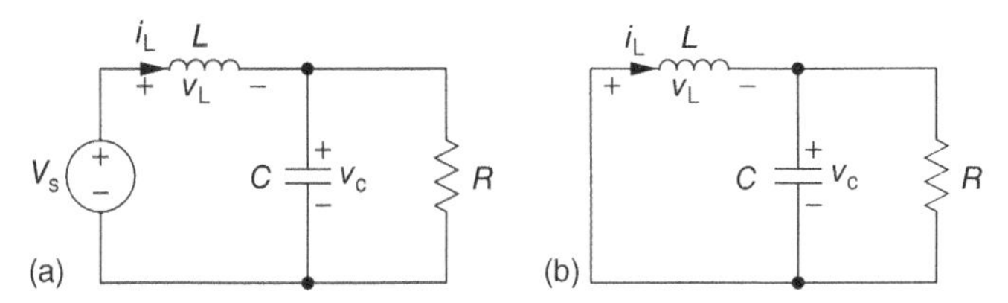

Figure 6.14 Equivalent circuits of the buck converter: (a) State I, that is, when the switch is closed and (b) State II, that is, when the switch is open.

We can combine these two equations into a more convenient form by using matrix notation. By choosing a vector **x** such that

$$\mathbf{x} = \begin{bmatrix} i_{L} \\ v_{C} \end{bmatrix}, \tag{6.24}$$

we can now write (6.22) and (6.23) in the form

$$\frac{d}{dt}\mathbf{x} = \mathbf{A_1}\,\mathbf{x} + \mathbf{B_1}\,V_{S} \tag{6.25}$$

as

$$\frac{d}{dt}\begin{bmatrix} i_{L} \\ v_{C} \end{bmatrix} = \begin{bmatrix} 0 & \frac{-1}{L} \\ \frac{1}{C} & \frac{-1}{RC} \end{bmatrix}\begin{bmatrix} i_{L} \\ v_{C} \end{bmatrix} + \begin{bmatrix} \frac{1}{L} \\ 0 \end{bmatrix} V_{S}. \tag{6.26}$$

6.5.2 State II

Using Fig. 6.14b as our basis, we can write

$$L\frac{di_{L}}{dt} = -v_{C} \tag{6.27}$$

$$C\frac{dv_{C}}{dt} = i_{L} - \frac{v_{C}}{R}. \tag{6.28}$$

Again we can combine these two equations into the form of (6.25) using

$$\frac{d}{dt}\mathbf{x} = \mathbf{A_2}\,\mathbf{x} + \mathbf{B_2}\,V_{S} \tag{6.29}$$

as

$$\frac{d}{dt}\begin{bmatrix} i_{L} \\ v_{C} \end{bmatrix} = \begin{bmatrix} 0 & \frac{-1}{L} \\ \frac{1}{C} & \frac{-1}{RC} \end{bmatrix}\begin{bmatrix} i_{L} \\ v_{C} \end{bmatrix} + \begin{bmatrix} 0 \\ 0 \end{bmatrix} V_{S}. \tag{6.30}$$

6.5.3 Combining the Two States

We can complete the mathematical model of the converter by combining the two states. Let us once again use the form of (6.25) to represent the circuit operation. We combine the two states by combining the matrices in (6.30) and (6.30) to yield the following equation.

$$\frac{d}{dt}\mathbf{x} = \mathbf{A}\,x + \mathbf{B}\,V_{S}, \tag{6.31}$$

where

$$\mathbf{A} = \begin{cases} \mathbf{A_1} & \text{when } s(t) = 1 \\ \mathbf{A_2} & \text{otherwise} \end{cases} \tag{6.32}$$

and

$$\mathbf{B} = \begin{cases} \mathbf{B}_1 & \text{when } s(t) = 1 \\ \mathbf{B}_2 & \text{otherwise.} \end{cases} \tag{6.33}$$

Although this is a complete description of the circuit operation, solving this system of differential equations requires some very advanced mathematics. That said, firstly, all is not lost, we can still rescue the situation by using a simplified analysis of the circuit by making certain assumptions and using the characteristics of the inductor and the capacitor as discussed in Section 6.4.2. Then, secondly, we have not wasted our time in developing these equations because we can make a couple of very important observations from them and can even use them to write small but accurate simulation programmes to simulate the converter operation using a numerical integration routines in MATLAB or Python. Formal theory has been developed to solve these matrices, for example, the state space averaging technique. We are following a simplified approach because this is a first-level book.

6.5.4 Simplified Analysis Approach

To simplify the analysis of the converter, let us make the following assumptions:

1. The system is in steady state, which implies that $\mathbf{x}(t) = \mathbf{x}(t + T_S)$ for time t long enough after converter turn-on that the system has reached steady state.
2. The capacitor is large enough so that the voltage across the capacitor is a constant, that is, $v_C(t) = V_C$. We will revisit this assumption a bit later to see if it was good and valid.
3. The inductor has a high enough inductance so that the change in current during the switching cycle is relatively low.
4. Let us assume, for now, that the current through the inductor is continuous and is never zero during a switching cycle.

As we assume that the system is in steady state, where $\mathbf{x}(t) = \mathbf{x}(t + T_S)$ and we have assumed that $v_C(t) = V_C$, the assumption of the circuit being in steady state boils down to the requirement that

$$i_L(t) = i_L(t + T_S). \tag{6.34}$$

Consider the circuit waveforms in Fig. 6.15. The current slope during State I can be calculated as

$$m_1 = \frac{di_L}{dt} = \frac{V_S - V_C}{L}. \tag{6.35}$$

Likewise, the current slope during State II is

$$m_2 = \frac{di_L}{dt} = \frac{-V_C}{L} \tag{6.36}$$

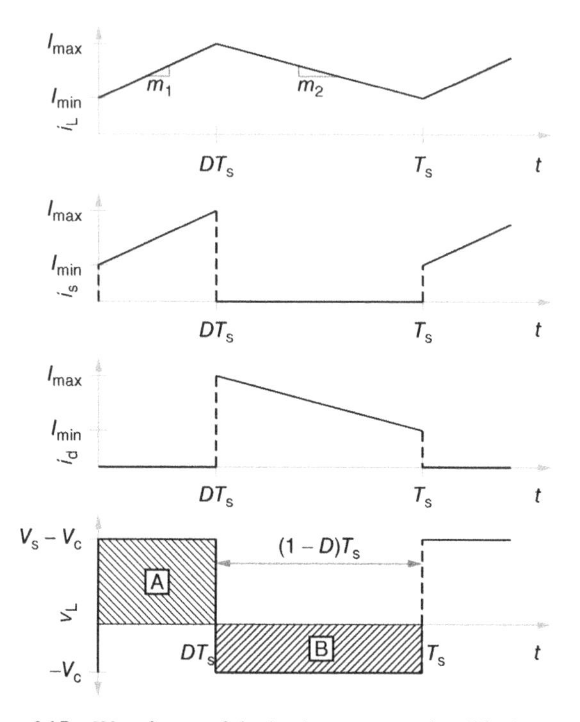

Figure 6.15 Waveforms of the buck converter, simplified approach.

We can now calculate the current ripple in the inductor current, denoted as Δi_L, as

$$\Delta i_L = I_{max} - I_{min} = (I_{min} + m_1 DT_s) - I_{min} = \frac{V_s - V_c}{L} DT_s, \tag{6.37}$$

but we can even go further by stating that because $i_L(t) = i_L(t + T_s)$, we know that

$$m_1 DT_s + m_2(1 - D)T_s = 0 \tag{6.38}$$

and therefore,

$$\frac{V_s - V_c}{L} DT_s + \frac{-V_c}{L}(1 - D)T_s = 0 \rightarrow V_c = DV_s. \tag{6.39}$$

We can also choose to represent this finding by expressing it in terms of the voltage across the inductor. As we know that

$$\Delta i_L = \frac{1}{L} \int_0^t v_L \, dt \tag{6.40}$$

we can now write

$$\frac{1}{L} \int_t^{t+DT_s} v_L \, dt + \frac{1}{L} \int_{t+DT_s}^{T_s} v_L \, dt = 0 \rightarrow V_c = DV_s. \tag{6.41}$$

We can visualise this equation by noting that it implies that for the circuit to be in steady state, that is, $i_L(t) = i_L(t + T_s)$, the areas A and B in Fig. 6.15 must be equal. The only time that these areas will be equal is when $v_c = DV_s$. The reason for representing the equation in terms of the voltage–time integral will become clear in Section 6.5.5.

We can complete the analysis of the waveforms shown in Fig. 6.15 by investigating the current through the source (i_s) and the current through the diode (i_d). We know from the characteristics of the inductor that the current through the inductor (I_L) cannot change instantaneously, and this result is shown in Fig. 6.15. That said, because of the switching behaviour of the circuit, there must be some abrupt changes in the current flow, and there are. We observe these abrupt changes in the waveforms of the switch and the diode current. While the switch is closed, State I, the inductor current flows through the switch and no current is conducted through the diode. This makes sense because when the switch is closed, the voltage across the diode (v_{pwm}) is equal to V_s and because of the characteristics of the diode, it cannot conduct current from cathode to anode. We also note that the voltage across the switch is equal to zero, which would make sense because the switch is closed.

In a similar vein, when the switch is open, State II, the diode conducts the current. At this point, the voltage across the diode (v_{pwm}) is equal to zero and the voltage across the switch is now equal to V_s. So what would happen if we did not add the freewheeling diode? Without the diode, the current in the inductor would have to change instantaneously from some value to zero when the switch is opened because there is no path for the current to flow along. When we investigate this hypothetical situation on paper, we might be tempted to think that the resulting infinitely large voltage across the inductor is not of major concern. However, paper is patient. Should we ever try this in real life, our experiment will end with a big bang, smoke and a broken switch.

6.5.5 What If $v_c(t) \neq V_c$?

Up to now in our analysis, we have assumed that the output filter capacitor is really big and the voltage across it is, therefore, nearly constant. But what would happen if the voltage is not constant and we have to write the voltage as

$$v_c(t) = V_c + \tilde{v}_c(t)? \tag{6.42}$$

Here $\tilde{v}_c(t)$ represents the small variations from steady state (or an average) value of the voltage $v_c(t)$ which we represent as V_c. Recall that for a capacitor

$$v_c(t) = \frac{1}{C}\int i_c\,dt. \tag{6.43}$$

In our circuit, as shown in Fig. 6.13, the capacitor current can be found as

$$i_c = i_L - \frac{v_c}{R} \tag{6.44}$$

but because this current depends on the voltage, it is difficult (however, by no stretch of the imagination impossible) to solve. Let us assume that the output current, which we will denote as i_o, is constant such that

$$i_c(t) = i_L(t) - I_o. \tag{6.45}$$

Let us represent this graphically as shown in Fig. 6.16. Again we will use the assumption that the circuit is in steady state and, therefore,

$$v_c(t) = v_c(t + T_s). \tag{6.46}$$

Remembering that the amount of charge delivered to the capacitor in a switching cycle can be found as

$$\Delta Q = \int_t^{t+T_s} i_c \, dt \tag{6.47}$$

but because

$$\Delta v_c = \frac{\Delta Q}{C} \tag{6.48}$$

and the system is in steady state where (6.46) holds, we know that

$$\Delta Q = 0 \rightarrow \int_t^{t+T_s} i_c \, dt = 0. \tag{6.49}$$

Now if we consider Fig. 6.16, we see two areas marked A and B. If the total charge delivered to the capacitor during the switching period is zero then area A = area B. We can now also represent the voltage ripple Δv_c in terms of the area A, which we will denote as A_A (or equivalently area B, denoted as A_B) as

$$\Delta v_c = \frac{1}{C} A_A = \frac{1}{C} A_B. \tag{6.50}$$

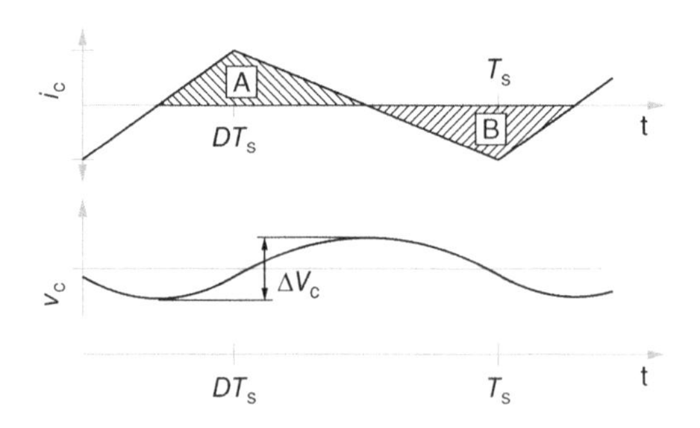

Figure 6.16 Waveforms of the buck converter capacitor, simplified approach.

In most power electronic converters, the capacitor voltage ripple is so small that the approach with which we assume that $v_c(t) \approx V_c$ is valid and the resulting error is very small. Let us investigate this statement with an example.

Example 6.3 The specifications of a buck converter, similar to that in Fig. 6.13, are tabulated in Table 6.1. Calculate the inductor current using the assumption that the capacitor voltage ripple is zero. Using this current, what would the capacitor voltage ripple be? Do you think the assumption of a negligible voltage ripple on the capacitor was valid?

Solution

The average output voltage is

$$V_o = DV_s = 24.75\ \text{V},$$

and the average output current and, therefore, by implication the average inductor current is

$$I_o = \bar{i}_L = \frac{V_o}{R} = 3\ \text{A}.$$

We can calculate the inductor ripple current by investigating the time when the switch is off, that is, State II. During this state, the voltage across the inductor is equal to the output voltage and

$$\begin{aligned}\Delta i_L &= \frac{1}{L}\int_{DT_s}^{T_s} v_L\,dt \\ &= \frac{1}{L}V_o(T_s - DT_s) \\ &= 0.446\ \text{A}.\end{aligned}$$

As the current changes in a linear manner, we can calculate the maximum and minimum values of the current, as defined in Fig. 6.15, as

$$I_{max} = I_o + \frac{1}{2}\Delta i_L = 3.22\ \text{A}$$

$$I_{min} = I_o - \frac{1}{2}\Delta i_L = 2.78\ \text{A}.$$

TABLE 6.1 Buck Converter Specifications, Example 6.3

Item	Value	Item	Value
Switching frequency, f_s	25 kHz	Duty cycle, D	0.55
Input voltage, V_s	45 V	Load resistance, R	8.25 Ω
Filter inductor, L	1.5 mH	Filter capacitor, C	220 μF

We can now calculate the voltage ripple on the capacitor. To do that, we will have to calculate either area A or area B in Fig. 6.16. For calculating area A let us choose a starting point in time where $t = 0$ at the instant when the switch closes. Owing to the linear rise in the current, and from the fact that the current starts to rise from a minimum value at $t = 0$ and reaches a maximum at $t = DT_s$, we know that the capacitor current will cross the zero line in Fig. 6.16 at $t = \frac{1}{2}DT_s$. Likewise, because the current decreases linearly from the maximum at $t = DT_s$ to the minimum value at $t = T_s$, the capacitor current will cross the zero line at $t = 0.5DT_s + 0.5(1 - D)T_s = 0.5(1 + D)T_s$. We can calculate area A by considering the two triangles separately. The area is

$$\begin{aligned} A_A &= \frac{1}{2}\frac{1}{2}\Delta i_L \frac{1}{2}DT_s + \frac{1}{2}\frac{1}{2}\Delta i_L \frac{1}{2}(1-D)T_s \\ &= \frac{1}{2}^3 \frac{\Delta i_L D}{f_s} + \frac{1}{2}^3 \frac{\Delta i_L (1-D)}{f_s} \\ &= \frac{1}{2}^3 \frac{\Delta i_L}{f_s} \\ &= 2.33\ \mu\text{A s}, \end{aligned}$$

and the voltage ripple is, therefore,

$$\Delta v_c = \frac{\Delta Q}{C} = \frac{A_A}{C} = 10.1\ \text{mV}.$$

The ripple component is only

$$\frac{\Delta v_c}{V_c} = 0.04\%$$

of the dc value; therefore, the original assumption of no voltage ripple will yield an acceptably accurate answer.

For the sake of interest, the LC filter discussed here has a cut off frequency of

$$\omega = \frac{1}{\sqrt{\text{LC}}} \rightarrow f = \frac{1}{2\pi\sqrt{\text{LC}}} = 277\ \text{Hz}.$$

Example 6.4 Choose the value of a filter inductor and filter capacitor for a buck converter, similar to that in Fig. 6.13. The desired operating conditions are tabulated in Table 6.2.

Solution

The average output voltage is

$$V_o = DV_s = 52.5\ \text{V}$$

TABLE 6.2 Buck Converter Specifications, Example 6.4

Item	Value	Item	Value
Switching frequency, f_s	10 kHz	Duty cycle, D	0.35
Input voltage, V_s	150 V	Output power, P_o	2 kW
Current ripple, Δi_L	20%	Output voltage ripple, Δv_c	< 0.25 V

and the average output current, and therefore by implication the average inductor current is

$$I_o = \bar{i}_L = \frac{P_o}{V_o} = 38.1 \text{ A}.$$

The allowable ripple current is, therefore,

$$\Delta i_L = 0.2\bar{i}_L = 7.62 \text{ A}.$$

We can calculate the inductor ripple current by investigating the time when the switch is off, that is, State II. During this state, the voltage across the inductor is equal to the output voltage and

$$\begin{aligned}
\Delta i_L &= \frac{1}{L}\int_{DT_s}^{T_s} v_L \, dt \\
&= \frac{1}{L} V_o (T_s - DT_s) \\
\therefore L &= \frac{1}{\Delta i_L} V_o (T_s - DT_s) \\
&= 450 \,\mu\text{H}.
\end{aligned}$$

As the current changes in a linear manner, we can calculate the maximum and minimum values of the current, as defined in Fig. 6.15, as

$$I_{max} = I_o + \frac{1}{2}\Delta i_L \approx 41.9 \text{ A}$$

$$I_{min} = I_o - \frac{1}{2}\Delta i_L \approx 34.3 \text{ A}.$$

Again, to calculate the voltage ripple on the capacitor, we will calculate either area A or area B in Fig. 6.16. The area is

$$\begin{aligned}
\Delta Q &= \frac{1}{2}\frac{1}{2}\Delta i_L \frac{1}{2} DT_s + \frac{1}{2}\frac{1}{2}\Delta i_L \frac{1}{2}(1 - D)T_s \\
&= \frac{1}{2}^3 \frac{\Delta i_L}{f_s} \\
&= 95.3 \,\mu\text{A s}
\end{aligned}$$

and using the voltage ripple requirement,

$$\Delta v_c = \frac{\Delta Q}{C}$$

$$\therefore C > \frac{\Delta Q}{\Delta v_c} = 381\ \mu\text{F}$$

$$\because \Delta v_c < 0.25\ \text{V}.$$

The ripple component is only

$$\frac{\Delta v_c}{V_c} = 0.04\%$$

of the dc value; therefore, the original assumption of no voltage ripple is accurate. Again, for interest, we can calculate the LC filter cutoff frequency as

$$\omega = \frac{1}{\sqrt{\text{LC}}} \rightarrow f = \frac{1}{2\pi\sqrt{\text{LC}}} = 295\ \text{Hz}.$$

6.6 DISCONTINIOUS CONDUCTION MODE

So far in our discussion, we have made the basic assumption that the current through the inductor is continuous. However, if we investigate Fig. 6.15, we see that if the average inductor current drops low enough, the lowest current point will tend toward zero. You should be able to verify the statement that we can typically reduce the inductor current by increasing the load resistance. If the average current is low enough, the inductor current will eventually reach the zero point, and because the inductor current can never be negative, there is no current flow path for a negative inductor current in the state where the switch is in the off-state so the current remains zero until the switch is switched to the on-state.

6.6.1 Boundary between CCM and DCM

Using Fig. 6.15 as our basis, we can define the boundary between the continuous conduction mode (CCM) and the discontinuous conduction mode (DCM) as the point where $I_{\text{min}} = 0$. The inductor current and voltage waveforms at this point are shown in Fig. 6.17a. Therefore, we can write that

$$I_{\text{max}} = m_1 DT_s = \frac{V_s - V_c}{L} DT_s \tag{6.51}$$

and the average value of the inductor current is, therefore,

$$\bar{\text{i}}_{\text{L}} = \frac{V_s - V_c}{2L} DT_s. \tag{6.52}$$

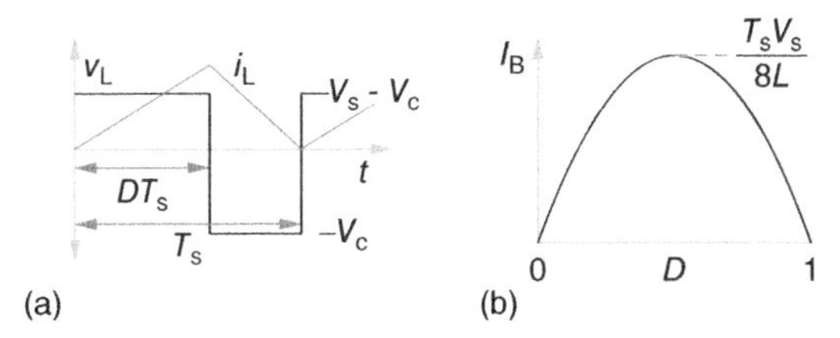

Figure 6.17 The buck converter boundary between the CCM and the DCM. (a) Inductor current and voltage waveforms at the boundary. (b) The boundary current as a function of duty cycle.

When the average inductor current reaches this value, we are at the boundary between the CCM and the DCM. Let us denote this by renaming the $\bar{i}_L$ term as I_B to denote the fact that this is the current that constitutes the boundary. At this boundary point we are technically still in the CCM; we know that $V_c = DV_s$, and therefore, we can rewrite the expression as

$$I_B = \frac{V_s(D - D^2)}{2Lf_s}. \tag{6.53}$$

If, for a specified converter, the average current is higher than this value (at the desired duty cycle), the converter will be operating in the CCM, and conversely, if the average inductor current is lower than this value, the converter will be operating in the DCM. The boundary current as a function of duty cycle is shown in Fig. 6.17b. We see that the boundary current reaches a maximum at $D = 0.5$.

Example 6.5 A buck converter, as shown in Fig. 6.13, switches at 100 kHz and uses a 45 μH inductor. If the input voltage is 35 V and the desired output voltage is 12 V, what is the maximum load resistance, R, that can be connected to the converter with the condition that the converter operates in the CCM?

Solution

As the converter still operates in the CCM, the duty cycle at this operating point can be calculated as

$$D = \frac{V_c}{V_s} = \frac{12}{35} = 34.29\%.$$

According to (6.53), the boundary current at this point is

$$I_B = \frac{V_s(D - D^2)}{2Lf_s} = \frac{35(0.3429 - 0.3429^2)}{2 \times 45 \times 10^{-6} \times 100 \times 10^3} = 0.88\ \text{A}.$$

The maximum load resistance is, therefore,

$$R = \frac{V_c}{I_B} = \frac{12}{0.88} = 13.7\ \Omega.$$

We can see that if the load resistance is higher than this value, the average inductor current will drop and the converter will enter the DCM. Remember that in the steady state, the average capacitor current is zero and, therefore, the average inductor current and the load current are equal to each other.

6.6.2 Relationship between V_s and V_c in DCM

The steady state waveforms of the buck converter operating in the DCM are shown in Fig. 6.18. At first glance, we see that the inductor current is not continuous for the whole time interval where the switch is in the off state. To describe the circuit operation, let us define three new variables, D_2, Δ_1 and Δ_2, as shown in Fig. 6.18. The terms are defined such that (as is clear from the definitions in the figure)

$$1 = D + \Delta_1 + \Delta_2 \tag{6.54}$$

$$D_2 = D + \Delta_1 = 1 - \Delta_2. \tag{6.55}$$

We know from the fundamental characteristics of the inductor, discussed in Section 6.4.2, that the average voltage over the inductor is zero in the steady state.

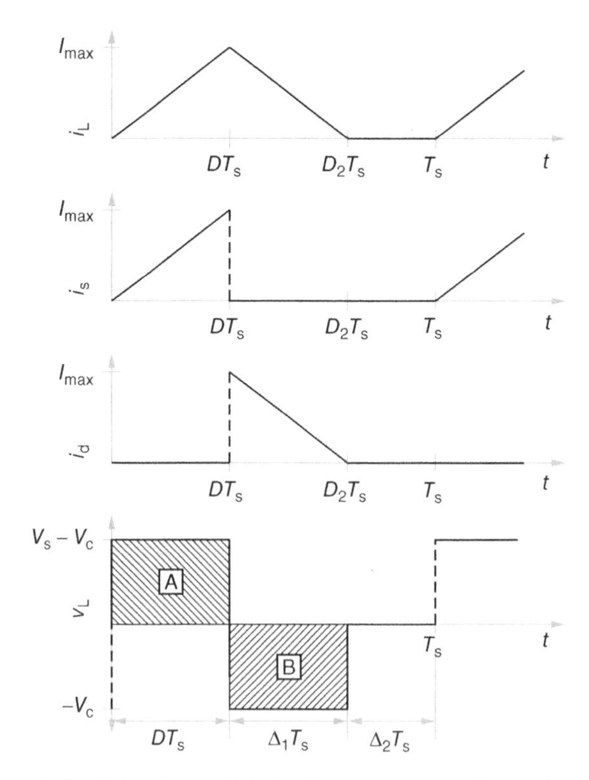

Figure 6.18 Waveforms of the buck converter operating in the DCM.

Therefore, with respect to Fig. 6.18, the areas marked 'A' and 'B' should be equal. Therefore, we can say that

$$(V_s - V_c)D = \Delta_1 V_c. \tag{6.56}$$

We can also quite easily calculate the average inductor current (and therefore, the output current because the average capacitor current in the steady state is zero). The average inductor current can be found as

$$I_o = \frac{1}{T_s} \int_{T_s} i_L(t)\, dt, \tag{6.57}$$

which basically boils down to finding the time–current area contained by the inductor current divided by the switching period. The average current is

$$\begin{aligned} I_o &= \frac{1}{2T_s} I_{max} D_2 T_s \\ &= \frac{I_{max}(D + \Delta_1)}{2}. \end{aligned} \tag{6.58}$$

Here, the term I_{max} is unknown, but it can be easily calculated by using either the up-slope or the down-slope of the current as

$$I_{max} = \frac{1}{L}(V_s - V_c)DT_s = \frac{1}{L}V_c\Delta_1 T_s. \tag{6.59}$$

We can now calculate the average output current by substituting (6.59) into (6.58)

$$\begin{aligned} I_o &= V_c\Delta_1 T_s \frac{(D + \Delta_1)}{2L} \\ &= \frac{V_c T_s}{2L}(D + \Delta_1)\Delta_1 \end{aligned} \tag{6.60}$$

At this point, we reach a cross-point. It is not possible to give an expression for Δ_1 that depends only on the known parameters of the buck converter. This statement should make sense when we realise that the circuit will be in DCM only if the load is small enough; the load, therefore, determines whether the converter is in the DCM or not. As we will now prove, while the circuit is in the DCM, the load influences the ratio of the input voltage and output voltage. Let us address the problem using two avenues: investigating the influence of the load resistance on the voltage transfer ratio and investigating the effect of the output current on the voltage transfer ratio.

6.6.2.1 *Effect of the Output Resistance* As we are modeling an ideal buck converter, that is, a converter without losses, we know that

$$P_{\text{in}} = P_{\text{out}}. \tag{6.61}$$

We can use this mathematical trick shot here. Remember that the average input power can be calculated using the average source current, as shown in Fig. 6.18. Therefore, (remembering from the definition in Fig. 6.13 that v_c is also the voltage across the output resistance which we denote here as v_o),

$$\begin{aligned} P_{\text{in}} &= P_{\text{out}} \\ V_s \frac{I_{\max} D T_s}{2 T_s} &= \frac{V_o^2}{R} \end{aligned} \tag{6.62}$$

or

$$V_o^2 = \frac{R I_{\max} D V_s}{2}. \tag{6.63}$$

Replacing the $I_{\max}$ term with (6.59) yields

$$\begin{aligned} V_o^2 &= V_s (V_s - V_o) \frac{R D^2 T_s}{2L} \\ 0 &= V_o^2 + V_o V_s \frac{R D^2}{2 L f_s} - V_s^2 \frac{R D^2}{2 L f_s}. \end{aligned} \tag{6.64}$$

For reasons that will become clear later, let us define a new variable, α, to describe the ratio of the resistance to the inductor impedance at the switching frequency. Let

$$\alpha = \frac{R}{2\pi f_{sL}}. \tag{6.65}$$

Solving (6.62) for the roots of V_o (with the condition that V_o is positive) and substituting the definition of α, (6.65), yields

$$\begin{aligned} V_o &= \frac{1}{2}\left(-V_s D^2 \pi\alpha + \sqrt{V_s^2 D^4 \pi^2 \alpha^2 + 4 V_s^2 D^2 \pi\alpha}\right) \\ &= \frac{1}{2} V_s D^2 \pi\alpha \left(-1 + \sqrt{1 + \frac{4}{D^2 \pi\alpha}}\right) \\ \therefore \frac{V_o}{V_s} &= \frac{1}{2} D^2 \pi\alpha \left(-1 + \sqrt{1 + \frac{4}{D^2 \pi\alpha}}\right) \end{aligned} \tag{6.66}$$

The relationship between the output voltage, V_o, and the source voltage, V_s, is clearly no longer a linear function. We can also see that it depends strongly on the

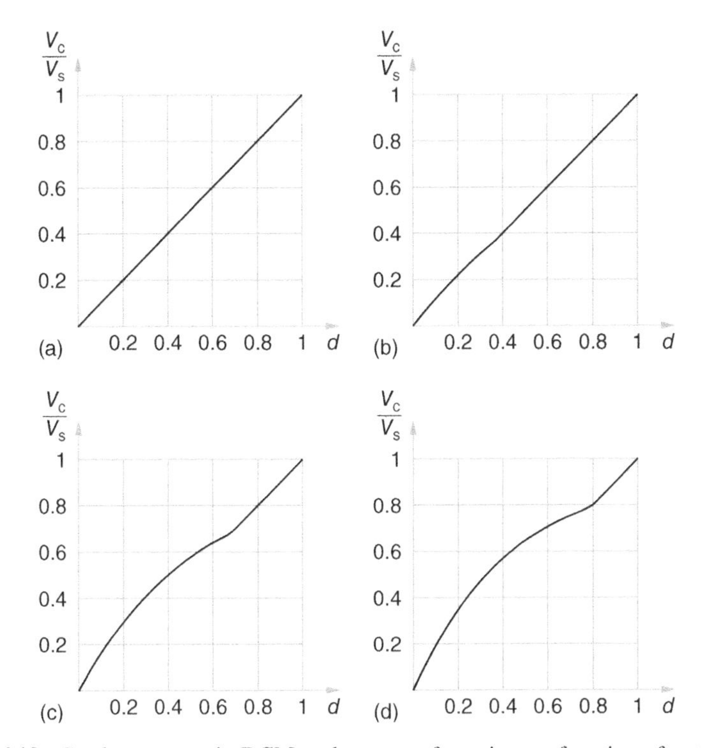

Figure 6.19 Buck converter in DCM: voltage transfer ratio as a function of output resistance (a) $\alpha = 0.1$, (b) $\alpha = 0.5$, (c) $\alpha = 1$, and (d) $\alpha = 1.5$.

relative size of the load and the filter inductor, the α term. In Fig. 6.19, the voltage transfer ratio of the buck converter is shown for values of $\alpha \in \{0.1, 0.5, 1, 1.5\}$.

We see that with small values of α, the converter is always in the CCM and the voltage transfer function is linear throughout the range of the duty cycle. However, when the load resistance increases, the converter is in the DCM for a larger span of duty cycles. For example, the converter is in the DCM for all values of duty cycle below 80% when $\alpha = 1.5$. To avoid the onset of the DCM in power electronic converters, the value of inductance is normally selected to be relatively large when compared to the expected load.

6.6.2.2 ***DCM as a Function of the Average Output Current*** Another method of investigating the behaviour of the buck converter while in the DCM is to look at the influence of the output current on the voltage transfer function. We have seen that the converter enters DCM when the output current is lower than the boundary current described by (6.53). This boundary current reaches a maximum value at $D = 0.5$, let us denote this maximum value by

$$I_{\mathrm{BM}} = \frac{T_{\mathrm{S}} V_{\mathrm{S}}}{8L}. \tag{6.67}$$

We have previously derived an expression for the average output current of the buck converter operating in the DCM. This expression is repeated from (6.60) as

$$I_o = \frac{V_o T_S}{2L}(D + \Delta_1)\Delta_1. \tag{6.68}$$

Let us substitute (6.56) into this expression to yield

$$I_o = \frac{V_S T_S}{2L} D\Delta_1, \tag{6.69}$$

solving for Δ_1 and using the definition for I_{BM} yields

$$\Delta_1 = \frac{I_o}{4 I_{BM} D}. \tag{6.70}$$

From (6.56), we know that

$$\frac{V_o}{V_s} = \frac{D}{D + \Delta_1}, \tag{6.71}$$

now substituting (6.70) yields

$$\frac{V_o}{V_s} = \frac{D^2}{D^2 + \frac{1}{4}\frac{I_o}{I_{LB}}}. \tag{6.72}$$

The voltage transfer ratio of the buck converter is shown in Fig. 6.20 as a function of the average output current. Once again, we see the highly non-linear behaviour of the converter while in the DCM.

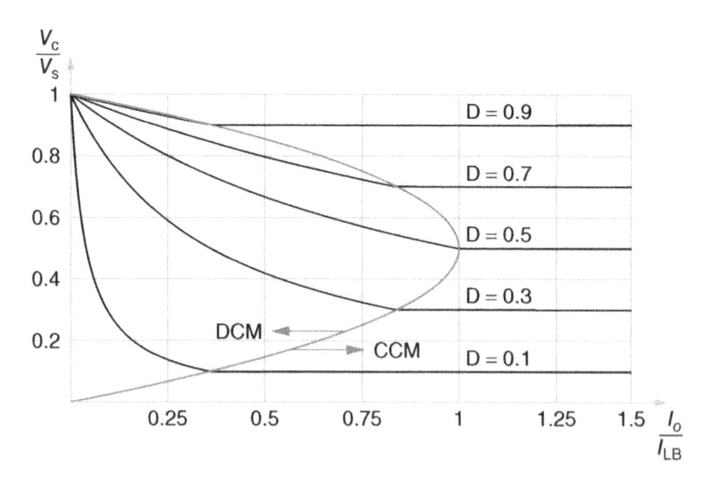

Figure 6.20 Buck converter in DCM: voltage transfer ratio as a function of output current.

6.7 OTHER BASIC CONVERTER STRUCTURES

Other basic converter structures apart from the basic buck converter exist. The buck converter discussed in Section 6.5 can only realise an output voltage that is lower than the source voltage, as is evident from (6.41). In this section, we will briefly discuss two other converters. The boost converter is only able to realise an output voltage higher than the source voltage. The buck–boost converter, on the other hand, is able to realise an output voltage that is either higher or lower than the source voltage.

In Sections 6.7.1 and 6.7.2, we will briefly discuss the steady-state operation of the boost and the buck–boost converters. Although these converters can also operate in the DCM if the load is light, we will not discuss the DCM boundary or the voltage transfer ratio of these converters in detail.

6.7.1 Boost Converter

The boost converter structure is shown in Fig. 6.21. If we compare it with the structure of the buck converter in Fig. 6.13, we see that the basic structure of the converter remains the same but the position of the switch, diode and inductor have changed. Let us investigate the converter operation in the steady state by analysing the converter waveforms shown in Fig. 6.22.

When the switch is in the closed position, the voltage across the inductor is equal to the source voltage and the diode is reverse biased. We can, therefore, conclude that $i_d = 0$ and that the inductor current slope

$$m_1 = \frac{V_s}{L}. \tag{6.73}$$

Conversely, when the switch is in the off state, the diode is forward biased and we know that $i_d = i_L$. The voltage across the inductor now becomes $V_s - V_c$, and the inductor current slope can be written as

$$m_2 = \frac{V_s - V_o}{L}. \tag{6.74}$$

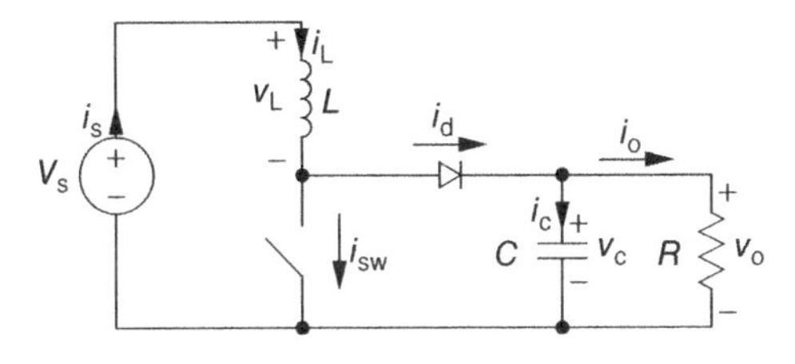

Figure 6.21 The boost converter.

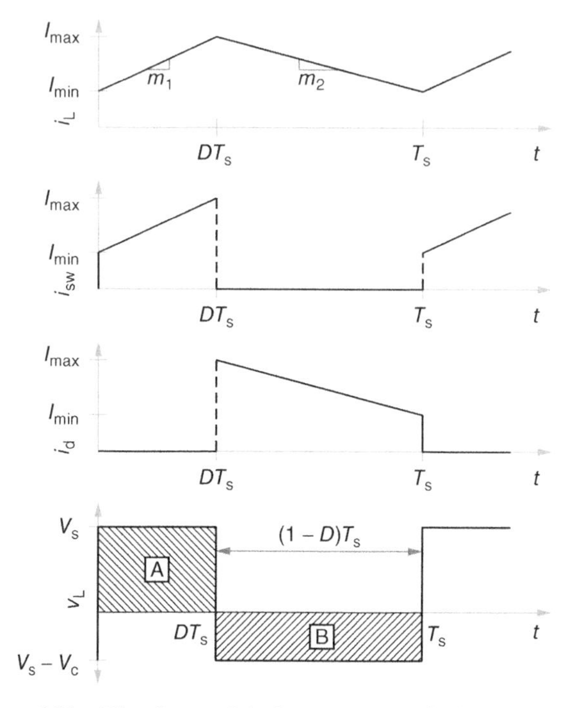

Figure 6.22 Waveforms of the boost converter in the steady state.

Now, because the converter is in steady state, and we know that

$$i_L(t) = i_L(t + T_s), \tag{6.75}$$

it is true that

$$m_1 DT_s = m_2(1 - D)T_s \tag{6.76}$$

and, therefore,

$$\frac{V_o}{V_s} = \frac{1}{1 - D}. \tag{6.77}$$

We can again assume that the circuit is lossless and, therefore,

$$\begin{aligned} P_{in} &= P_{out} \\ V_s I_s &= V_o I_o \\ \therefore \frac{I_o}{I_s} &= 1 - D. \end{aligned} \tag{6.78}$$

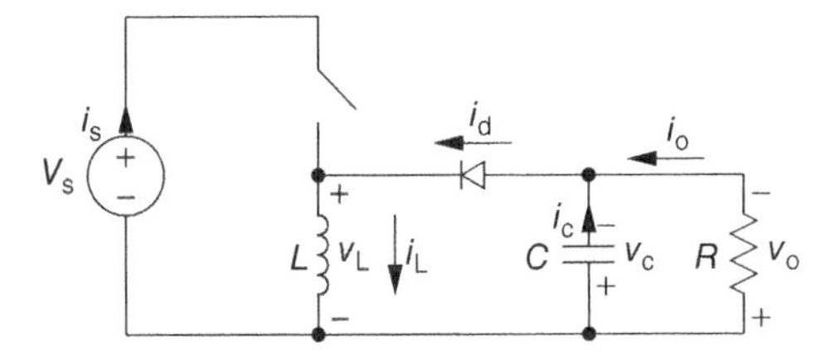

Figure 6.23 The buck–boost converter.

6.7.2 Buck–Boost Converter

The buck-boost converter structure is shown in Fig. 6.23. If we compare it with the structure of the buck converter in Fig. 6.13 or even the boost converter in Fig. 6.21, we will see that the basic structure of the converter remains the same but the position of the switch, diode and inductor have changed. Let us investigate the converter operation in the steady state by analysing the converter waveforms shown in Fig. 6.24.

When the switch is in the closed position, the voltage across the inductor is equal to the source voltage and the diode is reverse biased. We can, therefore, conclude

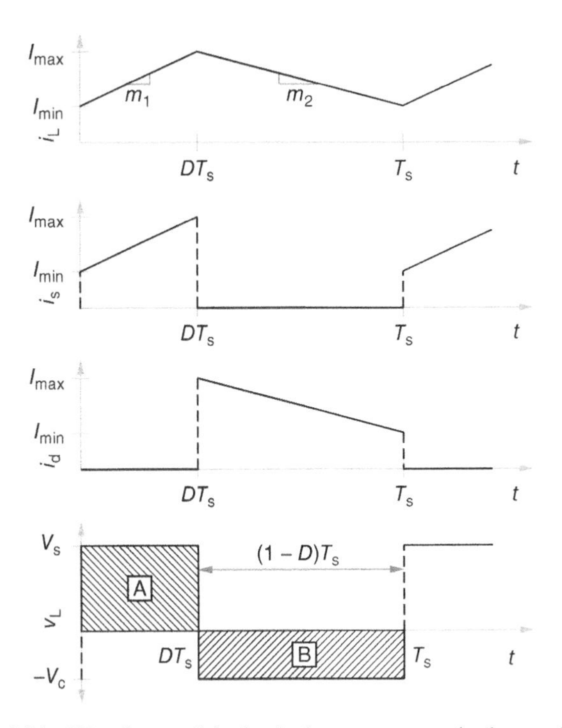

Figure 6.24 Waveforms of the buck–boost converter in the steady state.

that $i_d = 0$ and that the inductor current slope

$$m_1 = \frac{V_s}{L}. \tag{6.79}$$

Conversely, when the switch is in the off state, the diode is forward biased and we know that $i_d = i_L$. The voltage across the inductor now becomes $-V_o$ and the inductor current slope can be written as

$$m_2 = \frac{-V_o}{L}. \tag{6.80}$$

Now, because the converter is in the steady state, and we know that

$$i_L(t) = i_L(t + T_s), \tag{6.81}$$

it is true that

$$m_1 DT_s = m_2(1 - D)T_s \tag{6.82}$$

and, therefore,

$$\frac{V_o}{V_s} = \frac{D}{1 - D}. \tag{6.83}$$

We can again assume that the circuit is lossless and, therefore,

$$\begin{aligned} P_{in} &= P_{out} \\ V_s I_s &= V_o I_o \\ \therefore \frac{I_o}{I_s} &= \frac{1 - D}{D}. \end{aligned} \tag{6.84}$$

6.8 DC–DC CONVERTERS WITH ISOLATION

In many, if not most, applications where electric energy must be transformed from one voltage level to another, we have a need for galvanic isolation for safety purposes. Galvanic isolation means that we electrically isolate different parts of electrical systems to prevent current flowing between the sections. Often it is used to break ground loops to prevent accidental current from flowing to the ground through a person's body.

In Fig. 6.25, a 600 V/750 V dc/dc converter that is used to connect the power system of a ship to the land power system is shown. To prevent electrolysis of the hull of the ship, it is important that the two power supplies are electrically isolated. It uses a 25 kHz transformer and weighs only 10 kg! It is five hundred times more powerful at half the weight compared to the linear supply in Fig. 6.3.

Figure 6.25 A 50 kW galvanically isolated dc/dc converter.

In this section, we will discuss two converters with galvanic isolation. The flyback converter is a converter in which coupled inductors are used for isolation purposes. The energy is stored in the primary-side inductor and then released through the secondary-side inductor. The second converter, the half-bridge converter, uses a transformer as an isolation medium.

6.8.1 Coupled Inductor Isolation: Flyback

The flyback converter is derived from the boost converter by replacing the inductor of the boost converter with a coupled inductor. Recall that in the boost converter, energy is stored in the inductor while the switch is closed; this energy is then released to the load when the switch is opened. Now consider the flyback circuit in Fig. 6.26.

The key to understanding the flyback circuit operation is to realise that the switch in the primary side and the diode in the secondary side will conduct at different times. With the current direction definitions of the circuit in Fig. 6.26, it follows that when the switch is closed and $i_p > 0$ then $i_s = 0$ because the diode cannot reverse conduct. Likewise, when the switch is open, we always have the condition that $i_p = 0$. This decoupling of the primary-side and the secondary-side circuits simplifies the analysis of the converter considerably. That said, it is important to

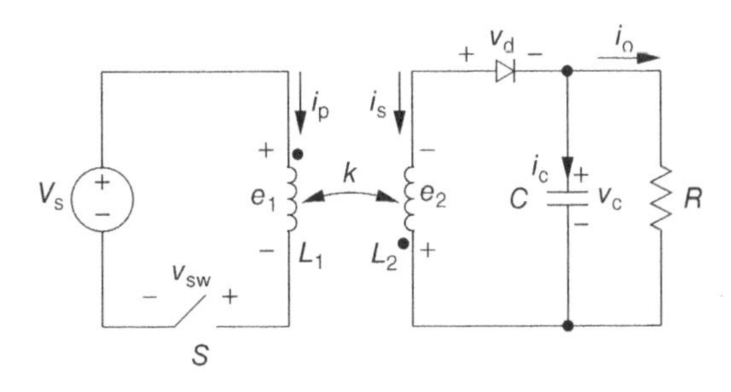

Figure 6.26 A flyback converter.

point out that this decoupling is strictly only possible in the ideal case where $k = 1$. If the coupling between the inductors is non-ideal, the leakage inductance on the primary and secondary sides somewhat complicates the analysis. Here we will, however, only analyse the ideal case. The waveforms are shown in Fig. 6.27 and Fig. 6.28.

As the flyback transformer is in essence a coupled inductor, this section should be read in conjunction with Section 4.3.2. In fact, many of the expressions derived there will be used here.

Let us start with the case where the switch, S, is closed. From 4.61, and substituting $e_1 = V_s$, $\frac{d}{dt}i_s = 0$ and the fact that $k = 1 \leftrightarrow M = \sqrt{L_1 L_2}$ yield

$$\begin{bmatrix} V_s \\ e_2 \end{bmatrix} = \begin{bmatrix} L_1 & -\sqrt{L_1 L_2} \\ \sqrt{L_1 L_2} & -L_2 \end{bmatrix} \frac{d}{dt} \begin{bmatrix} i_p \\ 0 \end{bmatrix}. \tag{6.85}$$

We can now rewrite this system of differential equations as (where N_1 and N_2 are the number of turns on the primary and secondary inductors, respectively)

$$e_2 = \sqrt{\frac{L_2}{L_1}} V_s = \frac{N_2}{N_1} V_s \tag{6.86}$$

and

$$\frac{d}{dt} i_p = \frac{V_s}{L_1}. \tag{6.87}$$

The voltage across the diode is, therefore,

$$v_d = -v_c - \frac{N_2}{N_1} V_s. \tag{6.88}$$

Now, let $I_{\min p}$ be the current in the primary inductor at the moment when the switch closes; with $t = 0$ defined as the instance when the switch closes, so the primary current can be written as

$$i_p(t) = I_{\min p} + \frac{1}{L_1} V_s t. \tag{6.89}$$

And, therefore, the maximum value of the primary current will be at the point where D is the duty cycle and T_s the switching period,

$$I_{\max p} = I_{\min p} + \frac{1}{L_1} V_s D T_s. \tag{6.90}$$

Meanwhile, on the secondary side, because $i_s = 0$, the system is described by the equation

$$-\frac{d}{dt} v_c = \frac{v_c}{R}. \tag{6.91}$$

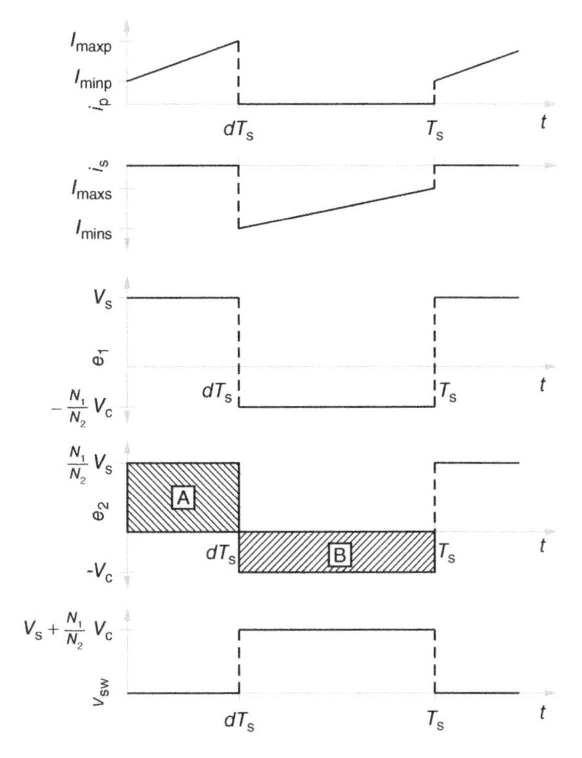

Figure 6.27 Waveforms of the flyback converter, simplified approach.

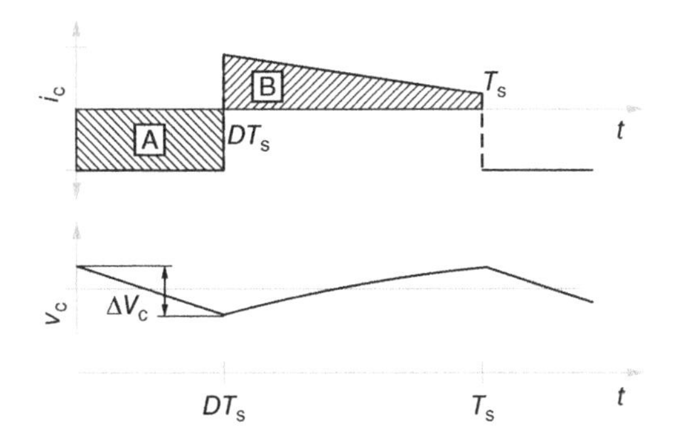

Figure 6.28 Waveforms of the flyback converter capacitor, simplified approach.

When the switch is open, two parameters of the circuit are known, namely, $i_p = 0$ and $e_2 = -v_c$. Again, substitution into (4.61) yields

$$\begin{bmatrix} e_1 \\ -v_c \end{bmatrix} = \begin{bmatrix} L_1 & -\sqrt{L_1L_2} \\ \sqrt{L_1L_2} & -L_2 \end{bmatrix} \frac{d}{dt} \begin{bmatrix} 0 \\ i_s \end{bmatrix}. \tag{6.92}$$

Therefore,

$$\frac{d}{dt}i_s = \frac{v_c}{L_2} \tag{6.93}$$

and

$$e_1 = -\sqrt{L_1L_2}\frac{d}{dt}i_s = -\sqrt{\frac{L_1}{L_2}}v_c = -\frac{N_1}{N_2}v_c. \tag{6.94}$$

So, the voltage across the open switch will be

$$v_{sw} = V_s + \frac{N_1}{N_2}v_c. \tag{6.95}$$

We can now write equations to describe the secondary current and the capacitor voltage. Recall that we defined $t = 0$ as the moment when the switch is closed; therefore, the switch opens at $t = DT_s$. If $I_{\min s}$ is the current at the secondary immediately after the switch is opened then

$$i_s = I_{\min s} + \frac{v_c}{L_2}(t - DT_s) \tag{6.96}$$

and

$$\frac{d}{dt}v_c = -i_s - \frac{v_c}{R}. \tag{6.97}$$

The secondary current at the point where the switch closes again will be, where we neglect the capacitor voltage ripple and assume that $v_c(t) \approx V_c$,

$$I_{\max s} = I_{\min s} + \frac{V_c}{L_2}(1 - D)T_s. \tag{6.98}$$

Finally, we have enough information to put everything together. The operation of the flyback converter can be explained by the fact that when the switch S is closed the source stores energy in the primary inductor L_1. However, it is important to note that because the inductors are coupled (and assuming $k = 1$), the energy stored in the secondary inductor is also the energy stored in the secondary-side inductor. When the switch opens, this energy stored in the inductor is released to the secondary side.

The maximum amount of energy stored in the inductor is

$$W_{\max} = \frac{1}{2}L_1I_{\max p}^2 \tag{6.99}$$

at the moment just before the switch opens. As soon as the switch has opened, because the energy cannot be destroyed or removed from the system, the secondary current must be

$$|I_{\text{min s}}| = \sqrt{\frac{2W_{\text{max}}}{L_2}} = \frac{N_1}{N_2} I_{\text{max p}}. \tag{6.100}$$

Likewise, at the point just before and after the switch closes, the primary and secondary currents are related by

$$|I_{\text{min p}}| = \frac{N_2}{N_1} I_{\text{max s}}. \tag{6.101}$$

Owing to the definition of the secondary current direction in Fig. 6.26, the secondary current is negative. *It is possible to derive all the equations presented here with an alternate definition of the secondary current direction; however, that would require a rewriting of the equations describing the coupled inductors developed in Section 4.3.2.*

In steady state, where $i_{\text{p}}(t + T_{\text{s}}) = i_{\text{p}}(t)$, we can relate these expressions as

$$\begin{aligned}
I_{\text{min p}} &= -\frac{N_2}{N_1}\left(I_{\text{min s}} + \frac{V_{\text{c}}}{L_2}(1-D)T_{\text{s}}\right)\\
&= -\frac{N_2}{N_1}\left(-\frac{N_1}{N_2}\left(I_{\text{min p}} + \frac{V_{\text{s}}}{L_1}DT_{\text{s}}\right) + \frac{V_{\text{c}}}{L_2}(1-D)T_{\text{s}}\right)\\
&= I_{\text{min p}} + \frac{V_{\text{s}}}{L_1}DT_{\text{s}} - \frac{N_2}{N_1}\frac{V_{\text{c}}}{L_2}(1-D)T_{\text{s}}.
\end{aligned}$$

As the system is in steady state, and using the fact that $L_1 = \mathscr{P}N_1^2$ and $L_2 = \mathscr{P}N_2^2$, we can write

$$\begin{aligned}
\frac{N_2}{N_1}\frac{V_{\text{c}}}{L_2}(1-D)T_{\text{s}} &= \frac{V_{\text{s}}}{L_1}DT_{\text{s}}\\
\frac{V_{\text{c}}}{V_{\text{s}}} &= \frac{N_1}{N_2}\frac{\mathscr{P}N_2^2}{\mathscr{P}N_1^2}\frac{D}{1-D}
\end{aligned}$$

and, therefore,

$$\boxed{\frac{V_{\text{c}}}{V_{\text{s}}} = \frac{N_2}{N_1}\frac{D}{1-D}.} \tag{6.102}$$

The voltage across the open switch, as derived in (6.95), can now be rewritten for steady state as

$$\boxed{v_{sw} = \frac{V_{\text{s}}}{1-D}.} \tag{6.103}$$

6.8.2 Transformer Isolation: Half-bridge

The half-bridge converter differs from the flyback in that it does not store energy in the isolation device. Rather it uses a transformer to directly deliver energy from the primary side to the secondary side.

Consider the circuit shown in Fig. 6.29. This consists of two switches and two capacitors and the transformer primary is connected to the mid-point of each leg. The source, with value V_s, that is connected to the two capacitors is not shown in this figure.

The first requirement of this circuit that we must address is the fact that the two switches in series can never be on at the same time. If both the switches are on, the capacitors and the source will be effectively shorted and the switches, and probably more, will fail. One way to generate the switching signals of the converter is shown in Fig. 6.30. Here we use two phase-shifted carrier waveforms, v_{c1} and v_{c2}. The two carriers are compared to the same reference value. The two switching signals are then found as

$$s_1(t) = \begin{cases} 1 & \text{if } v_r \geq v_{c1} \\ 0 & \text{if } v_r < v_{c1} \end{cases} \tag{6.104}$$

$$s_2(t) = \begin{cases} 1 & \text{if } v_r \geq v_{c2} \\ 0 & \text{if } v_r < v_{c2}. \end{cases} \tag{6.105}$$

The requirement that the two switches are never on at the same time now equates to the requirement that

$$\boxed{v_r < \frac{\hat{V}_c}{2},} \tag{6.106}$$

or equivalently that $D < \frac{1}{2}$.

When switch S_1 is closed, the voltage across the transformer primary winding is equal to $\frac{1}{2}V_s$. This will cause diode D_1 to be forward biased, and the rectified

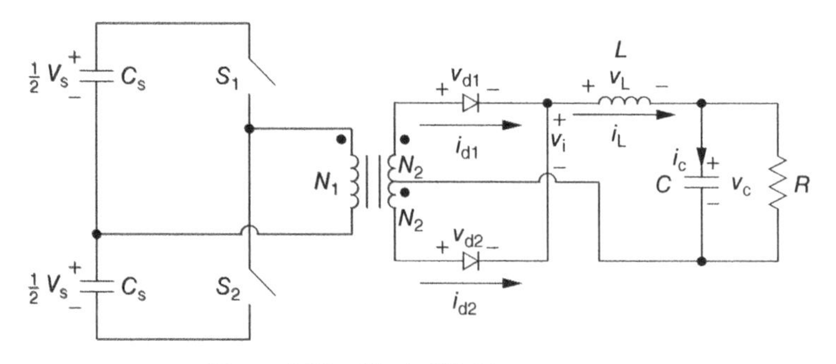

Figure 6.29 The half-bridge converter.

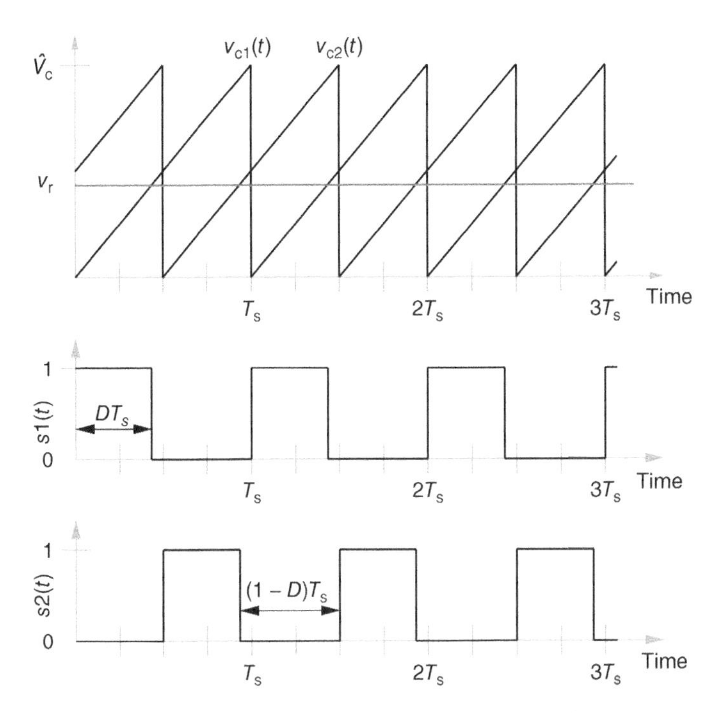

Figure 6.30 The half-bridge switching signals.

secondary voltage is

$$v_i = \frac{N_2}{N_1}\frac{V_s}{2}. \tag{6.107}$$

The voltage across the inductor is, therefore,

$$v_L = v_i - v_c = \frac{N_2}{N_1}\frac{V_s}{2} - v_c \tag{6.108}$$

$$\therefore \frac{d}{dt}i_L = \frac{1}{L}\left(\frac{N_2}{N_1}\frac{V_s}{2} - v_c\right). \tag{6.109}$$

As the diode D_1 is forward biased and, therefore, conducting while diode D_2 is reverse biased, we have

$$i_{d1} = i_L \text{ and } i_{d2} = 0. \tag{6.110}$$

When switch S_1 opens and before S_2 closes, the voltage across the transformer primary is equal to zero. During this interval, the voltage across the inductor is

$$v_L = -v_c \tag{6.111}$$

and, therefore,

$$\frac{d}{dt}i_L = -\frac{v_c}{L}. \tag{6.112}$$

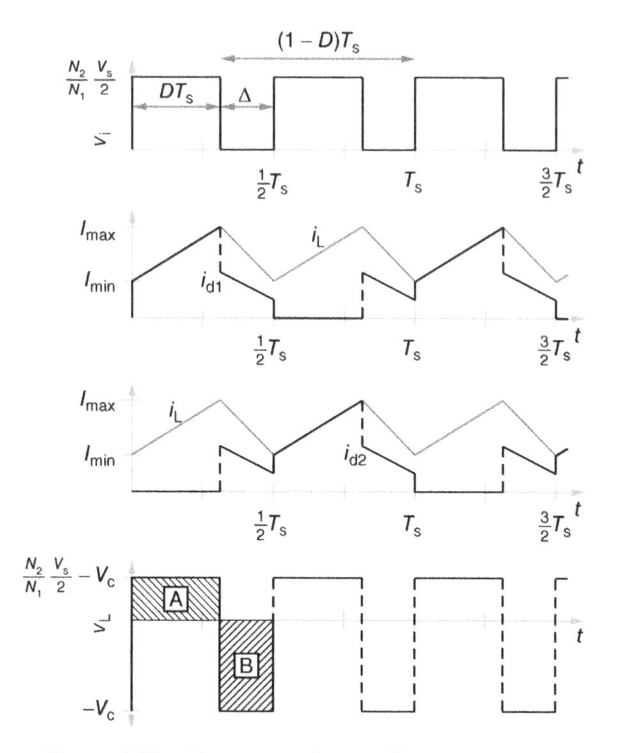

Figure 6.31 Waveforms of the half-bridge converter.

The inductor current splits equally between the two diodes.

At the point where switch S_2 closes, the process starts again, although at this point the diode D_2 will be conducting. The waveforms of the converter are shown in Fig. 6.31.

In steady state, because

$$i_L(t) = i_L(t + T_s) = i_L\left(t + \frac{T_s}{2}\right), \tag{6.113}$$

we have the requirement that (refer to Fig. 6.31) area A must be equal to area B. Therefore,

$$\left(\frac{N_2}{N_1}\frac{V_s}{2} - V_c\right) DT_s = V_c\left(\frac{1}{2}T_s - DT_s\right) \tag{6.114}$$

and, therefore,

$$\boxed{\frac{V_c}{V_s} = \frac{N_2}{N_1}D.} \tag{6.115}$$

As discussed in Section 6.5.5, if the output low pass filter (as defined by L and C) has a cutoff frequency such that

$$\frac{1}{LC} \ll 2\pi f_c, \tag{6.116}$$

we can assume that the output voltage of the converter is nearly constant, that is,

$$v_c \approx V_c \longrightarrow \tilde{v}_c \approx 0. \tag{6.117}$$

In this case, the load current can be written as (neglecting the small signal ripple components)

$$I_o = \frac{V_c}{R}. \tag{6.118}$$

The maximum current in the inductor, I_{max} in Fig. 6.31, can be calculated, while the converter is operating in the CCM, by applying the fact that the inductor current ripple is

$$\boxed{\Delta i_L = I_{max} - I_{min} = \frac{DT_s}{L}\left(\frac{V_s}{2}\frac{N_2}{N_1} - V_c\right) = \left(\frac{1}{2D} - 1\right)\frac{V_c DT_s}{L}.} \tag{6.119}$$

The peak inductor current is, therefore,

$$\boxed{I_{max} = I_o + \frac{1}{2}\Delta i_L = I_o + \left(\frac{1}{2D} - 1\right)\frac{V_c DT_s}{2L}} \tag{6.120}$$

and the minimum load current necessary to keep the converter in the CCM is found by the requirement that $I_{min} > 0$ and, therefore,

$$I_o > \left(\frac{1}{2D} - 1\right)\frac{V_c DT_s}{2L}. \tag{6.121}$$

Example 6.6 Calculate the value of the inductance for a half-bridge converter similar to that in Fig. 6.29 such that the ripple current is less than 10% under full load conditions. The desired operating conditions are tabulated in Table 6.3. With this inductance, what is the maximum load resistance that can be connected while the converter is in the CCM?

Solution

The converter duty cycle is

$$D = \frac{V_c}{V_s}\frac{N_1}{N_2} = 41.67\%,$$

TABLE 6.3 Half-Bridge Converter Specifications, Example 6.6

Item	Value	Item	Value
Switching frequency, f_s	25 kHz	Transformer turns ratio, $\frac{N_2}{N_1}$	0.75
Input voltage, V_s	48 V	Output voltage, V_c	15 V
Full load power, P_{max}	750 W	Filter capacitor, C	4700 μF

and the average output current, and therefore, by implication, the average inductor current, under full load conditions is

$$I_o = \frac{P_{max}}{V_c} = 50\,\text{A}.$$

The allowable ripple current is, therefore,

$$\Delta i_L = 0.1\bar{\imath}_L = 5\,\text{A}.$$

To keep the ripple current smaller than 5 A, the filter inductor should, therefore, be

$$\Delta i_L < 5\,\text{A} \rightarrow L > \left(\frac{1}{2D} - 1\right)\frac{V_c D T_s}{5}$$
$$\therefore L > \left(\frac{1}{D} - 1\right)\frac{V_c D}{5 f_s}$$
$$> 10\,\mu\text{H},$$

To keep the converter in the CCM, the load current must be larger than $\frac{1}{2}\Delta i_L$ and, therefore,

$$R > \frac{2V_c}{\Delta i_L} = 6\,\Omega,$$

which implies that the converter must deliver more than 37.5 W to remain in the CCM region.

6.8.3 Transformer Isolation: Full-bridge

The half-bridge converter does have the advantage that only two active switching elements are used. However, because of the fact that only a half of the input voltage is ever applied to the transformer primary, the current through the primary is larger than that for the full-bridge converter. Close inspection of the half-bridge circuit operation shows that the transformer primary current also flows through the two bus capacitors. At higher power levels, this current becomes problematic because the capacitance must increase significantly and an alternate solution should be found.

The full-bridge converter differs from the half-bridge not only in the increase of semiconductor devices but also in that the full bus voltage can be applied to the transformer primary. Although the same centre-tapped rectifier described in the operation of the half-bridge converter can also be used for the full-bridge converter, we will consider the use of a full-bridge rectifier here. The full-bridge rectifier can naturally be also used with the half-bridge converter.

Consider the circuit shown in Fig. 6.32. The primary side of the converter consists of four controllable switches. To make the discussion easier, let us define a switching function $s_1(t)$ that describes the switched position of S_1 such that

$$s_1(t) = \begin{cases} 1 & \text{if } S_1 \text{is on} \\ 0 & \text{if } S_1 \text{is off.} \end{cases} \tag{6.122}$$

We will use similar switching functions s_2, s_3 and s_4 to describe the switching positions of the other switches.

As with the half-bridge converter, it is clear that no two switches in series can ever be on at the same time; therefore, the combinations $s_1 = s_2 = 1$ and $s_3 = s_4 = 1$ are not allowed. In fact, the switch pairs in a leg are always switched as a complementary pair such that $s_1 + s_2 = 1$ and $s_3 + s_4 = 1$. Using the definitions shown, the voltage across the primary side of the transformer can be described as

$$v_p = \begin{cases} -V_s & \text{if } s_2 = s_3 = 1 \\ 0 & \text{if } s_1 = s_3 = 1 \text{ or } s_2 = s_4 = 1 \\ V_s & \text{if } s_1 = s_4 = 1. \end{cases} \tag{6.123}$$

There are several ways in which the switching functions for the different switches can be derived. Recall that for the previous converters we have discussed,

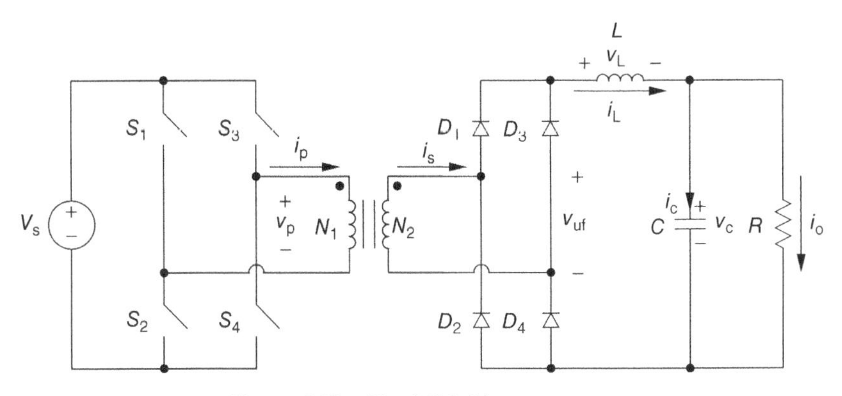

Figure 6.32 The full-bridge converter.

we compared a value with a triangular carrier and to change the effective duty cycle, we kept the carrier constant and changed the reference value. Although we can quite effectively do the same for the full-bridge converter, let us try something different.

Let us create two triangular carrier functions f_{c1} and f_{c2}, both with the same frequency and amplitude. However, let one carrier be shifted in time by a time period Δ such that

$$f_{c2}(t) = f_{c1}(t - \Delta). \tag{6.124}$$

One way of describing a triangular carrier is by using trigonometric functions

$$f_{c1} = \frac{2}{\pi} \arcsin\left(\sin\left(\frac{2\pi t}{T_s}\right)\right). \tag{6.125}$$

The switching signals s_1 through s_4 are now defined such that

$$s_1 = \begin{cases} 1 & = \text{if } f_{c1} \geq 0 \\ 0 & = \text{if } f_{c1} < 0 \end{cases} \tag{6.126}$$

$$s_2 = \begin{cases} 1 & = \text{if } f_{c1} < 0 \\ 0 & = \text{if } f_{c1} > 0 \end{cases} \tag{6.127}$$

$$s_3 = \begin{cases} 1 & = \text{if } f_{c2} \geq 0 \\ 0 & = \text{if } f_{c2} < 0 \end{cases} \tag{6.128}$$

$$s_1 = \begin{cases} 1 & = \text{if } f_{c2} < 0 \\ 0 & = \text{if } f_{c2} > 0. \end{cases} \tag{6.129}$$

The resulting switching waveforms are shown in Fig. 6.33.

Previously, it was convenient to describe the converter transfer behaviour by using the duty cycle as our measure. In the phase-shifted full-bridge, the transfer ratio is adjusted by phase shifting one carrier function with respect to another. To describe this behaviour, let us define a unitless constant, ϕ, such that

$$\phi = \frac{\Delta}{\frac{T_s}{2}}. \tag{6.130}$$

Using the same argumentation as for the half-bridge converter, we can draw the switching waveforms as shown in Fig. 6.34. However, in this case, during a half-cycle, the voltage across the transformer primary will be equal to zero for a $(1 - \phi)$

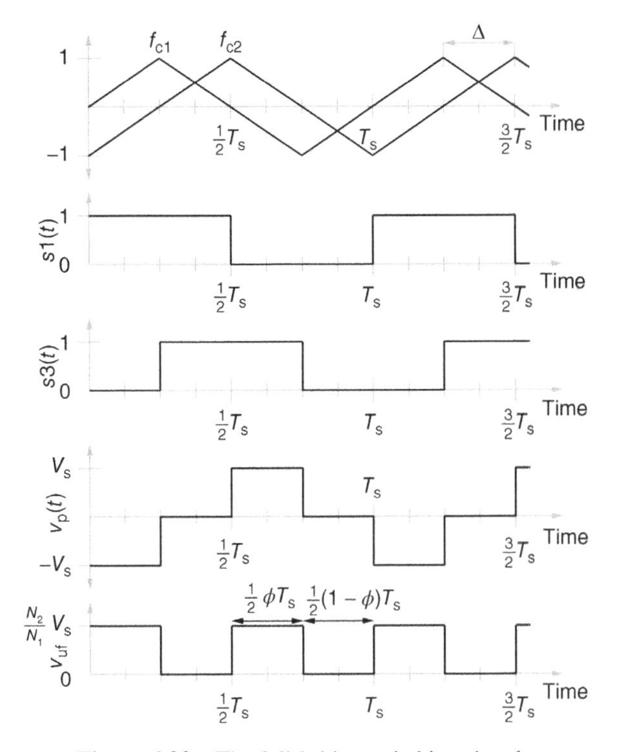

Figure 6.33 The full-bridge switching signals.

portion of time and equal to $\pm V_s$ for the remaining ϕ portion of time. As the areas A and B in Fig. 6.34 must be equal to one another because we have the requirement that $i_L(t) = i_L(t + T_s)$, it is true that

$$\left(\frac{N_2}{N_1}V_s - V_c\right)\phi\frac{T_s}{2} = V_c(1-\phi)\frac{T_s}{2} \tag{6.131}$$

and, therefore,

$$\boxed{\frac{V_c}{V_s} = \frac{N_2}{N_1}\phi.} \tag{6.132}$$

It is interesting to note that all four diodes conduct during the period when either S_1 and S_3 are on simultaneously or were S_2 and S_4 are on simultaneously. During these periods, the transformer primary is effectively clamped to 0 V and the current flowing through the output inductor forces all four diodes to conduct.

Example 6.7 Calculate ripple current in the inductor for a full-bridge converter, as shown in Fig. 6.32, when operated with the parameters indicated in Table 6.4.

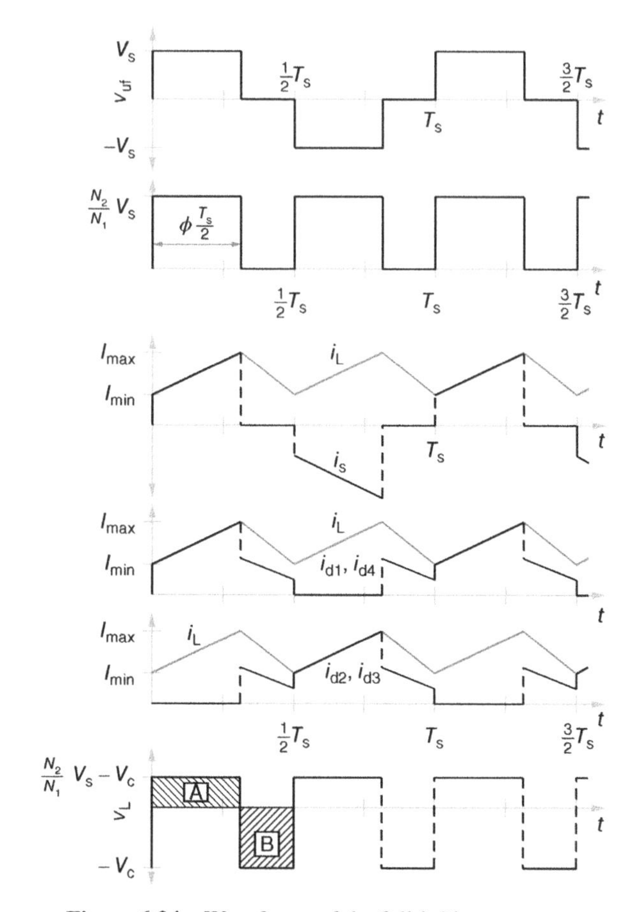

Figure 6.34 Waveforms of the full-bridge converter.

TABLE 6.4 Full-Bridge Converter Specifications, Example 6.7

Item	Value	Item	Value
Switching frequency, f_s	10 kHz	Transformer turns ratio, $N_2 : N_1$	1 : 6
Input voltage, V_s	420 V	Output voltage, V_c	48 V
Delivered load power, P	2.5 kW	Filter capacitor, C	4700 μF
Filter inductance, L	300 μH		

Solution

To deliver the required output voltage, the phase shift ratio must be

$$\phi = \frac{V_c}{V_s}\frac{N_1}{N_2} = 68.71\%,$$

and the average output current, and therefore, by implication, the average inductor current, is

$$I_o = \frac{P_{max}}{V_c} = 52\ \text{A}.$$

During the 'off' period, the voltage-seconds applied to the inductor is given by the area B as shown in Fig. 6.34

$$A_B = V_c(1 - \phi)\frac{T_s}{2} = 751\ \mu\text{Vs}$$

and the current ripple is, therefore,

$$\Delta i_L = \frac{751\ \mu\text{Vs}}{L} = 2.5\ \text{A}.$$

6.9 HIGHLIGHTS

- Efficiency is one of the main drivers for using switched mode operation in power electronics. Other considerations are weight and size reduction and the requirement for tightly controlled input and output parameters.
- The assumption made in the analysis of switchmode circuits is that the switch is ideal, which implies that
  1. when in the off state, the switch can handle any voltage across it and will not fail;
  2. when in the on state, the switch can handle any current through it without failing;
  3. the switch can switch instantaneously from the off to the on state and vice versa;
  4. there are no losses associated with the switching cycle;
  5. the switch only requires a signal to control the state of the switch.
- We assume that if the control signal is equal to '1', the switch is closed (i.e. 'on') and if the signal is equal to '0', the switch is 'off'. This signal is called $s(t)$ and by definition $s(t) \in \{0, 1\}$ or, in other words, the value of s can be either 1 or 0 for any and all points in time.
- In steady state, the average voltage across an inductor over one period is equal to zero

  $$\frac{1}{L}\int_{qT_s}^{(q+1)T_s} v_L\, dt = 0.$$

  Likewise, in steady state, the average current through a capacitor over one period is equal to zero

  $$\frac{1}{C}\int_{qT_s}^{(q+1)T_s} i_c\, dt = 0.$$

- The method for controlling the switch behaviour is through pulse-width modulation. The duty cycle, D, describes the duration of the on state of the switch. For steady-state operation, the voltage integral over the switching cycle is zero

$$\frac{1}{L}\int_{t}^{t+DT_s} v_L\, dt + \frac{1}{L}\int_{t+DT_s}^{T_s} v_L.$$

- To simplify the analysis of the converter, we first calculate the inductor current assuming that capacitor is large enough that the voltage across the capacitor is a constant, that is, $v_c(t) = V_c$. Using this inductor current, the charge ΔQ delivered to the capacitor in a switching cycle is found. The voltage ripple can then be calculated using the equation

$$\Delta v_c = \frac{\Delta Q}{C}$$

- A converter is in the CCM if the inductor current is continuous at all times and does not remain zero for periods of time. A converter that does not adhere to this requirement is said to be in the DCM. The DCM radically changes the transfer characteristic of a converter.
- The simplified analysis yields the following voltage transfer equations for the basic dc converters topologies under CCM:
  1. Buck
  $$\frac{V_o}{V_s} = D.$$
  2. Boost
  $$\frac{V_o}{V_s} = \frac{1}{1-D}.$$
  3. Buck–Boost
  $$\frac{V_o}{V_s} = \frac{D}{1-D}.$$
- The source and load currents are calculated by assuming that the circuit is lossless and, therefore,

$$P_{in} = P_{out}$$
$$V_s I_s = V_o I_o.$$

- DC–DC converters with isolation use coupled coils. If the number of turns on the source side is N_1 and on the load side is N_2 then the voltage transfer functions under CCM become
  1. Flyback
  $$\frac{V_c}{V_s} = \frac{N_2}{N_1}\frac{D}{1-D}.$$

2. Half-bridge

$$\frac{V_c}{V_s} = \frac{N_2}{N_1}D.$$

3. Full-bridge

$$\frac{V_c}{V_s} = \frac{N_2}{N_1}\phi,$$

where ϕ is the phase shift ratio.

PROBLEMS

6.1 Make a table for yourself showing equations for solving problems. Consider the basic converter types. Sometimes, the input voltage is constant and the output varies, while in other cases, the input may change, and the output is constant. There are many possibilities, and the challenge is to capture it all in a single table.

6.2 Under which conditions can the energy stored in a capacitor or an inductor destroy the semiconductor transistors in the power electronic converters?

6.3 In the simplified approach that we use to analyse the switch mode circuits of dc converters, we assume that the output capacitor voltage is dc when we calculate the inductor current. In a next step, we use this inductor current waveform to calculate the capacitor voltage ripple. Name the advantages and disadvantages of this method.

6.4 It is important that the magnetic cores in the inductors and transformers that we use in power electronics do not saturate. As the magnetic flux is determined by the integral of the voltage that is applied to the windings, we have to make sure that voltage averages out to zero over a cycle. For example, the 50 Hz ac of the power grid is a sinusoid, which by definition has an average of zero over a cycle. Explain how we ensure that inductors and transformers do not saturate in the power electronic circuits discussed in this chapter.

6.5 The input voltage of a linear regulator is variable between 10 and 14 V. $V_z = 6.8\,V$, $R_l = 10\,\Omega$, $V_{be} = 0.7\,V$ and $\beta = 100$.

1. If $I_z > 2\,mA$, find R_s.

2. Find the maximum power that can be dissipated in the transistor.

6.6 Which converter will have the largest peak current and which the smallest minimum current? The following values apply: $V_s = 150\,V$, $D = 0.4$, $f_s = 20\,kHz$, I_L (average) $= 10\,A$, $L = 1.2\,mH$, $C = 2\,mF$, and consider

1. buck converter;

2. boost converter;

3. buck–boost converter.

6.7 What is the output current of each converter? The following values apply $V_s = 150\,\text{V}$, $D = 0.4$, $f_s = 20\,\text{kHz}$, I_L (average) $= 10\,\text{A}$, $L = 1.2\,\text{mH}$, $C = 2\,\text{mF}$, and consider

1. buck converter;
2. boost converter;
3. buck–boost converter.

6.8 Which converter will have the largest voltage on the capacitor? The following values: apply $V_s = 150\,\text{V}$, $D = 0.4$, $f_s = 20\,\text{kHz}$, I_L (average) $= 10\,\text{A}$, $L = 1.2\,\text{mH}$, $C = 2\,\text{mF}$, and consider

1. Buck converter,
2. Boost converter,
3. Buck–boost converter.

6.9 What is the maximum voltage that the switch (transistor) has to block when it is turned off for each converter? The following values apply: $V_s = 150\,\text{V}$, $D = 0.4, f_s = 20\,\text{kHz}$, I_L (average) $= 10\,\text{A}, L = 1.2\,\text{mH}, C = 2\,\text{mF}$, and consider

1. Buck converter,
2. Boost converter,
3. Buck–boost converter.

6.10 What is the average current in the diode for each converter? The following values apply: $V_s = 150\,\text{V}$, $D = 0.4$, $f_s = 20\,\text{kHz}$, I_L (average) $= 10\,\text{A}$, $L = 1.2\,\text{mH}$, $C = 2\,\text{mF}$, and consider

1. buck converter;
2. boost converter;
3. buck–boost converter.

6.11 What is the voltage ripple on the output for each converter? The following values apply: $V_s = 150\,\text{V}$, $D = 0.4$, $f_s = 20\,\text{kHz}$, I_L(average) $= 10\,\text{A}$, $L = 1.2\,\text{mH}$, $C = 2\,\text{mF}$, and consider

1. buck converter;
2. boost converter;
3. buck–boost converter.

6.12 A buck converter supplies 12 V to a resistive load of 6 Ω from a 30 V source using the circuit in Fig. 6.13. The inductor current is continuous. The switching frequency is 5 kHz. Find

1. the value of the duty cycle;
2. the minimum value of inductance required;
3. the minimum and maximum values of i_L if $L = 1.5\,\text{mH}$;
4. power drawn from the source;
5. power delivered to the load.

6.13 A buck converter supplies an adjustable voltage and a resistive load with a variable resistance. The required output voltage range is 15–25 V. We know that the load resistance is $4\,\Omega \leq R \leq 10\,\Omega$. Any combination of the output voltage and the load resistance may occur. The input voltage is 40 V and the switching frequency is 8 kHz. Find

1. the range of duty cycle required;
2. the minimum value of inductance required to ensure a continuous inductor current under all operating conditions.

6.14 A buck converter, operating at 10 kHz, supplies 100 W at 12 V to a load from a 20 V source. Find

1. the inductance needed for a continuous inductor current;
2. the inductor current ripple with an inductance of 28.8 μH;
3. the size of output capacitance needed to keep the output voltage ripple below 0.1 V.

6.15 A buck converter requires an output power of 100 W at an output voltage of 24 V from a 60 V supply. The switching frequency is 60 kHz. The permitted value of ΔV_c is 0.15 V. Find values for D, L and C assuming L is three times as large as that required for continuous current.

6.16 A buck converter operates with a source that varies from 50 to 70 V. The output is 200 W at 30 V. Assuming a switching frequency of 60 kHz, design the remaining elements so that L is just large enough to maintain continuous current under all conditions and the output voltage ripple exceeds 0.1 V.

6.17 A boost converter with the configuration in Fig. 6.21 supplies 150 V to a 25 Ω load from a 40 V source. The capacitor is large enough that negligible load voltage ripple occurs; $L = 200\,\mu\text{H}$ and $T_s = 200\,\mu\text{s}$. Find

1. the value of the duty cycle;
2. the minimum value of the inductor current;
3. the maximum value of the inductor current;
4. the average diode current;
5. the output voltage ripple;
6. the critical inductance value that just permits continuous current in the inductor by using all the existing data, except the inductance value.

6.18 A boost converter has the following values: $V_s = 18\,V$, $T_s = 15\,\mu s$, $L = 20\,\mu\text{H}$, $D = 0.4$, $C = 200\,\mu\text{F}$ and $R = 6\,\Omega$. Find

1. the value of L necessary for the continuous current mode,
2. output voltage;
3. I_{max} and I_{min};
4. output voltage ripple.

6.19 A boost converter supplies a load of 40 V and 100 W from a 30 V source. The switching frequency is 25 kHz. Design values for L and C such that the peak switch current does not exceed 5 A and the voltage ripple is 1.0 V.

6.20 A boost chopper supplies a load of 50 V and 200 W from a source that varies from 20 to 30 V. The switching frequency is 60 kHz. Find an inductance value such that L is just sufficient to ensure that the inductor current is never in the discontinuous mode. Find C such that the voltage ripple is 0.3 V in the worst scenario.

6.21 An idealised flyback converter with a 1 : 1 ratio is to supply a variable voltage, ranging from 10 to 40 V, to a 5 Ω load resistor. The source voltage is 30 V. Assume continuous current mode of operation. If $T = 100\ \mu s$, find

1. the required duty cycle range;
2. the inductance value required for continuous current operation for all values of the duty cycle;
3. the peak switch current.

6.22 A flyback converter is to supply 100 A at 5 V from a 300 V dc source. The chopping frequency is 40 kHz. A step-down turn ratio of ($N_1/N_2 = 60$) is to be used in the actual converter.

1. Restatement of the problem for a 1 : 1 turn ratio.
2. Find the duty cycle that is required.
3. Find the value of L_1 for continuous current operation.
4. Find the actual value of L_2.
5. Find the voltage that the switch has to block.
6. The value of C for $\Delta VC = 0.05$ V if L_1 is large enough that the capacitor current is negative only during the time when the switch is closed.

6.23 A buck–boost supplies a load of 100 Ω from a 50 V source with $D = 0.45$, $T_s = 25\ \mu s$, $L = 375\ \mu H$ and $C = 20\ \mu F$. Find

1. the value of the minimum inductance that causes the continuous inductor current;
2. the value of the output voltage;
3. the values of I_{max} and I_{min};
4. the output voltage ripple value.

6.24 A flyback converter has the following data: $V_s = 200\ V$, $D = 0.4$, $L_1 = L_2 = 600\ \mu H$, $N_1/N_2 = 1$, $T_s = 15\ \mu s$, $R = 150\ \Omega$, $C = 60\ \mu F$. Find

1. the value of the minimum inductance that causes the continuous inductor current;
2. the value of the output voltage;
3. the values of I_{max} and I_{min};
4. the output voltage ripple value.

6.25 Design a flyback converter to provide 160 V from a 150 V source to a load of 200 Ω. The switching frequency is 40 kHz, the inductor current is just continuous and the output ripple may not exceed 0.5 V. Assume that the flyback transformer has a 1 : 1 turn ratio.

6.26 A full-bridge converter, operating at 100 kHz, supplies 100 W at 12 V to a load from a 300 V source. The number of turns ratio on the transformer is $N_1 = 150$ and $N_2 = 8$. Find

1. the phase shift ratio;
2. the minimum inductance for continuous inductor current when the output power is 50 W;
3. value of capacitance needed for $\Delta VC = 0.1$ V when the output power is 100 W and using the above inductance.

6.27 A galvanically isolated dc/dc converter has to supply an output of 48 V, the available input voltage is 600 V and continuous conduction exists. The switching frequency is 20 kHz, and the transistor conduction time is 10 μs. The number of turns on the secondary is $N_2 = 20$. Find

1. the value of N_1 if we would use a half-bridge;
2. the value of N_1 if we would use a full-bridge.

FURTHER READING

Erickson R.W., Maksimoviĉ D., *Fundamentals of Power Electronics*. 2nd edition, Kluwer Academic Press, Norwell, 2001.

Fisher M.J., *Power Electronics*. 1st edition, PWS-Kent, Boston, 1991.

Lappe R., Conrad H., Kronberg M., *Leistungelektronik*. 2nd edition, Verlag Technik GmbH, Berlin, 1991.

Mohan N., Undeland T.M., Robbins W.P., *Power Electronics: Converters, Applications, and Design*. 2nd edition, John Wiley and Sons, New York, 1995.

CHAPTER 7

SIMPLE ELECTRICAL MACHINES

7.1 INTRODUCTION

The physical principle of the production of mechanical force by the interaction between an electric current and a magnetic field was known as early as 1821. In electricity generation, an electric generator is a device that converts mechanical energy to electrical energy. The reverse conversion of electrical energy into mechanical energy is done by an electric motor, and motors and generators have many similarities. In fact, many motors can be mechanically driven to generate electricity and can often make acceptable generators.

Chapter 7 is the first of two chapters about electrical machines, the principle of operation is discussed and a simple dc electrical machine is introduced. The focus is on the principles of converting electrical power into mechanical power and on taking the initial steps for understanding how an electrical machine contributes to the operation of electrical energy conversion systems.

The dc machine made it possible to control the application of mechanical power. Older trains and trams used rheostats, another name for variable resistors, to control the current that flows through the armature and the field winding making it possible to accelerate smoothly and to control the top speed. Next time you go to a technical museum, take a look at the antique trains and trams and see how cleverly the old engineers designed the drivetrains.

The Principles of Electronic and Electromechanic Power Conversion: A Systems Approach, First Edition.
Braham Ferreira and Wim van der Merwe.

In this chapter, we do not discuss the various arrangements of field windings and the speed and torque characteristics of self-excited dc machines that many textbooks have as this is old fashioned. Nowadays, the control is done with power electronics.

7.2 MOTIONAL VOLTAGE AND ELECTROMAGNETIC FORCE

The following are the two basic phenomena that are responsible for electromechanical energy conversion and occur simultaneously in electrical machines.

1. Voltage is induced when a conductor moves in a magnetic field.
2. When a current-carrying conductor is placed in a magnetic field, the conductor experiences a mechanical force.

We will discuss the two energy conversion methods in Section 7.2.1 and 7.2.2, making an assumption that the magnetic field is uniform.

7.2.1 Conductor Moving in a Uniform Magnetic Field

When a conductor with length ℓ moves in a uniform magnetic field $\mathbf{B}$ with velocity $\mathbf{v}$, there will be a voltage induced in the conductor given by

$$\boxed{e = (\mathbf{v} \times \mathbf{B}) \cdot \mathbf{l}} \tag{7.1}$$

Here, the vector $\mathbf{l}$ is aligned with the orientation (or direction) of the wire and the vector end point coincides with the wire end point that is assumed to be positive. The choice of a positive terminal is completely arbitrary; if the initial assumption is 'wrong' then the voltage will simply have a negative value, indicating that the other terminal is in fact positive with respect to this terminal.

Notation Note 7.1

The symbol $\mathbf{a} \times \mathbf{b}$ is used to indicate the vector cross product of vectors $\mathbf{a}$ and $\mathbf{b}$. The dot symbol in $\mathbf{a} \cdot \mathbf{b}$ indicates the vector dot product of the vectors $\mathbf{a}$ and $\mathbf{b}$. The $\times$ symbol is also used to indicate scalar multiplication, for example, $3 \times 4 = 12$. However, because we denote vectors in a boldface typeface, we can distinguish between a vector cross product and a scalar multiplication.

The induced voltage is known as *motional voltage* or *speed voltage*. The above equation is a mathematical definition of Faraday's law of electromagnetic induction, which can also be written as

$$e = -\frac{d\lambda}{dt} = -N\frac{d\phi}{dt}, \tag{7.2}$$

where λ is the flux link linkage, ϕ is the flux and N is the number of turns.

Notation Note 7.2

Vectors are indicated by the notation $\langle a_x, a_y, a_z \rangle$ where the three components are in the direction of the unit vectors. For Cartesian systems, the unit vectors are indicated by the symbols $\hat{\mathbf{x}}$, $\hat{\mathbf{y}}$ and $\hat{\mathbf{z}}$, the ^ symbol is used to indicate that the vectors have a length of 1, that is, an unit vector. When no coordinate system is specified, the use of the Cartesian coordinate system is implied. However, the use of other coordinate systems such as the cylindrical $\langle \hat{\mathbf{r}}, \hat{\boldsymbol{\phi}}, \hat{\mathbf{z}} \rangle$ is of course possible, with the required indication that this coordinate system will be used.

Example 7.1 Using the template of Fig. 7.1 as the basis and taking the Cartesian coordinates, $\langle \hat{\mathbf{x}}, \hat{\mathbf{y}}, \hat{\mathbf{z}} \rangle$ as indicated, let

$$\mathbf{B} = \langle 0.1, 0, 0.75 \rangle \text{ T}$$
$$\mathbf{v} = \langle 6, 0.5, -2 \rangle \text{ m/s}$$
$$\mathbf{l} = \langle 0, 3, 0 \rangle \text{ m}.$$

Calculate the voltage induced in the wire. Also show which side will be positive.

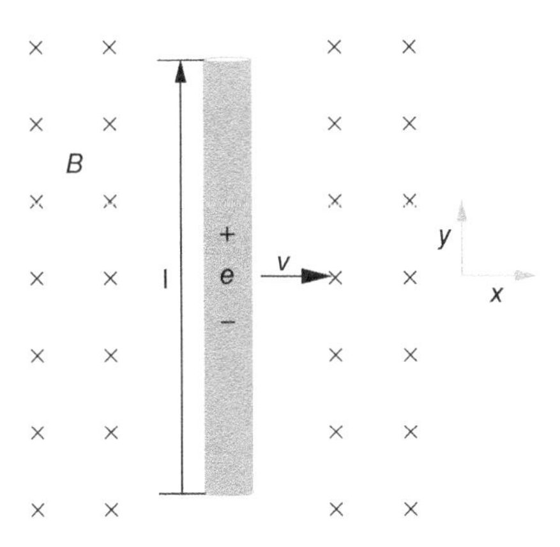

Figure 7.1 Conductor moving in the presence of a magnetic field.

Solution

The induced voltage can be found by

$$\begin{aligned} e &= (\langle 6, 0.5, -2\rangle \times \langle 0.1, 0, 0.75\rangle) \cdot \langle 0, 3, 0\rangle \\ &= \langle 0.375, -4.7, -0.05\rangle \cdot \langle 0, 3, 0\rangle \\ &= -4.7 \times 3 = -14.7 \text{ V}. \end{aligned}$$

As vector $\mathbf{l}$ is pointing towards the positive y-direction, the negative sign of the induced voltage implies that the end of the wire pointing towards the negative y-direction will be positive with respect to the other end.

Example 7.2 Let us use the same coordinate system, and let

$$\begin{aligned} \mathbf{B} &= \langle 0, 0, -0.75\rangle \text{ T} \\ \mathbf{v} &= \langle 8, 1, 0\rangle \text{ m/s} \\ \mathbf{l} &= \langle 1, 4, 0\rangle \text{ m}. \end{aligned}$$

Calculate the voltage induced in the wire. Also show which side will be positive.

Solution

We can use the alternate methods to calculate the result. The cross product between two vectors can also be expressed as

$$\mathbf{a} \times \mathbf{b} = |\mathbf{a}||\mathbf{b}| \sin(\theta)\hat{\mathbf{n}}$$

where θ is the smallest angle between the vectors and $\hat{\mathbf{n}}$ is a unit vector perpendicular to the plane spanned by $\mathbf{a}$ and $\mathbf{b}$. Firstly, because $\mathbf{B}$ is located in the z-direction and $\mathbf{v}$ is in the xy plane, we know that the vectors $\mathbf{B}$ and $\mathbf{v}$ are perpendicular to each other and, therefore, $\theta = 90°$. The unit vector $\hat{\mathbf{n}}$ must be in the xy plane to be perpendicular to $\mathbf{B}$ and can be found as

$$\hat{\mathbf{n}} = \langle -0.124, 0.9928, 0\rangle .$$

A graphical representation to aid in the calculation of $\hat{\mathbf{n}}$ is shown in Fig. 7.2. As the unit vector $\hat{\mathbf{n}}$ is calculated as $\hat{\mathbf{n}} = \langle -0.124, 0.9928, 0\rangle$, then

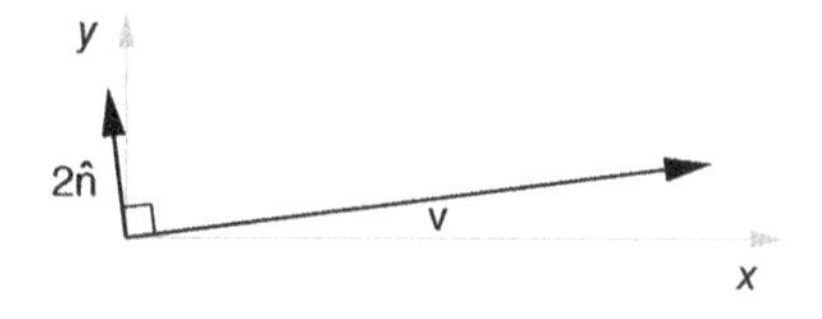

Figure 7.2 Graphical aid for Example 7.1.

$$\mathbf{v} \times \mathbf{B} = \left(\sqrt{8^2 + 1^2}\right)\left(\sqrt{-0.75^2}\right)\sin(90°)\hat{\mathbf{n}}$$
$$= 6.0467\,\langle -0.124, 0.9928, 0\rangle .$$

An alternate expression for the dot product is

$$\mathbf{a} \cdot \mathbf{b} = |\mathbf{a}||\mathbf{b}|\cos(\theta),$$

where θ is once again the smallest angle between the vectors. To calculate the angle between the vectors, consider Fig. 7.3. Let us define a vector $\mathbf{q}$ such that $\mathbf{l} + \mathbf{q} = \hat{\mathbf{n}}$. Now according to the cosine rule,

$$|\mathbf{q}|^2 = |\hat{\mathbf{n}}|^2 + |\mathbf{l}|^2 - 2|\hat{\mathbf{n}}||\mathbf{l}|\cos\theta,$$

so

$$\theta = \arccos\left(\frac{3.132^2 - 4.123^2 - 1^2}{-2 \times 4.123 \times 1}\right) = 6.7°.$$

The induced voltage can, therefore, be found by

$$e = 6.0467 \times 4.123 \times \cos\left(6.7°\right)$$
$$= 24.76 \text{ V}.$$

As vector $\mathbf{l}$ is pointing upwards, the top end of the wire will be positive with respect to the lower end.

7.2.1.1 *Ideal Configuration* We see from Example 7.1 that the vector calculations using cross and dot products can be very difficult and time consuming. The calculation is simplified when the flux and direction of movement are perpendicular to each other (if you wonder why this is the ideal case, remember the cross product term). Not only it is convenient when the equations are simple but also, more importantly, we then obtain the maximum voltage. We can write this

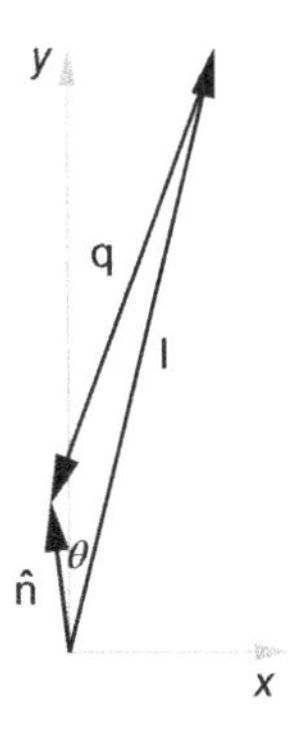

Figure 7.3 Graphical aid for Example 7.1.

requirement mathematically as $\mathbf{v} \perp \mathbf{B}$. For the conductor length ℓ, shown in Fig. 7.1, which is moving with a speed v (in the direction indicated) through a magnetic field with magnitude B that is directed into the plane of the paper (in a direction perpendicular to the velocity vector), the induced voltage is

$$
\begin{aligned}
e &= |\mathbf{B}||\mathbf{v}| \sin(90°)\hat{\mathbf{n}} \cdot \mathbf{l} \\
&= Bv\,(\hat{\mathbf{n}} \cdot \mathbf{l}). \qquad (7.3)
\end{aligned}
$$

If we now define ℓ to be the projected length of the conductor in the direction of $\hat{\mathbf{n}}$ then the expression for the induced voltage simply becomes

$$
\boxed{e = Bv\ell,} \qquad (7.4)
$$

where

$$
\ell = \hat{\mathbf{n}} \cdot \mathbf{l}. \qquad (7.5)
$$

Example 7.3 Consider the system shown in Fig. 7.1. It is true that $\mathbf{B} \perp \mathbf{v}$ and that the wire is orientated in parallel with the vector $\mathbf{v} \times \mathbf{B}$. If the magnetic field is 0.5 T, the conductor length is 3 m and the conductor is moving at 9.8 m/s what is the induced voltage?

Solution

The induced voltage is found as

$$
e = 0.5 \times 3 \times 9.8 = 14.7\,\text{V}.
$$

Example 7.4 Calculate the voltage induced in the conductor of Example 7.3 if the direction of movement is at an angle with the orientation of the wire, as shown in Fig. 7.4.

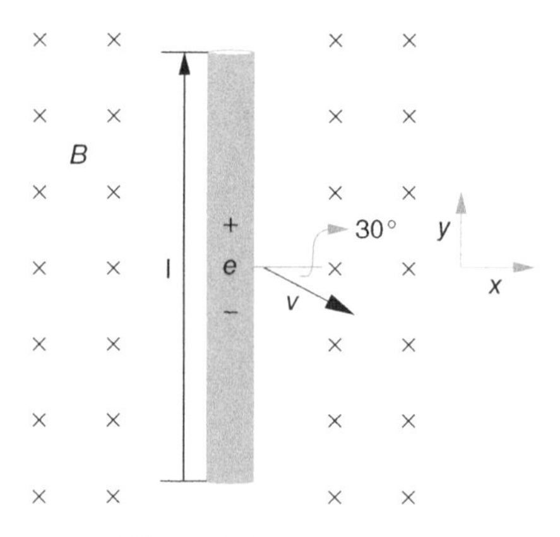

Figure 7.4 Example 7.4.

Solution

The length of the conductor projected onto the direction of movement is

$$\ell = |\mathbf{l}| \cos(30^\circ) = 2.598 \text{ m}.$$

The induced voltage is, therefore,

$$e = 0.5 \times 2.598 \times 9.8 = 12.73 \text{ V}.$$

7.2.2 Current-Carrying Conductor in a Uniform Magnetic Field

The converse of the conductor moving through a magnetic field is also true. If a current-carrying conductor is placed in a magnetic field, it would experience a force given by the Lorenz force equation. For a conductor, with length ℓ, carrying current i in a uniform magnetic field $\mathbf{B}$, this force is given by

$$\boxed{\mathbf{f} = i(\mathbf{l} \times \mathbf{B}),} \tag{7.6}$$

where the direction of the vector $\mathbf{l}$ indicates the direction of current flow.

It is interesting to note that this equation also adheres to Lenz's law, that was defined on p. 70, which states that '*An induced current is always in such a direction as to oppose the motion or change causing it.*' To explain this statement consider the following, let us take a conductor as shown in Fig. 7.1 and connect the end points with a wire that closes the loop in a hypothetical area where no magnetic field is present. Now, when the conductor is moved through the field, a voltage will be induced in the conductor as shown in Fig. 7.1 and a current will flow through the loop. With the direction of the voltage indicated, the current flow will be from the positive terminal through the connecting wire to the negative terminal of the conductor. Contrary to one's initial feeling, the current in the conductor moving through the magnetic field will, therefore, flow from the negative terminal to the positive terminal. This can also be understood by stating that this conductor is acting as a source; therefore, current will flow from the negative terminal to the positive because the source is delivering energy. In this case, according to Fig. 7.5, the force experienced by the conductor will be in the direction opposite to the direction of movement. In fact, in this case, the mechanical input power and the electrical output power would be equal, that is,

$$P_{\text{mech}} = vf = P_{\text{elec}} = vi. \tag{7.7}$$

Example 7.5 A 50 cm long conductor is placed in a uniform magnetic field. If the conductor carries a 10 A current and the B field and the conductor orientation ($\hat{\mathbf{n}}$) pointing to the direction of current flow are

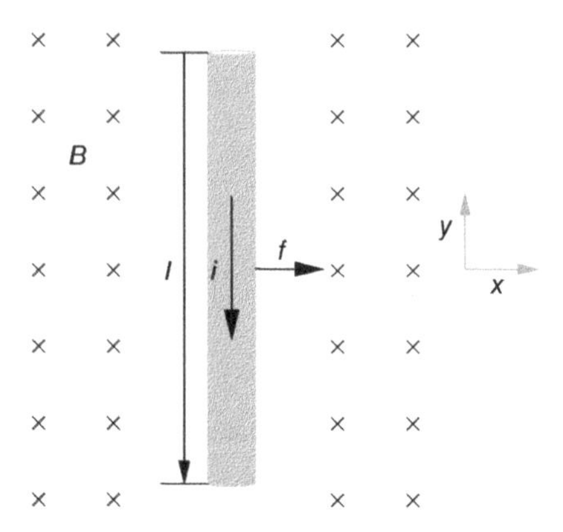

Figure 7.5 Current-carrying conductor in the presence of a magnetic field.

$$\mathbf{B} = \langle -0.2, 0.1, -0.6 \rangle$$
$$\hat{\mathbf{n}} = \langle 0.707, -0.707, 0 \rangle,$$

calculate the magnitude of the force experienced by the conductor.

Solution

Let us start by creating a vector $\mathbf{l}$ with the same length as the conductor and pointing in the direction of current flow. Therefore,

$$\mathbf{l} = \frac{1}{2}\hat{\mathbf{n}}.$$

The force can now be calculated as

$$\begin{aligned}\mathbf{f} &= i\,(\mathbf{l} \times \mathbf{B}) \\ &= 10\,(\langle 0.707, -0.707, 0 \rangle \times \langle -0.2, 0.1, -0.6 \rangle)\,\frac{1}{2} \\ &= 5\,\langle 0.4242, 0.4242, -0.707 \rangle \text{ N}.\end{aligned}$$

The total force is, therefore,

$$|\mathbf{f}| = 4.636 \text{ N}.$$

7.2.2.1 Ideal Configuration Again, we can consider the ideal case where $\mathbf{l} \perp \mathbf{B}$. This can be visualised by using the magnetic field established by a pair of magnets with a magnitude of B and a conductor of length ℓ carrying a current i. The conductor is positioned in such a manner in the magnetic field that $\mathbf{l} \perp \mathbf{B}$, an example of this configuration is shown in Fig. 7.5. In this ideal case, the force magnitude is found as

$$\boxed{f = Bil.} \tag{7.8}$$

Unfortunately, it is true that the direction of the force is not immediately clear when this expression is used.

Example 7.6 A conductor carrying a current of 5 A is placed in a uniform magnetic field given by

$$\mathbf{B} = \langle 0.37, -0.2, -.56 \rangle \text{ T}.$$

The conductor is 1 m long and is orientated so that the current if flowing in the direction $\hat{\mathbf{n}}$ is given as

$$\hat{\mathbf{n}} = \langle 0.6914, 0.6914, 0.21 \rangle.$$

Calculate the force exerted on the conductor.

Solution

The B-field and the direction of the current flow are perpendicular to each other because

$$\mathbf{B} \cdot \hat{\mathbf{n}} = \langle 0.37, -0.2, -.56 \rangle \cdot \langle 0.6914, 0.6914, 0.21 \rangle = 0.$$

The magnitude of the force can, therefore, be found as

$$f = \sqrt{0.37^2 + 0.2^2 + 0.56^2} \times 5 \times 1 = 3.5 \text{ N}.$$

You can verify this result (with the benefit of knowing the direction of the force) by using the full vector equation.

Example 7.7 In some cases, this simplified method is very easy to use, especially when the coordinate system is aligned to the current and magnetic field. Consider a conductor carrying a current of 5 A that is placed in a uniform magnetic field given by $\mathbf{B} = 0.7\,\hat{\mathbf{z}}$ T. The conductor is 1 m long and is orientated so that the current is flowing in the positive y-direction. Calculate the force exerted on the conductor.

Solution

The B-field and the direction of the current flow are perpendicular to each other because the z-axis and y-axis are, by definition, perpendicular. The magnitude of the force can, therefore, be found as

$$f = 0.7 \times 5 \times 1 = 3.5 \text{ N}.$$

As we know that

$$\hat{\mathbf{y}} \times \hat{\mathbf{z}} = \hat{\mathbf{x}},$$

we know that the force will be directed in the positive x-direction.

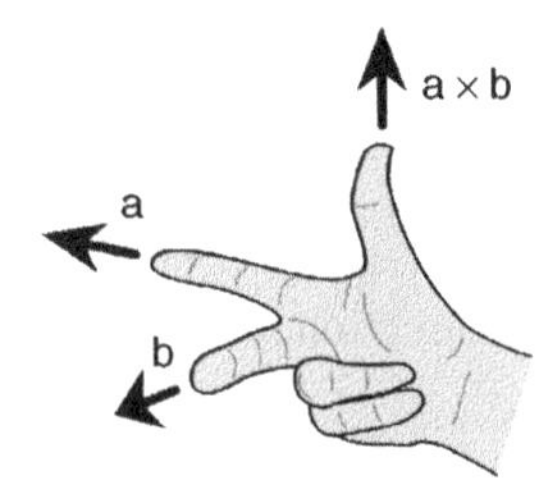

Figure 7.6 The cross product right-hand rule.

7.2.3 Right-Hand Rule

The right-hand rule as indicated in Fig. 7.6 is convenient for quickly calculating the direction of the cross product vector. As the induced voltage of a conductor moving in a magnetic field is given by

$$e = (\mathbf{v} \times \mathbf{B}) \cdot \mathbf{l},$$

the direction of the induced voltage can be found by using the right-hand rule. *If the index finger is pointing in the direction of movement, the middle finger in the direction of flux then the thumb of the right hand will point in the direction of the positive terminal.*

The right-hand rule can likewise be used to calculate the direction of the force experienced by a current-carrying conductor in a magnetic field. Recall that the force can be calculated by

$$\mathbf{f} = i(\mathbf{l} \times \mathbf{B}).$$

Therefore, *if the index finger is pointing in the direction of current flow, the middle finger in the direction of flux then the thumb of the right hand will point in the direction of the force.* You can verify this rule with the result of Example 7.7.

7.3 SIMPLE LINEAR dc MACHINE

We can begin the discussion of electromechanical energy conversion with the simplest form of energy conversion, the linear dc machine. A linear dc machine is shown in Fig. 7.7, and consists of a dc source, a switch, some connection wire (with a certain amount of resistance), two ideal smooth frictionless rails (and let us assume that the rails are superconducting, i.e. with zero resistance, to simplify the discussion), and, finally, a conducting bar (let us once again assume this bar to be superconducting) lying across the tracks. A uniform magnetic field is present in the vicinity of the tracks directed in a direction perpendicular to the track direction. In Fig. 7.7, the magnetic field is pointing directly into the page, whereas the track is running along the page to the right.

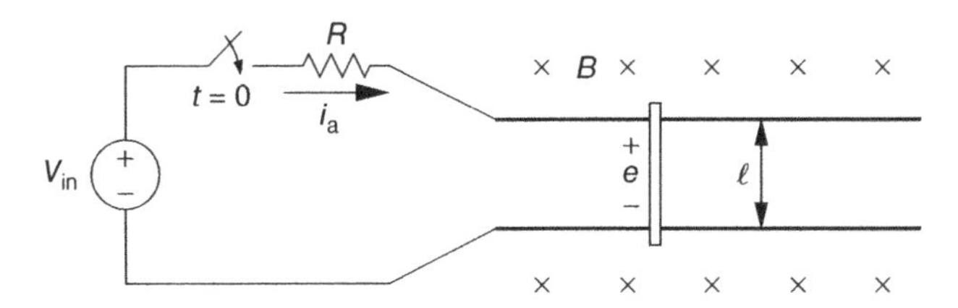

Figure 7.7 A linear dc machine.

We will make one further assumption. Let us assume that the magnetic field caused by the current in the conductors, especially in the rails and the conducting bar, is much smaller than the external magnetic field.

We can discuss the behaviour of our machine by investigating four different equations that stand central to the machine operation.

1. The force in the conducting bar due to the current in the presence of the magnetic field. Recall that the force is calculated as

$$\mathbf{f} = i\,(\mathbf{l} \times \mathbf{B})\,.$$

By application of the right-hand rule, we see that with respect to the system shown in Fig. 7.7, the induced force F_{ind} will be towards the right if the current is flowing downwards and towards the left if it flows upwards. As the current and the magnetic field are perpendicular to each other, we can write

$$f_{\text{ind}} = B i_{\text{a}} \ell, \tag{7.9}$$

where the directions of f and i_{a} are shown in Fig. 7.7.

2. The voltage induced by a conductor moving in a magnetic field is defined as

$$e = (\mathbf{v} \times \mathbf{B}) \cdot \mathbf{l}.$$

Once again, the direction of movement and the magnetic field are perpendicular to each other, and therefore, the induced voltage can be written as

$$e = B v \ell, \tag{7.10}$$

with the direction of e as indicated in Fig. 7.7. You can verify this statement with the right-hand rule.

3. Newton's second law of motion states that

$$f = ma, \tag{7.11}$$

where m is the mass and a is the linear acceleration. This law describes the motion of the conducting bar, with a the linear acceleration along the tracks and m the mass of the conducting bar.

4. Kirchoff's voltage law, which for the circuit in Fig. 7.7, states that

$$V_{\mathrm{in}} = i_{\mathrm{a}} R + e. \tag{7.12}$$

7.3.1 Starting of the Linear dc Motor

Let us use the four laws to discuss the start-up procedure of the linear dc machine. Let us assume that the conducting bar is at rest at $t < 0$. Therefore, the instant the switch closes, the induced voltage on the bar is zero and the current through the bar can be found as

$$i_{\mathrm{a}} = \frac{V_{\mathrm{in}} - e}{R} = \frac{V_{\mathrm{in}}}{R}. \tag{7.13}$$

This is the maximum current that the machine will take and is called the *starting current*. The current will flow in a downwards direction through the bar, so the force induced will be directed towards the right, as shown in Fig. 7.8. The magnitude of this force is

$$f_{\mathrm{ind}} = B\, i\, \ell = \frac{B\, V_{\mathrm{in}}\, \ell}{R}. \tag{7.14}$$

The induced force on the bar will now start to accelerate the bar towards the right according to Newton's second law. As the bar starts to move, it will gain speed and a voltage will be induced in the bar. For the circuit and definitions shown in Fig. 7.8, this induced voltage will be

$$e = B\, \ell\, v. \tag{7.15}$$

Therefore, as the induced voltage increases, the current will start to decrease according to Kirchoff's voltage law,

$$i_{\mathrm{a}} = \frac{V_{\mathrm{in}} - e}{R}.$$

The end result is that as the bar accelerates, the induced voltage will increase; hence, the current will decrease, the force will decrease and the acceleration will

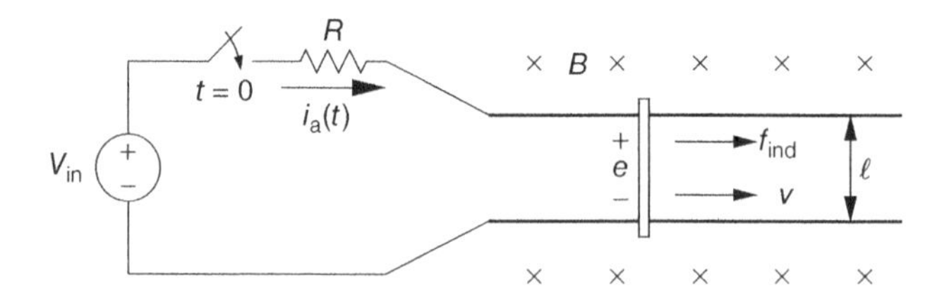

Figure 7.8 Starting of the linear dc machine.

slow. Eventually, the bar will reach *terminal velocity* where it cannot be accelerated further because

$$V_{\text{in}} = e = B\ell v.$$

At this point, the only way to accelerate the bar further is by increasing the applied voltage (V_{in}), by changing the geometry of the machine to change the length of the bar or, strangely enough, by decreasing the flux. This is a well-known phenomena which is observed in electrical machines, where the speed can be increased by *flux weakening*.

7.3.2 Linear dc Machine Operating as a Motor

Now we are in a position to discuss what will happen when the machine is mechanically loaded. Consider the machine, with the definitions of the relevant quantities in Fig. 7.9. We will initially assume that the bar is moving at the terminal velocity v_{tv} and that it is unloaded.

As the bar is moving at the terminal velocity, we know that

$$V_{\text{in}} = e \leftrightarrow i_{\text{a}} = 0. \tag{7.16}$$

As the current is zero, no force is induced. When an external force f_{load} is applied, the bar will start to decelerate according to Newton's second law. The acceleration is found as

$$a = \frac{f_{\text{ind}} - f_{\text{load}}}{m}. \tag{7.17}$$

The bar will decelerate until the induced force is equal to the applied force. This will naturally imply that the velocity at which this occurs is lower than the terminal velocity. The speed at this equilibrium point can be calculated as

$$f_{\text{load}} = -f_{\text{ind}} \tag{7.18}$$

$$\therefore i_{\text{a}} = \frac{f_{\text{load}}}{B\ell} \tag{7.19}$$

$$\therefore v = \frac{V_{\text{in}}}{B\ell} - \frac{f_{\text{load}}R}{B^2\ell^2}. \tag{7.20}$$

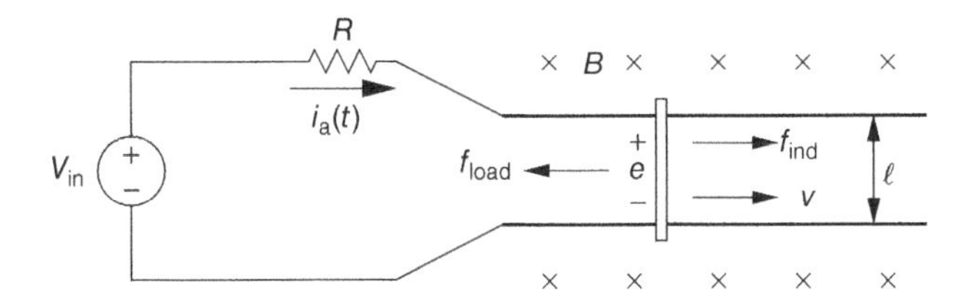

Figure 7.9 The linear dc machine operating as a motor.

We also know that according to the conservation of energy principle, the input power and the output power must be equal to one another. If we investigate the system operation in terms of power dissipation, at the equilibrium point, we find

$$P_{\text{mech}} = v f_{\text{load}} \tag{7.21}$$

$$P_{\text{elec}} = V_{\text{in}}\, i_{\text{a}} \tag{7.22}$$

$$P_{\text{loss}} = i_{\text{a}}^2 R \tag{7.23}$$

and of course we know that

$$P_{\text{elec}} - P_{\text{loss}} = P_{\text{mech}} \tag{7.24}$$

and, therefore,

$$V_{\text{in}}\, i_{\text{a}} - i_{\text{a}}^2 R = v f_{\text{load}}. \tag{7.25}$$

Let us substitute the equation

$$i_{\text{a}} = \frac{f_{\text{ind}}}{B\ell}$$

with the requirement that $f_{\text{ind}} = f_{\text{load}}$ into this result and rearrange for v. We find

$$v = \frac{V_{\text{in}}}{B\ell} - \frac{f_{\text{load}} R}{B^2\ell^2}, \tag{7.26}$$

which is the same as our previous result!

7.3.3 Linear dc Machine Operating as a Generator

Finally, let us consider the case where mechanical energy is delivered to our linear dc machine. Let the bar be initially unloaded and travelling at terminal velocity. Now let us apply a force, f_{app}, to the bar in the same direction as the direction of movement. This force will, according to Newton's law, accelerate the bar and the speed will increase. As the speed increases, the voltage induced in the bar will also increase, and it is, therefore, true that $e > V_{\text{in}}$. The current flow will, therefore, reverse (relative to the previously discussed motor operation) as shown in Fig. 7.10.

As the current direction has reversed, the induced force on the bar will now be directed to the left, acting to counteract the effects of the externally applied force. You can verify this statement by application of the right-hand rule. The bar will accelerate until the force induced in the bar is equal (but opposite) to the externally applied force.

Now, because the current direction is reversed, the dc source is now absorbing energy. The conservation of energy principle implies that the external mechanical force must be delivering energy to the system. As the applied force is in the direction

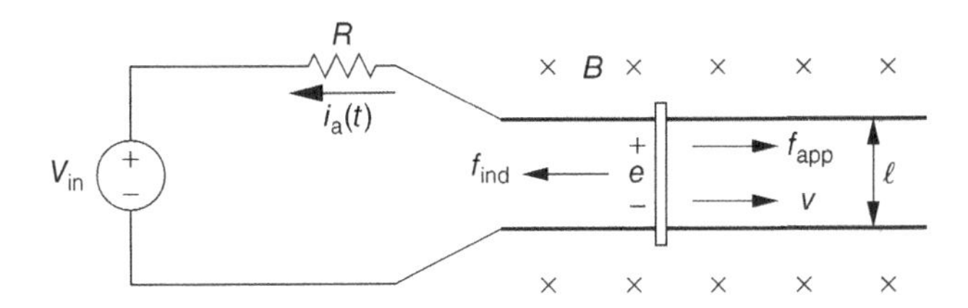

Figure 7.10 The linear dc machine operating as a generator.

of movement, this is indeed the case. We can calculate the equilibrium point speed by using the conservation of energy principle. Let

$$P_{mech} = v f_{app} \tag{7.27}$$

$$P_{elec} = V_{in}\, i_a \tag{7.28}$$

$$P_{loss} = i_a^2 R \tag{7.29}$$

and

$$P_{mech} = P_{elec} + P_{loss}, \tag{7.30}$$

therefore,

$$v f_{app} = V_{in}\, i_a + i_a^2 R. \tag{7.31}$$

Let us again substitute the fact that

$$i_a = \frac{f_{app}}{B\ell}$$

at the equilibrium point. Then

$$v = \frac{V_{in}}{B\ell} + \frac{f_{app} R}{B^2 \ell^2}. \tag{7.32}$$

You can verify this equation using the balance of force principle.

7.3.4 Electrical Equivalent Circuit of the Linear dc Machine

As the mechanical to electrical energy conversion stage is, in essence, lossless we can represent the conversion stage simply as an electrical source. Consider the equivalent circuit in Fig. 7.11, it is in all respects equivalent to our linear dc machine, with the exception that many of the mechanical and magnetic details are now hidden.

To simplify our analysis, we can now linearise the electrical model of the machine. This concept is possibly best illustrated by an example.

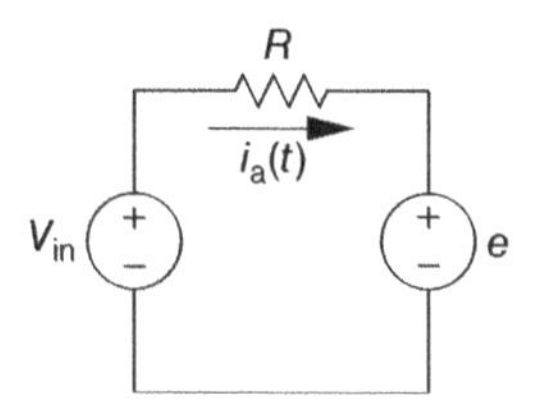

Figure 7.11 Equivalent electrical circuit of a linear dc machine.

Example 7.8 A linear dc machine has a resistance of 0.3 Ω. When connected to a 50 V dc source, the machine takes 4 A when the bar is moving at 10 m/s. Calculate the following:

1. The induced voltage on the bar and the power delivered.
2. The current and the power delivered should the speed change to 8 m/s.

Solution

1. The induced voltage can be calculated using Kirchoff's law,

$$e = V_{\text{in}} - i_{\text{a}}R = 48.8\,\text{V}.$$

The mechanical power delivered is, therefore,

$$P_{\text{mech}} = e\,i_{\text{a}} = 195.2\,\text{W}.$$

2. When the linear speed drops to 8 m/s, the induced voltage will change. With the assumption that the geometry and the magnetic flux remain constant, we can write

$$\frac{e_1}{e_2} = \frac{B\ell v_1}{B\ell v_2} = \frac{v_1}{v_2}.$$

The induced voltage at this speed is, therefore,

$$e_2 = \frac{v_2}{v_1}e_1 = 39.04\,\text{V}.$$

It follows that

$$i_{\text{a}} = \frac{V_{\text{in}} - e}{R} = 36.53\,\text{A}$$

$$P_{\text{mech}} = e\,i_{\text{a}} = 1\,426.3\,\text{W}.$$

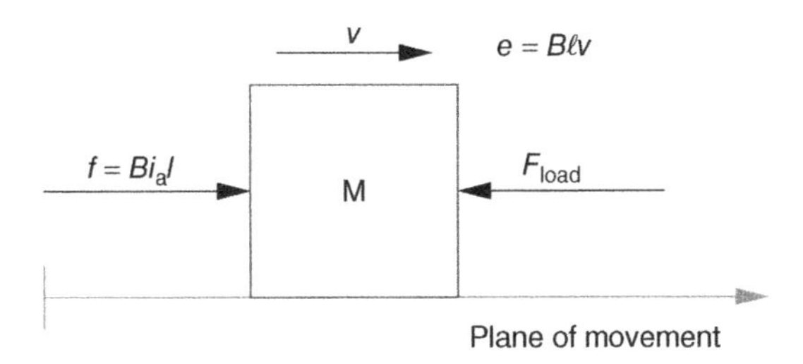

Figure 7.12 Equivalent mechanical circuit of a linear dc machine.

7.3.5 Mechanical Equivalent Circuit of the Linear dc Machine

Having discussed the electrical equivalent circuit, we can now represent the ideal source e used to model the energy conversion stage with an equivalent mechanical circuit, as shown in Fig. 7.12. Using this figure in conjunction with the electrical equivalent circuit, it is possible to describe the operation of the linear machine.

7.3.6 A Practical Example: The Railgun

All considered, the linear dc machine does not have many uses, bar maybe one, the railgun or the electrical linear accelerator. A railgun operates exactly as our linear dc machine with the exception that the conducting bar is now a projectile. Let us take a little time to investigate the possibility of replacing a system used to chemically propel a 9 mm projectile with a railgun. The basic characteristics of the chemical system is tabulated in Table 7.1.

Let us make the following assumptions for our railgun:

1. We have a 500 V dc source available.
2. The resistance of the interconnecting conductors, the conducting rails and the projectile combined is equal to 0.1 Ω.
3. We will connect the rails across two edges of the projectile; therefore, with respect to Fig. 7.9, the length $\ell = 0.009$ m.
4. We will use the strongest magnets available to create the external magnetic field. We will use Neodymium rare earth magnets with a flux density of 1.2 T.
5. We will ignore any inductive effects of the rails and all other conductors.

TABLE 7.1 Characteristics of the Chemical Propulsion System

Description	Value	Description	Value
Projectile weight	7.45 g	Exit velocity	381 m/s
Chamber length	125 mm	Projectile diameter	9 mm

We want the final velocity of the projectile to be 381 m/s. Let us calculate how much time we will need to accelerate the projectile to this speed. Let us use the equivalent circuit of Fig. 7.8. At the instant we pull the trigger, the projectile will be at a standstill; therefore, we can write the following equation:

$$V_{in} = e(t) + i_a(t)R \text{ where } e(0) = 0. \tag{7.33}$$

We can now substitute the following

$$i(t) = \frac{f_{ind}(t)}{B\ell}$$
$$e(t) = v(t)\,B\,\ell$$
$$\text{and } f_{ind}(t) = m\,a(t)$$

into this equation to yield

$$\begin{aligned} V_{in} &= v(t)\,B\,\ell + \frac{f_{ind}(t)\,R}{B\,\ell} \\ &= v(t)\,B\,\ell + \frac{m\,a(t)R}{B\,\ell}. \end{aligned} \tag{7.34}$$

Let us rearrange

$$V_{in} = v(t)\,B\,\ell + \frac{d\,v(t)}{dt}\frac{m\,R}{B\,\ell}$$
$$\frac{d\,v(t)}{dt} + v(t)\frac{B^2\ell^2}{m\,R} = \frac{V_{in}B\,\ell}{m\,R}, \tag{7.35}$$

which is a first-order linear differential equation of the form

$$\dot{x} + Ax = B$$

with solution

$$x(t) = C_1 e^{-At} + C_2.$$

The solution to (7.35) is, therefore,

$$v(t) = C_1 e^{-t/\tau} + C_2, \tag{7.36}$$

where τ is the time constant given as

$$\tau = \frac{m\,R}{B^2\,\ell^2}. \tag{7.37}$$

The constants C_1 and C_2 can be calculated by looking at the boundary conditions at $t = 0^+$ and $t \to \infty$. The steady-state speed where $e = V_{\text{in}}$ is

$$v(\infty) = \frac{V_{\text{in}}}{B\ell} \tag{7.38}$$

and the speed at the initial condition is zero, or

$$v(0^+) = 0. \tag{7.39}$$

Therefore, the constants can be found as

$$v(0^+) = C_1 + C_2 = 0$$

$$v(\infty) = C_2 = \frac{V_{\text{in}}}{B\ell}$$

$$\therefore C_2 = \frac{V_{\text{in}}}{B\ell} \tag{7.40}$$

$$\text{and } C_1 = -\frac{V_{\text{in}}}{B\ell}, \tag{7.41}$$

and the equation for the speed is, therefore,

$$v(t) = \frac{V_{\text{in}}}{B\ell}\left(1 - e^{-t/\tau}\right). \tag{7.42}$$

We can now calculate how much time we need to accelerate the projectile to the desired muzzle velocity of 381 m/s. Let us denote this desired speed by v_{mv}. Now

$$\frac{v_{\text{mv}} B\ell}{V_{\text{in}}} = 1 - e^{-t/\tau}$$

$$t_{\text{r}} = -\tau \ln\left(1 - \frac{v_{\text{mv}} B\ell}{V_{\text{in}}}\right) = 52.78\,\text{ms} \tag{7.43}$$

The distance travelled by the projectile during this time period defines the minimum barrel length of the railgun. The length, ℓ, can be found as

$$\begin{aligned}\ell &= \int_0^{t_{\text{r}}} v(t)dt \\ &= \left.\frac{V_{\text{in}}}{B\ell}\right|_0^{t_{\text{r}}} + \tau \left.\frac{V_{\text{in}}}{B\ell} e^{-t/\tau}\right|_0^{t_{\text{r}}} \\ &= 10.07\,\text{m}.\end{aligned} \tag{7.44}$$

We can see that this is not really feasible; apart from the fact that the barrel of our gun will need to be about 10 m long and we will also have to carry around a

500 V battery pack that is able to deliver about 5 kA for 50 ms. For this reason, the railgun concept is only of concern for very large artillery pieces or for use on ships. In these guns, large capacitors are used rather than a battery pack and the gun is also normally composed of several smaller guns in series to reduce the impedance of the conductors.

As a side note, it is important to point out that in this example we have ignored the inductive effects of the rails and other conductors and also the magnetic field induced by the current. As an interesting side note, consider the following. The magnetic field contribution by the current in one of the rails in the vicinity of the projectile can be calculated using (4.1)

$$B_c = \mu \frac{i}{2\pi \frac{1}{2}\ell} \approx 0.22\,\text{T}. \tag{7.45}$$

The total magnetic field is, therefore, in the region of $1.2 + 2(0.22) = 1.64$ T, which will reduce the required length of the barrel.

Another factor we will need to consider is that the rails will also be subjected to the same force as the projectile. The force experienced by one of the rails can be calculated as, with a starting current of $i_s = 5$ kA,

$$f_R = B i_s \ell = 1.2 \times 5000 \times 10 = 60\,\text{kN}. \tag{7.46}$$

That is a huge force trying to blow the railgun apart!

7.4 BASIC OPERATION OF THE DC MACHINE

The benefit of the linear dc machine discussed in Section 7.3 is that it is very easy to understand the operation of the machine. However, because of the fact that the machine can only deliver power linearly over a limited range, the machine is of very little practical use. But what if we could 'bend' the machine such that the start point and the end point become the same point? If this is possible then the machine can deliver continuous power. Let us now consider the simplest form of a rotating dc machine: a single loop, free to rotate about an axis, placed in a magnetic field. Such a setup is shown in Fig. 7.13.

7.4.1 Induced Voltage

Consider Fig. 7.13, the flux is directed in the negative x-direction and the rotation axis is aligned with the z-axis. Let us define a number of terms. We will refer to the one-turn winding that is free to rotate about the axis as the rotor, and the stationary part of the machine, which for this simple machine consists of the external magnetic field as the stator. The magnetic field in the vicinity of the rotor is known as the *field flux*. Finally, let us define a rotation angle θ. We will define θ as the angle between the line from point a to b and x-axis. In Fig. 7.13, the angle is $\theta = 0°$.

We can calculate the voltage induced in the coil using two methods. Let us first use the more versatile, and more complex, vector-based approach.

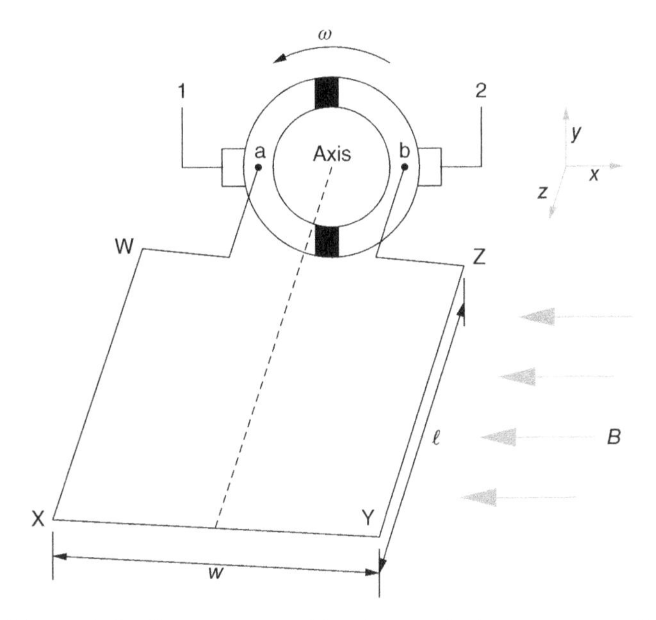

Figure 7.13 Commutators and brushes attached to a one-turn coil.

7.4.1.1 The Vector-Based Approach Assume that the flux is uniform, and let us consider each segment of the loop separately. We will use the Cartesian components with vectors written as $\langle \hat{\mathbf{x}}, \hat{\mathbf{y}}, \hat{\mathbf{z}} \rangle$.

1. *Section* W–X In this section, the velocity component of the wire can be written as

$$\mathbf{v}_{\mathrm{WX}}(\theta) = \frac{w}{2}\omega \langle \sin\theta, -\cos\theta, 0 \rangle . \tag{7.47}$$

   Let us choose the length vector of this segment as $\mathbf{l} = \ell \langle 0, 0, 1 \rangle$. The voltage induced in this segment can, therefore, be calculated as

$$\begin{aligned} e_{\mathrm{WX}}(\theta) &= \left(\mathbf{v}_{\mathrm{WX}} \times \mathbf{B}\right) \cdot \mathbf{l} \\ &= \frac{w}{2}\omega\ell B \left(\langle \sin\theta, -\cos\theta, 0 \rangle \times \langle -1, 0, 0 \rangle\right) \cdot \langle 0, 0, 1 \rangle \\ &= \frac{w}{2}\omega\ell B \langle 0, 0, -\cos\theta \rangle \cdot \langle 0, 0, 1 \rangle \\ &= -\frac{w}{2}\omega\ell B \cos\theta . \end{aligned} \tag{7.48}$$

   We see that the point W is, therefore, more positive than the point X.

2. *Section* X–Y In this section, the velocity component of the wire cannot be written as easily as the velocity vector for the section W–X. However, we know that the velocity vector (although being a function of position along

the segment) will be in the x–y plane and, therefore, the vector $\mathbf{v} \times \mathbf{B}$ will only have a component in the z-direction. As vector $\mathbf{l}$ is also directed in the x–y plane and, therefore, perpendicular to the vector $\mathbf{v} \times \mathbf{B}$, it follows that $(\mathbf{v} \times \mathbf{B}) \cdot \mathbf{l} = 0$. Therefore,

$$e_{\mathrm{XY}} = 0. \tag{7.49}$$

This segment is called the *end-turn*, and although it does not contribute to the induced voltage in the loop, it serves the very important function of completing the loop.

3. *Section* Y–Z In this section, the velocity component of the wire can be written as

$$\mathbf{v}_{\mathrm{YZ}}(\theta) = \frac{w}{2}\omega \langle -\sin\theta, \cos\theta, 0\rangle . \tag{7.50}$$

Let us choose the length vector of this segment as $\mathbf{l} = \ell \langle 0, 0, 1\rangle$. The voltage induced in this segment can, therefore, be calculated as

$$\begin{aligned} e_{\mathrm{YZ}}(\theta) &= \left(\mathbf{v}_{\mathrm{YZ}} \times \mathbf{B}\right) \cdot \mathbf{l} \\ &= \frac{w}{2}\omega\ell B \left(\langle -\sin\theta, \cos\theta, 0\rangle \times \langle -1, 0, 0\rangle\right) \cdot \langle 0, 0, -1\rangle \\ &= \frac{w}{2}\omega\ell B \langle 0, 0, \cos\theta\rangle \cdot \langle 0, 0, -1\rangle \\ &= -\frac{w}{2}\omega\ell B \cos\theta. \end{aligned} \tag{7.51}$$

We see that the point Y is, therefore, more positive than the point Z.

4. *Parts of Section* Z–W We can follow the same argument with the parts of this segment as with section X–Y. The voltage induced in these wire segments will always be directed in the z-axis, and because the segments are aligned in the x–y plane, no voltage is induced between the end points of the wire segments. Therefore, for all sub-segments of section Z–W,

$$e = 0. \tag{7.52}$$

We can now calculate the induced voltage between points a and b by adding the induced voltages in the different sections of the loop. Let us choose point b as our reference potential. The voltage from point b to Z is zero, from Z to Y is $\frac{w}{2}\omega\ell B \cos\theta$, zero again from Y to X, from X to W the induced voltage is $\frac{w}{2}\omega\ell B \cos\theta$. Finally, no voltage is induced from W to a. The total induced voltage is, therefore,

$$\boxed{e_{\mathrm{ab}}(\theta) = \omega\, \ell\, B\, w \cos(\theta).} \tag{7.53}$$

7.4.1.2 *Simplified Total Flux Linkage Approach*

Let us again assume that the flux in the vicinity of the loop is uniform. We can calculate the flux linked by the loop, or said in a different way, the total flux contained by the loop, as (where A_e is the equivalent area of the loop)

$$\begin{aligned}\phi(\theta) &= BA_e(\theta) \\ &= B\ell w \sin(\theta) \end{aligned} \tag{7.54}$$

From (7.2), we can write the induced voltage as

$$\begin{aligned} e_{ab}(\theta) &= \frac{d\phi}{dt} \\ &= B\ell w \frac{d\theta}{dt} \cos(\theta) \\ &= \omega B \ell w \cos(\theta) \end{aligned} \tag{7.55}$$

It is true that this method is easy to use and intuitive for this simple motor. However, this method does not tell us the polarity of the induced voltage. We can determine the polarity of the voltage by considering the loop at a rotation of $\theta = 90°$. At this point, the coil is turned vertically and point b will be right above point a. According to Lenz's law, the voltage induced in the coil will serve to counteract the effects of the external magnetic field. Therefore, using Amperè's law and the flux-current right-hand rule, we see that the current flowing in the coil because of the induced voltage must be flowing from point b to point a. Therefore, at $\theta = 90°$, point a must be more positive than point b and (7.55) as stated is correct. Remember that the current is closed by an external circuit between points a and b. In this external circuit, current will flow from point a to point b, but inside the machine, which for our purposes is acting as a source, the current is flowing from the negative terminal to the positive, or then from point b to point a even though point a is positive with respect to point b.

7.4.2 Mechanical Voltage Rectification

We have seen in the previous section that the voltage induced in the coil is sinusoidal. To convert this ac voltage to dc, some rectification must take place. Although it is possible to rectify the ac voltage with the use of a diode (bridge), the resulting circuit can then only be used as a generator because current can never flow in the opposite direction through a diode bridge.

A mechanical rectifier is used in dc machines to convert the ac to dc while allowing current to flow in both directions. This mechanical rectifier in a dc machine is called a *commutator*. Consider the circuit in Fig. 7.13, we see that points a and b are located on a circular segment. In fact, the two circular segments can actually be manufactured from one conducting disc with a hole in it (or more mathematically correct an *annulus*) that is cut into three parts. The middle segment is thrown away keeping only the outer two parts. We can now construct the commutator by taking the two outer segments of the disc and replacing the centre portion with a non-conducting material, the black part in Fig. 7.13, to reconstruct the original disc. The commutator is completed by adding two stationary brushes that establish contact between the conducting parts of the disc and the outside world. We can see, therefore, that as the coil rotates. point a is sometimes connected to point 1 and sometimes to point 2. In this way, we can now rectify the ac voltage. The resulting induced voltages are indicated in Fig. 7.14, with the relevant definitions in Fig. 7.13 and Section 7.4.1.

The voltage between points 1 and 2 will be

$$\boxed{e_{12} = \omega B \ell w \,|\cos(\theta)|.} \tag{7.56}$$

The average dc voltage can be calculated from (7.55) as

$$\boxed{\begin{aligned} e_{dc} &= \frac{1}{\pi}\int_0^{\pi} e_{12} d\theta \\ &= \frac{\omega B \ell w}{\pi}\int_0^{\pi} \cos\theta d\theta \\ &= \omega B \frac{2\ell w}{\pi}. \end{aligned}} \tag{7.57}$$

We see, therefore, that the induced dc voltage depends on three factors. Firstly, the machine geometry, the $\frac{2\ell w}{\pi}$ term, which cannot be altered. Secondly, the flux in the area of the coil, B, which cannot be altered in permanent magnet machines, but in some machines, the flux is created using an electromagnet, and in these instances, the flux can be changed. Finally, the induced dc voltage depends on the rotational speed of the machine, ω. The faster the rotation of the machine, the higher the induced voltage.

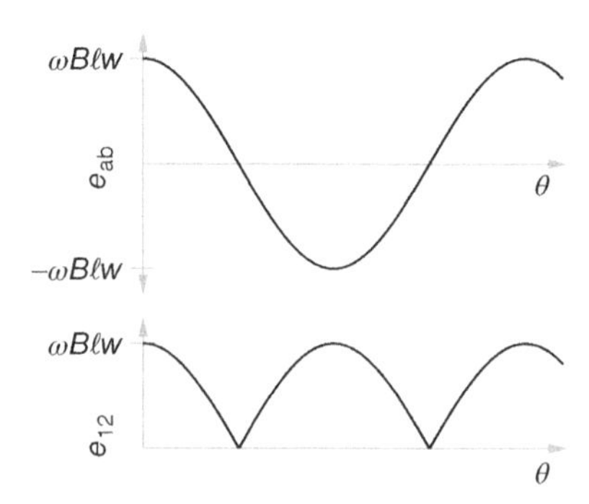

Figure 7.14 The induced voltage rectified by the commutator.

According to the definitions in Fig. 7.13 and the discussion in Section 7.4.1, the rectified voltage on points 1 and 2 will have a polarity such that point 1 is positive with respect to point 2 when the rotation direction and flux direction is as indicated in Fig. 7.13. Should either the flux direction or the rotation direction change then the voltage will change polarity. However, should both the flux direction and the direction of rotation change, the induced voltage polarity will still be the same.

7.4.3 Force and Torque

We know that any conductor carrying a current in a magnetic field will be subjected to an induced force. Consider the circuit shown in Fig. 7.13. We have proven in the previous discussion that an ac voltage will be induced between terminals a and b should the coil rotate about the axis. This voltage is then rectified to yield a dc voltage between points 1 and 2. What will happen if we use this developed electrical energy? Let us assume that the circuit between points 1 and 2 in Fig. 7.13 is completed and that there is a current I flowing from point 1 to point 2. For simplicity, we will assume that the current is a constant.

Definition 7.1

A counter-clockwise direction of rotation is defined as being positive. This definition stems from the definition of the cross product. Likewise, a torque that acts in a counter-clockwise direction is defined as positive.

In Section 7.2.2, we have seen that a conductor carrying a current in a magnetic field is subjected to an induced force

$$\mathbf{f} = i(\mathbf{l} \times \mathbf{B}).$$

We have also seen in Chapter 5 and specifically in Fig. 5.4 that the torque on an axis is determined by the tangential force acting on a distance from the axis point. Let us now calculate the torque on the coil, using the same method as in Section 7.4.1 and using the definitions of Figs. 7.13 and 7.15. Note that we have chosen the current direction to be from point 2 to point 1, consistent with the dc machine delivering energy to the outside world. In Fig. 7.15, the rotation angle is $\theta = 30°$.

1. *Section* W–X In this section, the current direction vector and the length of the conductor can be combined to yield the length vector as

$$\mathbf{l}_{\mathrm{wx}} = \ell \langle 0, 0, -1 \rangle . \tag{7.58}$$

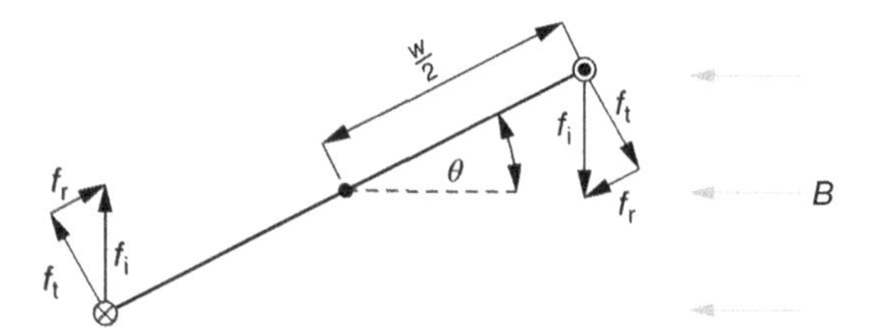

Figure 7.15 Forces experienced by the current-carrying conductors in the magnetic field and the resulting torque.

The induced force on this segment can, therefore, be calculated, where the flux vector $\mathbf{B} = B\langle -1, 0, 0\rangle$, as

$$\begin{aligned} \mathbf{f}_{\mathrm{WX}} &= i\left(\mathbf{l}_{\mathrm{WX}} \times \mathbf{B}\right) \\ &= BI\ell\,\langle 0, 1, 0\rangle . \end{aligned} \tag{7.59}$$

We now need to convert this force into its tangential and radial components to enable us to calculate the torque on the coil. Consider the definitions in Fig. 7.15, and let us define the induced force, which we can define as a scalar with the definitions in Fig. 7.15, as

$$f_{\mathrm{i}} = BI\ell. \tag{7.60}$$

Therefore, the tangential component of the force is

$$f_{\mathrm{t}}(\theta) = f_{\mathrm{i}} \cos\theta \tag{7.61}$$

and the torque is

$$t_{\mathrm{WX}}(\theta) = -\frac{1}{2} BI\ell w \cos\theta. \tag{7.62}$$

The sign of the clockwise-directed torque is defined as negative because it is acting in a clockwise direction, consider Definition 7.4.3.

2. *Section* X–Y In this section, the current direction component of the wire cannot be written as easily as that for the section W–X. However, we know that the current direction vector (although being a function of rotation) will be in the x–y plane and, therefore, the vector $\mathbf{I} \times \mathbf{B}$ will only have a component in the z-direction. Any force directed along the z-axis will not contribute to the torque experienced by the coil (as measured along the axis of rotation). Therefore,

$$t_{\mathrm{XY}} = 0. \tag{7.63}$$

3. *Section* Y–Z In this section, the current direction in the wire can be written as

$$\mathbf{l}_{YZ} = \ell \langle 0, 1, 0 \rangle . \tag{7.64}$$

The induced force is, therefore,

$$\begin{aligned} f_{YZ} &= i \left(\mathbf{l}_{YZ} \times \mathbf{B} \right) \\ &= B I \ell \langle 0, -1, 0 \rangle . \end{aligned} \tag{7.65}$$

Using the same method as for the section W–X and the definitions in Fig. 7.15, we see that

$$f_t(\theta) = B I \ell \cos \theta \tag{7.66}$$

and

$$t_{YZ}(\theta) = -\frac{1}{2} B I \ell w \cos \theta . \tag{7.67}$$

4. *Parts of Section* Z–W We can follow the same argument with the parts of this segment as with section X–Y. The force induced on these wire segments will always be directed in the z-axis and, therefore, the torque contribution of these segments (around the axis of rotation) is zero and

$$t_{ZW} = 0. \tag{7.68}$$

Adding the torque contributions together, the total torque can be calculated as

$$\boxed{t(\theta) = -B I \ell w \cos \theta .} \tag{7.69}$$

However, this equation does not include the effects of the commutator. The current will always flow from point b to point a as a result of the commutation and, therefore,

$$\boxed{t(\theta) = -B I \ell w | \cos \theta | .} \tag{7.70}$$

7.4.4 Power Balance between Mechanical and Electrical Power

Consider the example discussed in Sections 7.4.1 and 7.4.3. When the machine is rotated as shown in Fig. 7.13 and a constant current I flows from point 1 to point 2 then the electrical power delivered by the machine is calculated from $p = vi$ as

$$p_{elec}(\theta) = -\omega B I \ell w | \cos(\theta) | . \tag{7.71}$$

In turn, the mechanical power is calculated from $p = t\omega$ as

$$p_{mech}(\theta) = B I \ell w | \cos \theta | \omega . \tag{7.72}$$

Since the machine is delivering electrical power, the electrical power is negative, and because the machine is taking mechanical power (the torque direction is in an opposite direction to the direction of rotation), the mechanical power is positive. We see, therefore, that

$$p_{\text{elec}} + p_{\text{mech}} = 0 \tag{7.73}$$

and that the principle of energy conservation holds.

Example 7.9 In general, a single-turn single-pole machine is of no real benefit to anyone. To illustrate this point, let us construct a dc machine with the strongest permanent magnets available. The dimensions and requirements of the machine are listed in Table 7.2. Let us make the cylindrical rotor of a very special material (that does not exist) with the same density as aluminium but assuming that it has a high permeability (unlike aluminium). How long will it take for the rotor to accelerate from standstill to the required speed?

Solution

We know that the output torque of the machine is

$$t(\theta) = B I \ell w |\cos(\theta)|$$

and the average torque is, therefore,

$$t_{\text{ave}} = \frac{2 B I \ell w}{\pi}.$$

Therefore, with a 5 A current, the machine can deliver an average torque of 1.15 Nm. The volume of the rotor can be calculated as

$$V_{\text{rotor}} = \pi r^2 \ell = 0.094\,\text{m}^3$$

and the weight is, therefore, 254.5 kg. From Chapter 5, we know that the inertia of a cylinder is

$$J = \frac{1}{2} m r^2$$

TABLE 7.2 Design Data of the Single-Turn DC Machine of Example 7.9

Description	Value	Description	Value
Magnetic field	1.2 T	Machine length	750 mm
Rotor radius	200 mm	Maximum current	5 A
Winding resistance	1 Ω	Required speed	1 000 rpm
Density of aluminium	2 700 kg/m^3		

and the inertia of the rotor is, therefore,

$$J_{\text{rot}} = \frac{1}{2}254.5 \times 0.2^2 = 5.09\,\text{kg m}^2.$$

Assuming that the machine is connected to an ideal current source that will always deliver 5 A, the required time is

$$t = \frac{\Delta\omega}{\alpha} = \frac{\Delta\omega J}{T} = \frac{1\,000 \times 2\pi \times 5.09}{60 \times 1.15} = 463.5\,\text{s}$$

that is almost $7\frac{3}{4}$ minutes! Clearly, this machine is of no real use to anyone. The acceleration time will be even longer if magnetic steel with a density of around $7\,600\,\text{kg/m}^3$ is used for the rotor.

7.4.5 The benefit of a Uniform Air Gap

The basic dc machine shown in Fig. 7.13 has a variable air gap and, therefore, the flux is always uniform and unidirectional. What will happen if we can change the geometry of the dc machine slightly to ensure that the air gap is uniform?

Let us alter the geometry of Fig. 7.13 slightly and use two magnets with curved faces to create the magnetic field. Let us also place the rotating coil in a cylindrical container with a relatively high permeability. Such a machine is shown in Fig. 7.16. The most important difference between the magnetic circuit of Figs. 7.13 and 7.16 is that the uniform air gap that causes the magnetic field lines to bend in the region of the air gap. Apart from the regions at $\theta = 90°$ and $\theta = 270°$, the magnetic field lines are always perpendicular to the faces of the cylindrical container.

We can calculate the induced voltage using the same methods as in Section 7.4.1. However, because of the change in direction of the field lines, the induced voltage is no longer sinusoidal but tends to have a flat top as shown in Fig. 7.17. The induced voltage is now a lot more uniform and the magnitude of the rectified dc voltage is

$$\boxed{e_{\text{dc}} \approx \omega B \ell w.} \tag{7.74}$$

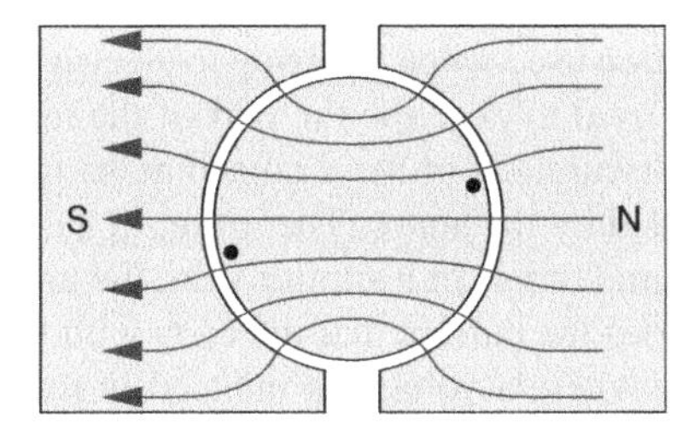

Figure 7.16 The induced voltages in a machine with a uniform air gap.

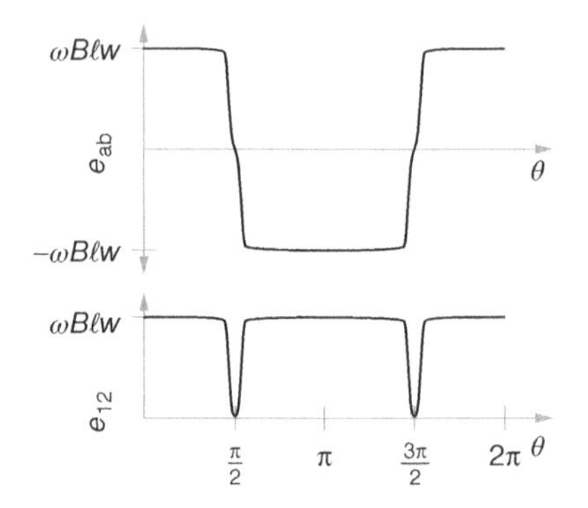

Figure 7.17 The effect of a uniform air gap.

The torque delivered by the machine is likewise changed by the alteration in the magnetic flux distribution. The torque developed can now be approximated, for the definitions in Section 7.4.3, as

$$T \approx \begin{cases} BI\ell w & \text{under the pole faces} \\ 0 & \text{beyond the pole edges,} \end{cases} \tag{7.75}$$

where *under the pole faces* implies that the current-carrying conductors are in the region of the uniform magnetic field. This implies naturally that *beyond the pole edges* implies that the current-carrying conductors are in the region where no magnetic flux is present, that is, $\theta \approx 90°$ or $\theta \approx 270°$.

7.5 PRACTICAL DC MACHINE CONSTRUCTION

The basic parts of any rotating electrical machine are shown in Fig. 7.18. As we can see, the machine consists of a stator, rotor, bearing shaft and some sort of base. More often than not, the base is equipped with bolt holes, so that the machine can be secured to a stable platform.

Two terms that are often used when referring to electrical machines are *field* and *armature*. The field is used to describe the part of the machine that provides the magnetic field. In our discussions of the dc machine so far, we have seen that the magnetic field is provided by the stator. Therefore, in a dc machine, the field and the stator refer to the same thing. In a similar vein, the armature refers to the part of the machine that carries the current that interacts with the field or, equivalently, the part of the machine where a voltage is induced in response to the movement of conductors through the magnetic field. Therefore, in a dc machine, the armature and the rotor refer to the same thing.

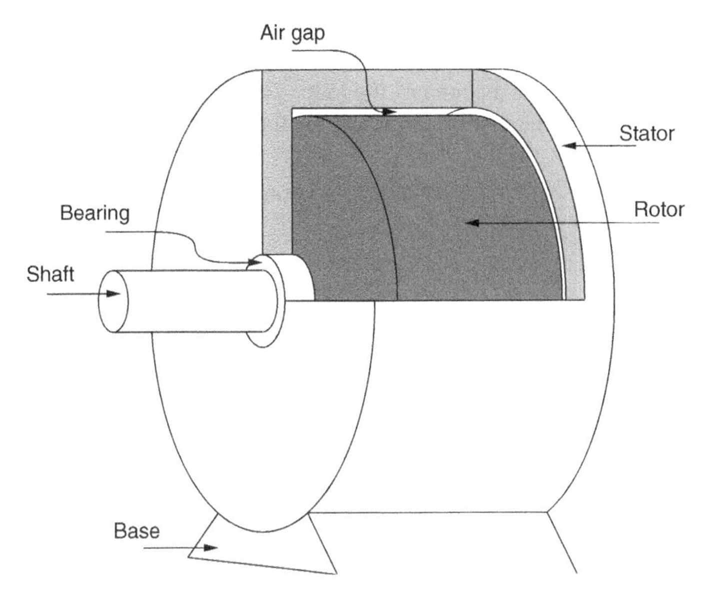

Figure 7.18 Basic parts of an electrical machine.

The simple machine of Fig.7.13 and also that of Fig. 7.16 are shown to have only two poles, one N-pole and one S-pole. In practical machines, there can be more than two poles, but they are always in pairs, for better utilisation of the space around the stator and rotor circumference, and also for the different torque–speed characteristics of the machine. In general, higher torque machines will have a larger number of pole pairs.

In addition, the machine geometry is better utilised when there are multiple turns in a coil instead of the single turn shown in Fig. 7.13 for the simple machine. The multiple turns will increase the induced voltage. A multi-turn arrangement is known as a *winding*. In machines, many conductors are arranged radially along the rotor surface and connected in series or parallel combinations depending on the number of poles and type of winding scheme. In dc machines, the conductor connections are arranged in one of two winding patterns, known as *lap* or *wave windings*.

How does the addition of more pole and turns per winding affect the torque and induced voltage produced by a real machine? Let us start with the induced voltage.

7.5.1 Induced Voltage in a Real dc Machine

We have seen in Section 7.4.1 that the induced voltage in the dc machine, which is often called the *back-emf*, depends on the following.

1. The rotational speed, ω, of the machines rotor.
2. A constant that refers to the construction of the machine, this statement will become clearer later in this discussion.

3. The flux, ϕ, in the machine. Recall from Chapter 4 that $\phi = B A_e$ where A_e is the equivalent area. In (7.55), we see the term $B \ell w$ that refers to the multiplication of B and the area of the loop. This term, therefore, represents the maximum flux that can be 'captured' by the loop.

We have seen in Section 7.4.1 that the maximum voltage induced in a coil (right under the pole face) is equal to

$$e_{max} = \omega B \ell w = v B \ell, \tag{7.76}$$

where v is the linear speed of the conductor. If we now manufacture our coil with N turns, in other words, all these turns are in series, then this voltage will be induced in each turn. As the turns are connected in series, the total maximum voltage now becomes

$$e_{max} = \omega N B \ell w. \tag{7.77}$$

However, in many machines, we need the capability to carry large currents in the windings, and therefore, we might create a number of coils and place them in parallel to form one winding. Considering this, let a winding consist of Z conductors. If the winding consists of a coils connected in parallel then each coil will have Z/a turns. For practical dc machines, it is not always easy to see how the different coils are connected in parallel and we speak of the number of current paths rather the number of parallel coils. With this definition, the maximum induced voltage becomes

$$e_{max} = \frac{Z \omega B \ell w}{a}. \tag{7.78}$$

Now that we have dealt with an increase in the number of conductors from one to any number, what will happen when we increase the number of poles? Let us denote the number of poles by P. From (7.78), we note again the term $B \ell w$ that denotes the flux cut by the coil. Let us now define the amount of flux per pole as ϕ_P. The rotor of a machine is shaped like a cylinder; the area of a cylinder is

$$A = 2\pi r \ell = \pi w \ell. \tag{7.79}$$

If there are P poles, the area associated with each pole becomes the total area divided by the number of poles or

$$A_P = \frac{2\pi r \ell}{P} = \frac{\pi \ell w}{P}. \tag{7.80}$$

The *total flux per pole* is, therefore,

$$\phi_P = B A_P = \frac{2\pi r \ell B}{P} = \frac{\pi \ell w B}{P}. \tag{7.81}$$

Using this expression for the total flux per pole to rewrite the expression for the maximum induced voltage,

$$\begin{aligned} e_{\text{max}} &= \frac{Z\omega B\ell w}{a} \\ &= \frac{ZP}{\pi a}\frac{B\pi\ell w}{P}\omega \\ &= \frac{ZP}{\pi a}\phi_{\text{P}}\,\omega \end{aligned} \tag{7.82}$$

we can rewrite this using the constant

$$K_{\text{m}} = \frac{ZP}{\pi a} \tag{7.83}$$

as

$$e_{\text{max}} = K_{\text{m}}\,\phi_{\text{P}}\,\omega. \tag{7.84}$$

Up until now in our discussion we have worked exclusively with the peak-induced voltage. However, in real machines, the air gap is always uniform (the field pole faces are curved), so according to the discussion in Section 7.4.5, the induced voltage is approximately equal to the peak-induced voltage. In fact, for dc machines with many different coils and poles, the induced emf can be written as

$$\boxed{e = K_{\text{m}}\,\phi_{\text{P}}\,\omega.} \tag{7.85}$$

7.5.2 Torque Produced in a Real dc Machine

We have seen in Section 7.4.3 that the produced torque depends on

1. the current in the magnetic field;
2. some constant that refers to the construction of the machine, this statement will become clearer later in this discussion;
3. once again the flux, ϕ, in the machine.

We have seen in Section 7.4.3 that the maximum voltage induced in a coil (right under the pole face) is equal to

$$T_{\text{max}} = B I \ell w, \tag{7.86}$$

where I is the current in the conductor. If we now manufacture our coil with N turns, in other words, all these turns are in series, then each conductor will contribute the same amount of tangential force and the total maximum torque now becomes

$$T_{\text{max}} = N B I \ell w. \tag{7.87}$$

However, again because many machines need the capability of carrying large currents, we often use a number of coils and place them in parallel to form one winding. If, therefore, a winding consist of Z conductors and consists of a parallel current paths then each coil will carry a current of I/a. With this definition, the maximum produced torque becomes

$$T_{max} = \frac{ZBI\ell w}{a}. \tag{7.88}$$

We can again rewrite this equation, using the definition of ϕ_P, as

$$\begin{aligned} T_{max} &= \frac{ZBI\ell w}{a} \\ &= \frac{ZP}{\pi a}\frac{B\pi\ell w}{P} I \\ &= \frac{ZP}{\pi a}\phi_P I \\ &= K_m \phi_P I, \end{aligned} \tag{7.89}$$

where

$$K_m = \frac{ZP}{\pi a}. \tag{7.90}$$

Once again, in a practical machine where the air gap is uniform, as discussed in Section 7.4.5, the produced torque is approximately equal to the peak value calculated here. In fact, for dc machines with many different coils and poles, the produced torque can be written as

$$\boxed{T = K_m \phi_P I.} \tag{7.91}$$

In these expressions for K_m, the term a represents the number of parallel paths in the winding, although it is not important now to know exactly why for wave windings $a = 2$ but for lap windings $a = P$. Wave windings are, therefore, often used for low current high voltage applications and lap windings for high current low voltage applications.

Example 7.10 An armature that consists of 72 coils each with 12 turns is used in a six-pole dc machine. There are 12 parallel current paths. The flux per pole in the dc machine is 40 mWb. How much back emf is present when the machine turns at 400 rpm?

Solution

As there are twelve parallel current paths,

$$a = 12.$$

The number of conductors is

$$Z = 72 \times 12 = 864.$$

We can now calculate the machine constant as

$$K_{\mathrm{m}} = \frac{ZP}{\pi a} = \frac{864 \times 6}{\pi \times 12} = 137.51$$

and the back emf as

$$\begin{aligned} E_{\mathrm{a}} &= K_{\mathrm{m}}\, \phi_{\mathrm{P}}\, \omega \\ &= 137.51 \times 0.04 \times \frac{400 \times 2\pi}{60} \\ &= 230.4\,\mathrm{V}. \end{aligned}$$

As an interesting side note, we see that for this machine, $a = 12 = 2P$; therefore, we know that a double lap winding is used for this machine. In other words, the armature is wound twice with a wave winding. Recall that for a 'normal' wave winding $a = P$.

Example 7.11 A 12-pole wave wound machine contains 144 coils with 10 turns each. If the machine produces 2 900 V when rotating at 350 rpm, what is the flux per pole?

Solution

The machine has a wave winding, therefore,

$$a = 2.$$

The number of conductors is

$$Z = 144 \times 10 = 1\,440$$

and the machine constant is, therefore,

$$K_{\mathrm{m}} = \frac{ZP}{\pi a} = \frac{1\,440 \times 12}{2\pi} = 2\,750.2.$$

We can now calculate the flux per pole as

$$\begin{aligned} \phi_{\mathrm{P}} &= \frac{E}{K_{\mathrm{m}}\, \omega} \\ &= \frac{60 \times 2\,900}{2\pi \times 350 \times 2\,750.2} \\ &= 28.77\,\mathrm{mWb}. \end{aligned}$$

Example 7.12 We want to design a dc machine for use in a hybrid-electric bicycle. The dc machine is connected to the hub of the front wheel. The wheel has a diameter of 76 cm. The bicycle is equipped with a 25 V battery and we have to deliver 200 W to the bicycle when the bicycle speed is 15 kmh. To allow for losses in the system, we want the maximum current of the machine at this operating point to be 9 A. We have permanent magnets available that can deliver 20 mWb per pole. The size of the dc machine must be small and we have determined that we have space for only 110 conductors. Calculate how many poles the machine should have if a wave winding is used.

Solution

The rotational speed of the wheel when the bicycle speed is 15 kmh can be calculated as

$$\omega = \frac{v}{r} = \frac{15\,000}{60^2 \times \frac{1}{2}0.76} = 10.965\ \mathrm{rad/s}.$$

To produce 200 W at this point, the required torque is

$$T = \frac{P}{\omega} = \frac{200}{10.965} = 18.24\ \mathrm{Nm}.$$

To produce this torque with a current of 9 A, the machine constant should be

$$K_\mathrm{m} = \frac{T}{\phi_\mathrm{m} I} = \frac{18.24}{0.02 \times 9} = 101.33. \tag{7.92}$$

If we use a wave winding then

$$a = 2$$

and

$$P = \frac{K_\mathrm{m}\, \pi a}{Z} = \frac{101.33 \times 2\pi}{110} = 5.79,$$

therefore, a machine with six poles should be used.

Example 7.13 How many conductors would we need if we design the machine of Example 7.12 with six poles but using a lap winding?

Solution

If we use a lap winding then

$$a = P = 6.$$

As the design constants are the same as in Example 7.12, we know that the machine constant should still be

$$K_\mathrm{m} = 101.33 = \frac{ZP}{\pi a}$$

and, therefore,

$$Z = \frac{K_{\mathrm{m}}\,\pi a}{P} = K_{\mathrm{m}}\pi = 318.34.$$

However, the total number of turns should be a whole number and also a multiple of the number of poles; therefore,

$$Z = 324.$$

Although for all practical reasons, three hundred and eighteen conductors should also work because we will then need 9.01 A!

7.6 PRACTICAL DC MACHINE CONFIGURATIONS

Definition 7.2

An electrical machine that takes electrical power and converts it to mechanical power is referred to as being in the *motoring* mode or even simply referred to as a *motor*. Conversely, an electrical machine that takes mechanical power and converts it to electrical power is said to be in the *generating* mode or is referred to as a *generator*.

The dc machine discussed so far used permanent magnets to establish the flux. We will discuss this machine and its limitations in more detail in the next section. There is another alternative for generating the magnetic field; it could be generated by a second set of windings that is placed on the stator of the machine. The construction of a machine with an external field winding is shown in Fig. 7.19. Here the names field winding and armature winding come to their full right. The winding in the rotor is called the *armature winding*, while the winding in the stationary part is called the *field winding*. Both the armature and field windings are supplied with dc currents. The armature winding carries the main current while the field windings carry a small excitation current. The flux per pole is directly proportional to the field current; using the theory discussed in Chapter 4, we know that

$$\begin{aligned} &\phi_{\mathrm{P}} \propto i_{\mathrm{f}} \\ &\because \phi_{\mathrm{P}} = \mu \frac{N i_{\mathrm{f}} A}{\ell} \\ &\therefore \phi_{\mathrm{P}} = K_{\phi} i_{\mathrm{f}}, \end{aligned} \tag{7.93}$$

where K_{ϕ} is a constant that is inherent to the construction of the machine. The practical example of a dc machine in Fig. 7.20 demonstrates the main working

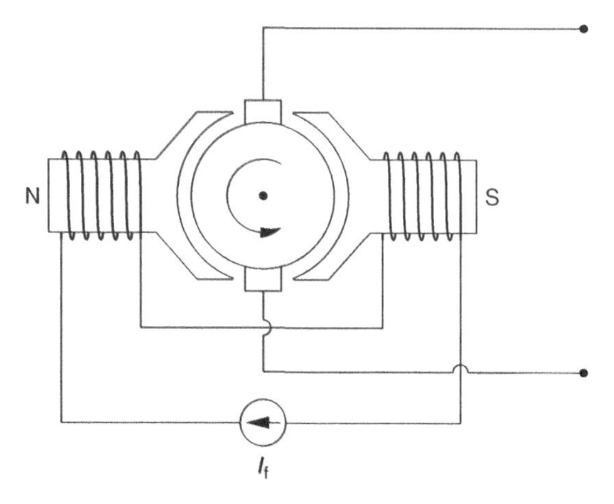

Figure 7.19 Field of the dc machine provided by means of an electromagnet (field winding).

Figure 7.20 A hundred year old dc machine.

parts. In the foreground the commutator and brushes can be seen. This machine has four poles, hence the four field windings on the stator.

It is important to note that the flux produced by the field winding and that produced by the armature winding are orthogonal to each other. In a dc machine, this orthogonality is ensured by the construction of the machine and the operation of the commutator and brushes. As we will see later, maximum torque is produced if the field and armature flux are at right angles to one another. The dc machine operation in this case is very simple because we do not have to control the armature flux and field flux in a manner to ensure orthogonality.

For sake of simplicity, we will introduce the following schematic definitions. The armature and brushes (and per implication the commutator) are shown on the right-hand side of Fig. 7.21. The field winding is shown on the left-hand side. In both windings, it is normal to include the winding resistance in the model. In some instances, depending on the required accuracy of the model, the power lost in the brushes can also be included in the model of the dc machine. Although there are different ways to model this loss, we will make the choice here to model the brush loss as a resistance in series with the armature winding.

It is possible to implicitly control both the field and the armature winding currents by using two dc sources, as shown in Fig. 7.26. Although this has many advantages, we need two controllable dc sources that increase the cost of such a system. It is also possible to operate the dc machine using only a single source, by connecting the field winding either in shunt or in series with the armature winding. The series connection is a popular configuration and has two very useful features. Firstly, it generates a very high starting torque which is why it is still used today as a starter motor for automobiles. Secondly, because both the magnetic fields generated by the armature and field windings are reversed when current polarity changes, a series dc motor can also be powered by alternating voltage; if designed for this purpose, the series-connected dc machine is sometimes referred to as the *universal machine*. For this reason, the series dc motor is often used in household appliances, for example, vacuum cleaners.

An electrical machine is designed for a maximum voltage and a maximum continuous current. Neither of these limits can be exceeded when the machine is used without compromising the lifetime of the machine. The maximum voltage limit is set by the amount of electrical insulation in the machine; if the maximum voltage limit is exceeded, the insulation could fail and the machine windings short circuited resulting in machine failure. The maximum continuous current limit is also connected to the insulation of the machine. As the machine windings are non-ideal, they experience resistive heating because of the power lost in them when used. The power lost in the winding, or equivalently the heating power, corresponds to i^2R and, therefore, increases with the square of the current. When the heat generated by the windings exceeds the temperature limit of the insulation, irreversible

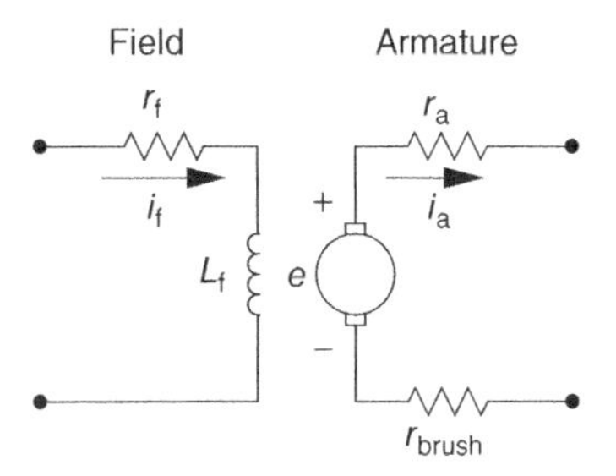

Figure 7.21 Schematic representation of a DC machine with field winding.

damage to the insulation would occur and the machine could fail because of short-circuited windings. However, unlike the maximum voltage limit where the limit cannot be safely exceeded even for short periods of time, the maximum current can be exceeded for short periods of time. This is because the heat generated by the windings takes some time to permeate through the machine and all the metal components of the machine takes a large amount of energy to heat up sufficiently to cause any damage. In many applications, electrical machines experiences over currents during starting. That said it is important to ensure that the current limit is adhered to during steady-state operation.

7.6.1 Permanent Magnet DC Machine

Permanent magnets are used to create the field magnetic flux in a permanent magnet dc machine. This has some advantages. Notably that no electrical power is used to create the field and also that the use of strong rare earth magnets can decrease the size of the machine. However, because the field cannot be controlled, the complete system does suffer some drawbacks. This will be explored in the following examples.

Example 7.14 A four-pole permanent magnet dc machine has two hundred and forty eight windings configured in a lap winding configuration. The machine is rated for 200 V and 30 A and the armature resistance is equal to 20 mΩ. The permanent magnets deliver 35 mWb per pole. Calculate the maximum speed and the maximum torque of the machine. Also calculate the combination of speed and torque at the maximum power point.

Solution

As a lap winding is used, $a = P$ and the machine constant can be found as

$$K_{\mathrm{m}} = \frac{ZP}{\pi a} = \frac{Z}{\pi} = 78.95;$$

the maximum torque is, therefore,

$$T_{\max} = K_{\mathrm{m}}\,\phi_{\mathrm{P}}\,I_{\max} = 82.89\ \mathrm{Nm}$$

and the maximum speed is

$$\omega = \frac{e_{\max}}{K_{\mathrm{m}}\phi_{\mathrm{P}}} = 72.39\,\mathrm{rad/s},$$

which is equal to 691.25 rpm. At the maximum power point, the machine is operating at the rated voltage and the rated current. The induced voltage at this point is

$$e = V_{\mathrm{in}} - i_{\mathrm{a}}\,r_{\mathrm{a}} = 199.4\,\mathrm{V},$$

which corresponds to a speed of

$$\omega = \frac{e}{K_m \phi_P} = 72.17\,\text{rad/s}$$

or 689.2 rpm. For the sake of interest, this corresponds to 5.98 kW of output power at the shaft, the remainder is dissipated in the armature winding as heat.

Example 7.15 The machine discussed in Example 7.14 is used as a hoist where the required torque is constant at all speeds. The torque requirement of the system is 40 Nm. Calculate the speed at which the motor will run when connected to this load. If the required machine speed is 220 rpm, what voltage should be applied to the machine?

Solution

When the machine is delivering 40 Nm to the load, the armature current will be

$$i_a = \frac{T}{K_m \phi_P} = 14.48\,\text{A}.$$

This implies that the induced emf will be

$$e = V_{in} - i_a r_a = 199.7\,\text{V},$$

which implies that the shaft speed will be

$$\omega = \frac{e}{K_m \phi_P} = 72.27\,\text{rad/s}$$

or 690 rpm. This operational point can be seen in Fig. 7.22, where the results of Example 7.14 were used to construct the graphical representation.

If we decrease the machine speed to 220 rpm, the machine is still delivering the same torque and the armature current will still be 14.48 A. However, the armature emf will be

$$e = K_m \phi_P \omega = K_m \phi_P \frac{220 \times 2\pi}{60} = 63.65\,\text{V}.$$

The terminal (input) voltage should, therefore, be

$$V_{in} = e + i_a r_a = 63.94\,\text{V}.$$

Example 7.16 The machine used in Examples 7.14 and 7.15 is used to drive a centrifugal pump that has to deliver exactly 4 kl of water per minute to a process. The total head height (the equivalent height against which the water must be pumped, which includes the resistance of the piping) is 25 m. The load profile of the pump can be found as $T = 0.014\,\omega^2$, and the pump is 55% efficient over its whole operating range. Calculate the efficiency of the system for the following.

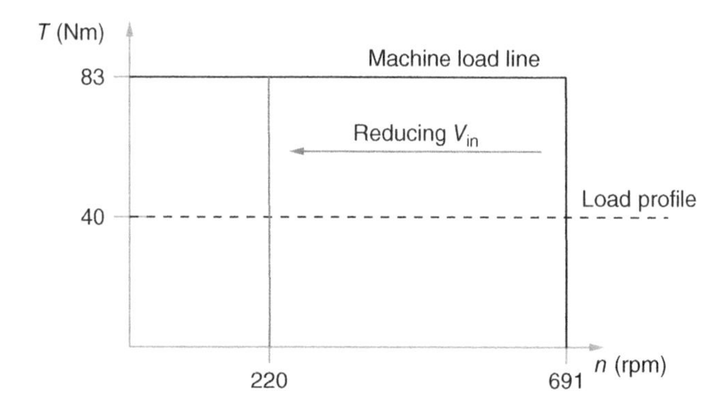

Figure 7.22 The machine load line and the load profile, Example 7.15.

1. If a valve is used to restrict the amount of water flowing through the pipe: in this case, the valve simply increases the total friction of the pipe and, hence, increases the effective head height of the water delivery system. There is, therefore, a considerable amount of losses in the valve system.
2. If the speed of the dc machine is regulated to a point where the pump delivers only the required amount of water.

Solution

When the pump is operating at the maximum speed of the dc machine, it will require

$$T = 0.014\,\omega^2 = 73.34 \text{ Nm},$$

which is less than the maximum torque of the machine. We see, therefore, that the system is limited by the maximum speed of the machine. The system will operate at the point where the torque delivered by the machine and that required by the pump are equal. We have an expression of the torque–speed characteristic of the pump; however, we need to create an expression for the torque–speed characteristic of the dc machine in the speed-limiting region. We can find the torque of the machine as

$$T_{\mathrm{m}} = K_{\mathrm{m}}\,\phi_{\mathrm{P}}\,i_{\mathrm{a}},$$

which can be rewritten using

$$i_{\mathrm{a}} = \frac{V_{\mathrm{in}} - e}{r_{\mathrm{a}}}$$

as

$$T_{\mathrm{m}} = K_{\mathrm{m}}\,\phi_{\mathrm{P}}\,\frac{V_{\mathrm{in}} - e}{r_{\mathrm{a}}}.$$

Finally, by substituting the expression

$$e = K_{\mathrm{m}}\,\phi_{\mathrm{P}}\,\omega$$

in this expression we get

$$T_{\mathrm{m}} = \frac{K_{\mathrm{m}}\,\phi_{\mathrm{P}}\,V_{\mathrm{in}} - K_{\mathrm{m}}^2\,\phi_{\mathrm{P}}^2\,\omega}{r_{\mathrm{a}}}.$$

To find the point where the torque developed by the dc machine and that required by the pump is equal, we can equate the torque–speed expressions and solve for ω:

$$T_{\mathrm{m}} = T_{\mathrm{p}}$$

$$\frac{K_{\mathrm{m}}\phi_{\mathrm{P}}V_{\mathrm{in}} - K_{\mathrm{m}}^2\phi_{\mathrm{P}}^2\,\omega}{r_{\mathrm{a}}} = 0.014\,\omega^2$$

$$\frac{K_{\mathrm{m}}\phi_{\mathrm{P}}V_{\mathrm{in}}}{r_{\mathrm{a}}} - \frac{K_{\mathrm{m}}^2\phi_{\mathrm{P}}^2\,\omega}{r_{\mathrm{a}}} - 0.014\,\omega^2 = 0$$

$$\omega = -27\,335.7\ \mathrm{rad/s}\ \text{or}\ \omega = 72.2\ \mathrm{rad/s}.$$

The viable speed of rotation is 72.2 rad/s, which is equal to 689.5 rpm. The load profile of the pump and the machine load line are shown in Fig. 7.23. At this point, the machine would deliver

$$T = 0.014\,\omega^2 = 73\ \mathrm{Nm}$$

to the pump. This corresponds to a total power of

$$P = T\omega = 5.2\,\mathrm{kW}$$

delivered to the pump. As the pump is only 55% efficient, the mechanical power produced is

$$P_{\mathrm{mech}} = \eta\,P = 2.86\,\mathrm{kW},$$

which implies that the pump will deliver

$$\frac{P_{\mathrm{mech}}}{h} = 114.4\ \mathrm{l/s}$$

or 6.86 kl/min.

However, we need a flow rate of 4 kl/min. If a valve is used to bleed the excess power from the system, the mechanical power will be equal to the flow rate multiplied by the *unrestricted* head height, remember that the extra restriction caused by the semi-closed valve is an additional loss. Therefore,

$$P_{\mathrm{mech}} = \frac{4000}{60}25 = 1.67\,\mathrm{kW}.$$

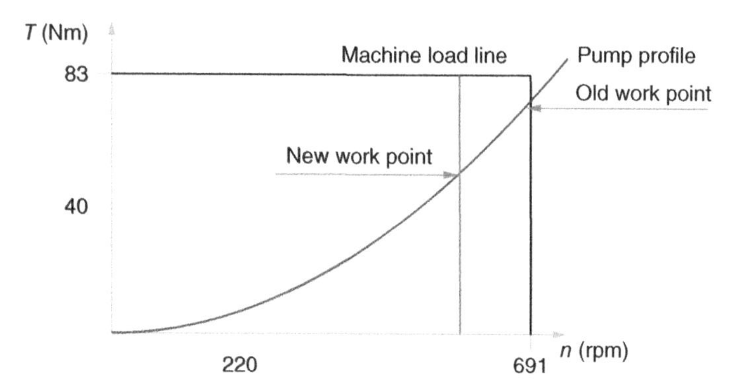

Figure 7.23 The machine load line and the load profile, Example 7.16.

The dc machine delivers 73 Nm to the pump; therefore, the armature current must be

$$i_a = \frac{T}{K_m \phi_P} = 26.42\,\text{A}$$

and the total input power to the machine is

$$P_{in} = V_{in}\, i_a = 5.28\,\text{kW}.$$

The total system efficiency is, therefore,

$$\eta = \frac{P_{mech}}{P_{in}} = 31.54\%.$$

If we reduce the voltage input to the machine to reduce the speed, we can alter the work point of the system to the exact operating point. To pump the required amount of water, the pump requires

$$P_{pump} = \frac{P_{mech}}{\eta_{pump}} = 3.03\,\text{kW}.$$

The speed of the pump at this work point can be calculated using the pump's load profile

$$P_{pump} = T\omega = 0.014\,\omega^3$$

$$\therefore \omega = \sqrt[3]{\frac{P}{0.014}} = 60\,\text{rad/s}$$

and the torque is, therefore,

$$T = 0.014\,\omega^2 = 50.47\text{ Nm}.$$

The armature current is, therefore,

$$i_a = \frac{T}{K_m \phi_P} = 18.27\,\text{A}$$

and the induced emf is

$$e = K_m \phi_P \omega = 165.78\,\text{V}.$$

The voltage source should, therefore, be adjusted to

$$V_{in} = e + i_a r_a = 166.14\,\text{V}.$$

The efficiency now becomes

$$\eta = \frac{P_{mech}}{V_{in}\, i_a} = 54.9\%$$

close to the efficiency of the pump.

Example 7.17 Consider the system in Example 7.16. If another more efficient pump with a load profile of $T = 0.023\,\omega^2$ is found, can it be used if there is no way to reduce the voltage? In other words, if only a fixed 200 V supply is available.

Solution

The machine–pump combination will run at a speed where the torque delivered by the machine and the torque required by the pump is equal. This point can either occur where the machine is delivering the maximum possible torque or where the machine is running at the maximum possible speed, as seen in Example 7.15. Let us investigate at what speed the pump will run when taking the maximum current from the machine. From the load profile of the pump, we know that at a torque of 82.89 Nm, the pump will run at

$$\omega = \sqrt{\frac{T}{0.023}} = 60\,\text{rad/s},$$

which is less than the maximum speed of the machine. Therefore, the system is current, or equivalently torque, limited. It is important to note, however, that the machine does not limit the current to the safe value, the operator of the machine would have to do that by limiting the input voltage to the machine. If left unrestricted, the machine will simply take more current from the supply until the system becomes speed limited. This process is shown in Fig. 7.24. In this case, we see clearly that the machine is overloaded because the operating point falls outside of the safe operating area as described by the machine load line.

Using the same methodology as in Example 7.16, we see that the stable work point of the unrestricted machine is $\omega = 72.07$ rad/s, which corresponds to an input

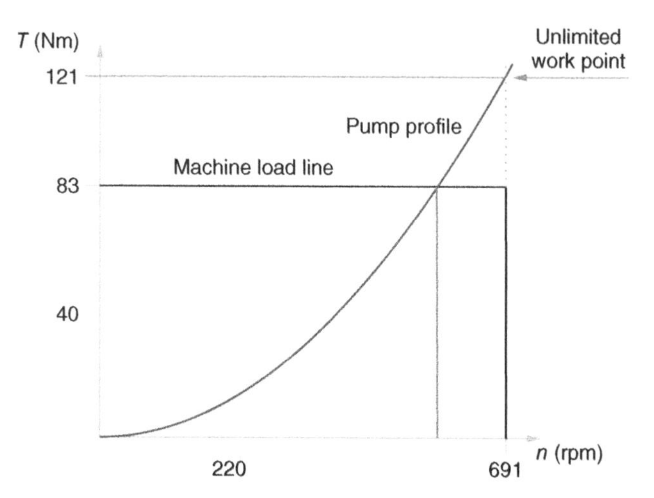

Figure 7.24 The machine load line and the pump load profile, Example 7.17.

current of 43.8 A, clearly more than the 30 A, for which the machine was designed. The only way in which this pump and machine can be safely used together is by reducing the terminal voltage and operating the system at the intersection of the pump load profile and the altered machine load profile, as shown in Fig. 7.24 by the second grey line. You can verify that the input voltage should be about 166 V for the machine to operate safely.

7.6.2 Field Winding DC Machines

In the permanent magnet dc machine, we have seen that we can decrease the speed of the machine by varying the input voltage to the machine. However, we have no real control over the current drawn by the machine, as this is set by the load characteristics, although we can force the machine to operate at a range of points by decreasing the input voltage. This process is visualised in Fig. 7.25.

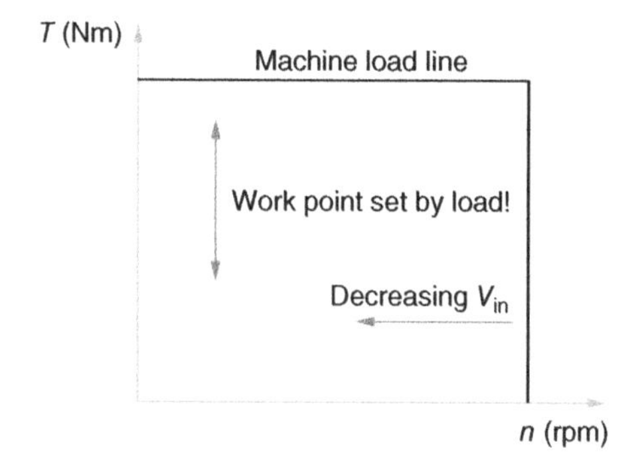

Figure 7.25 The control we have over the permanent magnet dc machine load line.

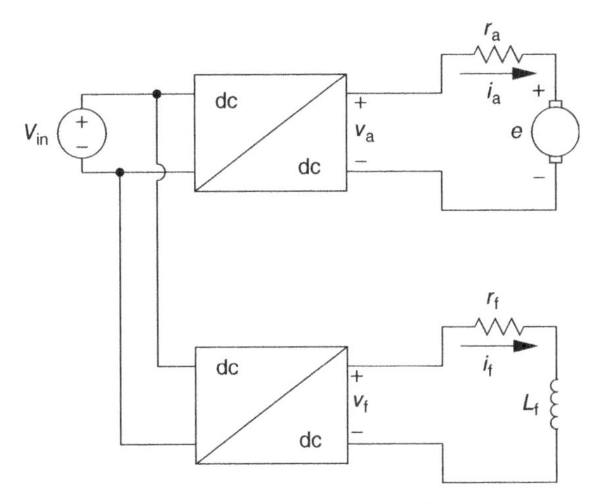

Figure 7.26 A separately excited dc machine shown with a dc drive that can control both the armature current and the field current.

This changes dramatically in the separately excited dc machine because we now have control over the field winding and, therefore, the amount of flux per pole. This allows us a lot of flexibility to control both the torque and speed by varying both the field current and the input voltage. By varying the input voltage, we can 'control' the armature current by using the interaction between the machine and the load. Such a system is shown in Fig. 7.26. At this point it is worthwhile pointing out that the power rating of the converter supplying the armature windings is much larger than that of the converter supplying the field winding. This is because the field winding is only used to create the flux in the region of the air gap, while the armature winding is delivering the electrical power that is converted into mechanical energy.

From the earlier discussion of the permanent magnet dc machine we know that we have two limitations on the operating conditions of the machine, namely, the maximum armature current and the maximum applied voltage. The armature current limit determines the maximum torque that the machine can deliver, and the maximum applied voltage determines the maximum speed of the machine. These two limits are often referred to as the *rated torque* and *rated speed* of the machine.

Definition 7.3

The nameplate rating of electrical equipment such as electrical machines and transformers refers to the maximum safe operating conditions of the equipment. Typical information included on the nameplate is rated voltage, rated current, rated torque, rated speed, rated frequency and sometimes internal connection diagrams. In the case of electrical machines, the rated speed refers to the maximum speed of the machine at rated flux.

The separately excited dc machine does, however, give us a method of safely exceeding the rated speed of the machine without exceeding the rated voltage of the machine. We can achieve this feat by a technique called *flux weakening*. Remembering that the induced emf of the dc machine is related to the speed of the machine as

$$e = K_m \phi_P \, \omega \tag{7.94}$$

this effectively sets the maximum speed of the machine since when the machine is delivering no power to the load (at no load), it is true that

$$T = 0 \longleftrightarrow i_a = 0 \tag{7.95}$$

$$\therefore e = V_{in} \tag{7.96}$$

$$\text{and } \omega_{rated} = \frac{V_{in}}{K_m \, \phi_{P-rated}}. \tag{7.97}$$

However, by decreasing the flux, we can increase the speed of the machine without increasing the applied voltage as can be seen in (7.97) when $\phi_P < \phi_{P-rated}$ so it must be true that $\omega > \omega_{rated}$. We call this region of operation the flux weakening region.

Although we can increase the speed of the machine above the rated speed by using flux weakening, this comes at a price. We know that the torque delivered by the machine is related to the flux

$$T = K_m \, \phi_P \, i_a; \tag{7.98}$$

therefore, when we decrease the flux, we also decrease the amount of torque that can be produced by the machine per unit of armature current. Therefore, we see that the maximum torque that the machine can deliver lowers proportionally with the weakening of the flux, while meanwhile the maximum possible speed of the machine increases. As both characteristics change linearly with the change in flux and we know that

$$P = T\omega, \tag{7.99}$$

we see that the output power remains constant. For this reason, the flux weakening area is also called the *constant power region*. The load line of the separately excited dc machine can be visualised for these two distinct regions as shown in Fig. 7.27.

Example 7.18 A separately excited machine has the following nameplate characteristics: $V_{in} = 250$ V, $I_{max} = 30$ A and rated speed 890 rpm. We have measured the armature winding resistance as 30 m Ω and the field resistance as 750 Ω. The nameplate states that the field winding is rated for 250 V. What should the field current be to enable the machine to deliver maximum power to a constant torque load of 55 Nm?

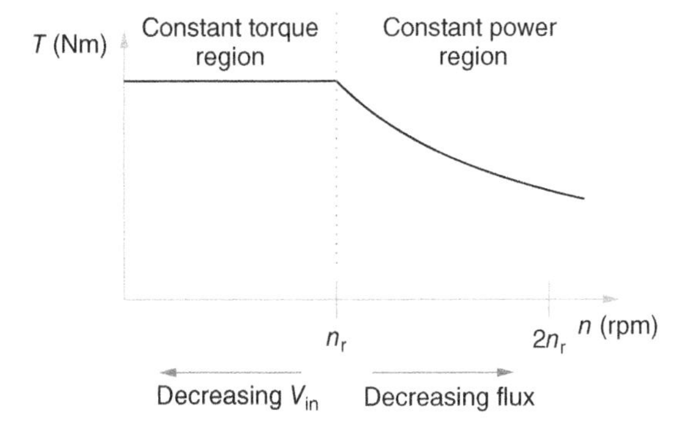

Figure 7.27 The load line of the separately excited dc machine in both the constant torque and the constant power regions.

Solution

We first need to calculate as many of the machine's parameters as possible. We know that the rated field current is

$$i_{\text{f-rated}} = \frac{250}{750} = \frac{1}{3}\,\text{A}.$$

From here, we can calculate the machine's constant. Remember that the flux of the machine is related to the field current by

$$\phi_{\text{P}} = K_{\phi}\, i_{\text{f}};$$

therefore, we know that

$$e = K_{\text{m}}\, \phi_{\text{P}}\, \omega = K_{\text{m}}\, K_{\phi}\, i_{\text{f}}\, \omega.$$

At the rated current and speed, the induced emf is

$$e = V_{\text{in}} - i_{\text{a}} r_{\text{a}} = 249.1\,\text{V}.$$

Therefore, we can calculate the product of the two constants as

$$K_{\text{m}}\, K_{\phi} = \frac{e}{\omega\, i_{\text{f}}} = \frac{249.1 \times 60 \times 3}{890 \times 2\pi} = 8.0182$$

and the maximum torque rating of the machine is, therefore,

$$T = K_{\text{m}}\, \phi_{\text{P}}\, i_{\text{a}} = K_{\text{m}}\, K_{\phi}\, i_{\text{f}}\, i_{\text{a}} = 80.18\,\text{Nm}.$$

Although the machine is rated for

$$P = T\,\omega \approx 7.5\,\text{kW}$$

when connected to the 55 Nm load, it will only be able to deliver

$$55\frac{2\pi 890}{60} = 5.13\,\text{kW}$$

to the load without the use of flux weakening. To deliver the full 7.5 kW, the rotational speed should be

$$\omega' = \frac{P_{\text{max}}}{T} = \frac{7500}{55} = 136.4\,\text{rad/s}.$$

As we are still delivering rated power, the armature current will still be 30 A and the induced emf will still be 249.1 V. We can, therefore, calculate the required field current as

$$i_{\text{f}} = \frac{e}{K_{\text{m}}\,K_{\phi}\,\omega} = 227.8\,\text{mA},$$

and the voltage applied to the field winding is, therefore,

$$v_{\text{f}} = i_{\text{f}}\,r_{\text{f}} = 170.9\,\text{V}.$$

7.6.3 Losses

On a system level, (7.85) and (7.91) fulfil an important function by coupling electrical and mechanical systems. Electrical power (EI) is converted into mechanical power ($T\omega$), or vice versa, in the air gap of the machine. However, the air gap power is not available at the electrical terminals and the shaft of the machine because of losses. As with any conversion component some of the energy is lost as heat. On the electrical side, resistive heating takes place in the copper windings and the process of magnetisation also has some losses associated with it. On the mechanical side, the rotor and shaft experience energy loss due to bearing friction and air friction. This power flow process is visualised in Fig. 7.28, for both motor and generator operation.

We see that for motor operation, we can write the equations

$$P_{\text{elec}} = P_{\text{field}} + P_{\text{brush}} + P_{\text{cu\_a}} + P_{\text{dev}} \tag{7.100}$$

$$P_{\text{dev}} = P_{\text{shaft}} + P_{\text{wf}}. \tag{7.101}$$

Naturally, some of the terms in these equations could fall away, depending on the complexity of the model or the machine used. For example, when a dc machine is used that uses permanent magnets to establish the field in the area of the armature, the field losses will be zero. It is also true that very often, the losses in the brushes are deemed to be small enough to be ignored and are often left out of the model. It is important to note in the model that the term P_{dev} refers to the power that is converted from electrical power to mechanical power.

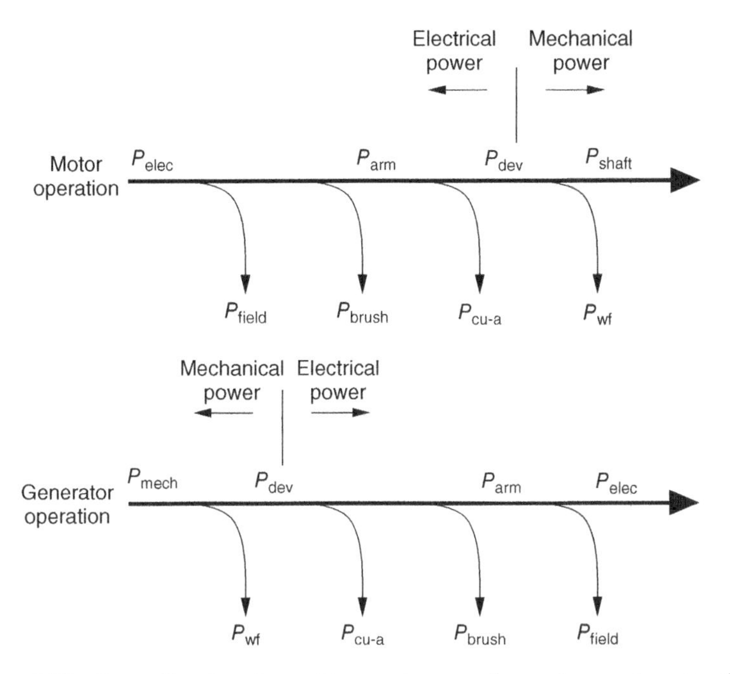

Figure 7.28 Power flow in a dc machine: motor operation and generator operation.

We can write the same set of equations for the dc machine operating as a generator. In this case, the input power is in the form of mechanical energy and it delivers electrical energy. We can express the power equations as

$$P_{mech} = P_{wf} + P_{dev} \tag{7.102}$$

$$P_{dev} = P_{cu\_a} + P_{brush} + P_{field} + P_{elec}. \tag{7.103}$$

This is in essence the same set of equations as those derived for the motoring operation; however, the order of the terms is slightly adjusted to better represent the operation of the system. Here, we have chosen to represent the input power as P_{mech} and not as P_{shaft}, but both terms are correct. We have chosen to name the input power P_{mech} to emphasise the fact that for generator operation, the input power is mechanical energy; for all practical purposes (and for the remainder of this book), the terms P_{mech} and P_{shaft} will be used interchangeably in the context of electrical machinery.

We can now define the efficiency of the machine as the ratio between the output power and the input power. As with all systems, efficiency is defined as

$$\eta = \frac{P_{out}}{P_{in}}. \tag{7.104}$$

However, for the dc machine operating as a motor, with the definitions of Fig. 7.28, we can also write

$$\eta = \frac{P_{\text{shaft}}}{P_{\text{elec}}} = \frac{P_{\text{mech}}}{P_{\text{elec}}}, \tag{7.105}$$

while for the machine operating as a generator, the efficiency becomes

$$\eta = \frac{P_{\text{elec}}}{P_{\text{mech}}} = \frac{P_{\text{elec}}}{P_{\text{shaft}}}. \tag{7.106}$$

7.7 DC MACHINE AS A COMPONENT IN A SYSTEM

In many applications where machines are used, we are not as interested in the precise operation of the machine but would prefer to treat the machine as a sub-system of the whole. By combining some of the equations discussed and with our understanding of the basic operation of the machine, we can now represent the machine as shown in Fig. 7.29.

As discussed earlier in this chapter, an electric machine is an electromechanical device used for energy conversion from electrical to mechanical and vice versa. In addition, it is also possible to reverse the rotation direction of a machine. In the case of a simple dc machine, the machine will turn in the other direction if the voltage polarity on the terminals is reversed. This gives us four possibilities for operating the same machine, which can be represented as four quadrants, as shown in Fig 7.30.

In an electric vehicle propulsion system, the electrical machine is operated in all four quadrants. The electrical machine can be designed to process supplied energy and to process power or torque to the transaxle driving forward and reversing the vehicle. The machine also processes the power flow in the reverse direction during regeneration when the vehicle is braking converting mechanical energy from the wheels into electrical energy. The term motor is used for the electric machine when energy is converted from electrical to mechanical, while the term generator is used when the power flow is in the opposite direction with the machine converting mechanical energy into electric energy. The breaking mode in electric machines is referred to as *regenerative breaking*. There are electrical, mechanical and magnetic losses during the energy conversion process in either direction in an electrical machine, which affects the conversion efficiency. Some energy is always lost from the system for any energy conversion process. However, the conversion efficiency of electric machines is typically high compared to that of other types of energy conversion devices.

To understand four quadrant operation, consider a permanent magnet machine that is powered by a buck converter as shown in Fig 7.31. We are connecting two voltage sources to each other. On the one hand we have the ideal dc voltage source, while the induced voltage of the dc motor is the load. By changing the duty cycle, the average voltage and the speed are adjusted. Because of the diode, the current can only flow from left to right.

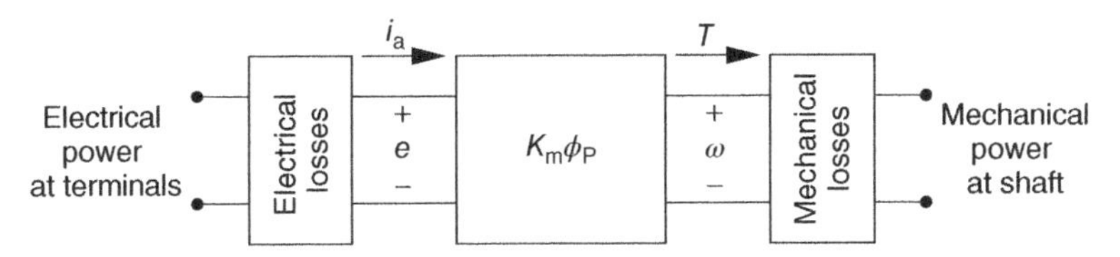

Figure 7.29 A dc machine as a component in a system.

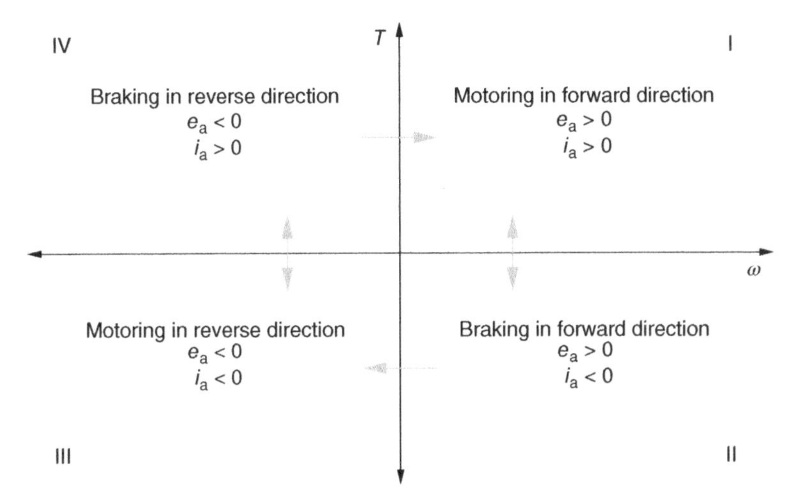

Figure 7.30 The four quadrants of machine operation.

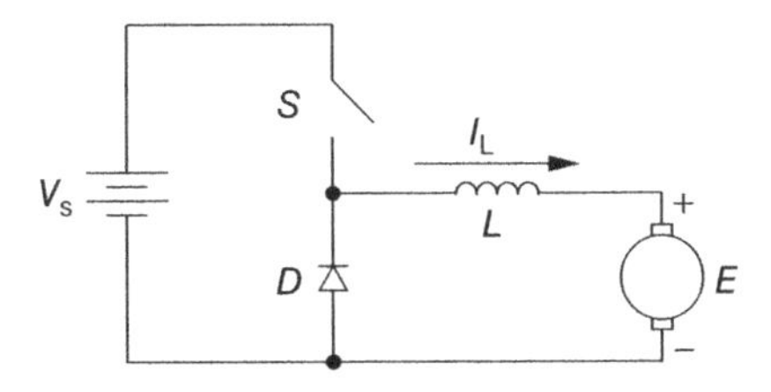

Figure 7.31 Buck converter delivering power to a dc motor.

If we can change the direction of the current then the power flow will be reversed. Then the dc machine would function as a generator that delivers power to the ideal voltage source. When the positions of the diode and switch are interchanged, the dc machine delivers energy to a boost converter that has a battery as a load as is shown in Fig 7.32

The buck and the boost converters cover two quadrants of operation. The two functions can be combined in one converter by using two switches with anti-parallel diodes. Changing the direction of the current changes the torque polarity, and it

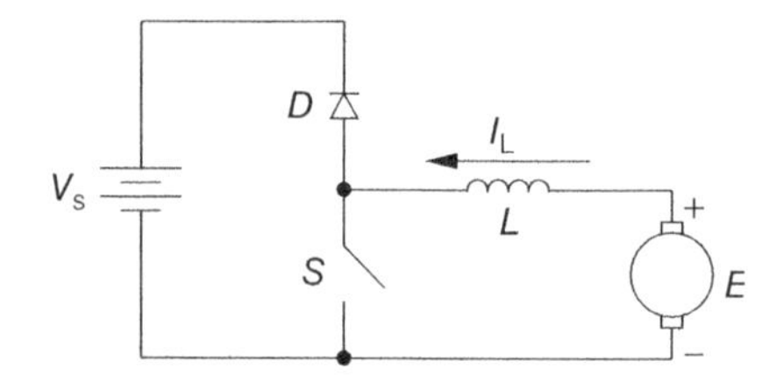

Figure 7.32 dc generator delivering power to a boost converter.

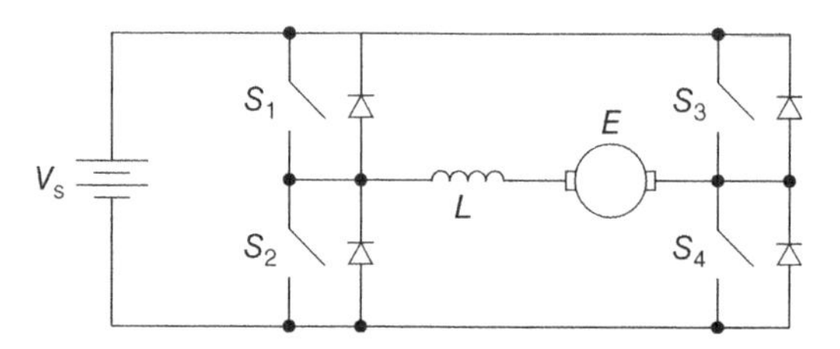

Figure 7.33 Four quadrant dc drive.

becomes possible to alternate between motor and generator operation. One can use this buck–boost converter in an electric vehicle. When the vehicle accelerates, the dc machine functions as a motor and becomes a generator during breaking. One drawback of the buck–boost converter is that the polarity of the voltage on the dc machine is always the same. If it is a permanent magnet machine then if would only rotate in one direction and the electric vehicle would not be able to reverse. If we use the double buck–boost converter in Fig 7.33 then we can reverse the voltage polarity. It seems that for every quadrant, we need one switch and one diode.

7.8 HIGHLIGHTS

- The voltage induced in a conductor moving through an uniform magnetic field can be described as

$$e = (\mathbf{v} \times \mathbf{B}) \cdot \mathbf{l}.$$

- In an ideal case, where $\mathbf{v} \perp \mathbf{B}$, the induced voltage can be written as

$$e = B\,v\,\ell.$$

where ℓ is the length of the conductor in the direction of the vector $\mathbf{v} \times \mathbf{B}$.

- The force on a current-carrying conductor can be described as

$$\mathbf{f} = i\,(\mathbf{l} \times \mathbf{B}).$$

- In an ideal case where $\mathbf{l} \perp \mathbf{B}$, the expression can be simplified to

$$f = B\,i\,\ell.$$

- The direction of the cross product result, and therefore the force on a current-carrying conductor in a magnetic field or the voltage induced in a conductor moving through a magnetic field, can be found with the right-hand rule.
- Motor operation of the simple dc machine is described by the following equation

$$P_{\text{elec}} - P_{\text{loss}} = P_{\text{mech}} \quad \longrightarrow \quad V_{\text{in}} i_{\text{a}} - i_{\text{a}}^{2} R = Fv,$$

  and during generator operation

$$P_{\text{elec}} + P_{\text{loss}} = P_{\text{mech}} \quad \longrightarrow \quad V_{\text{in}} i_{\text{a}} + i_{\text{a}}^{2} R = Fv.$$

- A single loop electrical machine operating with two magnets as field and a simple commutator will produce an induced voltage

$$e(\theta) = \omega\,B\,\ell\,w|\cos(\theta)|$$

  and a torque

$$T = -B\,I\,\ell\,w|\cos(\theta)|.$$

- The operation of a dc machine is much improved by using an uniform air gap, many conductor turns, as well as more than two magnetic poles.
- The induced voltage of a dc machine can be calculated as

$$e = K_{\text{m}}\,\phi_{\text{p}}\,\omega,$$

  where

$$K_{\text{m}} = \frac{ZP}{\pi a},$$

  here Z is the total number of conductors, P the number of magnetic poles and a the number of parallel current paths. The difference between the emf of the motor and the applied external voltage is the voltage drop across the armature resistance.

$$e_{\text{motor}} = V_{\text{in}} - i_{\text{a}} R_{\text{a}}.$$

- Likewise, the torque produced can be calculated as

$$T = K_{\text{m}}\,\phi_{\text{p}}\,I.$$

- dc machines with permanent magnets have a constant value for ϕ_p. The field is controllable when an electromagnet is used and the equations become

$$e_{\text{motor}} = K_m K_\phi i_f \,\omega$$

and

$$T_{\text{motor}} = K_m K_\phi i_f \,\omega,$$

where K_ϕ is a constant and i_f is the field winding current.
- The speed of a dc machine can be increased beyond the rated speed by using flux weakening.

PROBLEMS

7.1 Explain how the speed of the conductor bar determines whether the simple linear machine operates as a motor or as a generator.

7.2 Why is it possible to get a higher speed on the shaft when one uses a dc motor with field control instead of a permanent magnet dc motor?

7.3 Given the following parameters for the linear motor shown in Fig. 7.9: $V_{in} = 10$ V, $i_a = 1$ A, $B = 1$ T, $R = 10\,\Omega$ and $\ell = 0.5$ m, at what speed does the bar move?

7.4 Given the following parameters for the linear motor shown in Fig. 7.9: $V_{in} = 10$ V, $B = 1$ T, $R = 10\,\Omega$ and $\ell = 0.5$ m: if a 100 g bar accelerates at 1 m/s^2, at what speed does the bar move?

7.5 Given the following parameters for the linear motor shown in Fig. 7.9: $V_{in} = 10$ V, $B = 1$ T, $R = 10\,\Omega$ and $\ell = 0.5$ m, what is the terminal velocity of the bar?

7.6 Use the following parameters for the linear generator shown in Fig. 7.10: $V_{in} = 10$ V, $B = 1$ T, $R = 10\,\Omega$ and $\ell = 1$ m. Owing to the movement of the bar 2 W of power is delivered into the voltage source. At what speed does the bar move?

7.7 Calculate the full load current of a 580 V, 300 kW (mechanical power), four-pole DC motor at full load. The machine is 97% efficient.

7.8 A DC generator delivers 120 A at 300 V to an electrical load. If the machine is 92% efficient and the shaft speed is 1300 rpm, calculate the torque.

7.9 Given the following parameters for the single-turn rotating machine: $n = 1000$ rpm, $B = 1.5$ T, $w = 10$ cm, $\ell = 15$ cm. Calculate the average voltage between terminals 1 and 2.

7.10 Given the following parameters for the above single-turn rotating machine: $B = 1.5$ T, $w = 10$ cm and $\ell = 15$ cm. If a 10 Ω resistor is connected between terminals 1 and 2, how fast should the machine rotate (in rpm) to generate 1 W of power?

7.11 A permanent magnet motor has a machine constant $K_m = 50$, the magnet flux $\phi_P = 50$ mWb and an armature resistance $r_a = 1$ Ω and is fed by a 200 V supply. A 40 Nm load is connected to the motor. What is the maximum speed at which the machine can turn?

7.12 A dc motor with a separately excited field has a machine constant $K_m = 50$ and an armature resistance $r_a = 1$ Ω and is fed by a 40 A supply.

1. The maximum field of the separately excited machine also needs to be 30 mWb as stated previously. Calculate the value of the field constant K_ϕ if the field current is 1.5 A at this value.
2. What should the field current be to run the motor at a speed of 1000 rpm?

7.13 A permanent magnet dc machine runs? with? no load at 4000 rpm when connected to a 500 V supply drawing a 3 A current.

1. The shaft of the machine is clamped and is, therefore, stationary. A dc converter supplies 20 A to generate 100 Nm of torque. What is the voltage that the dc converter has to supply?
2. At what speed should the machine run if it would function as a generator supplying 10 A into a 50 V supply?

7.14 An electric vehicle needs 50 kW to maintain the designed top speed of 160 km/h. The dc electric motor is connected to a differential that reduces the speed by a ratio of 3 to the wheels that have a 50 cm diameter.

1. Find the maximum speed (in rpm) at which the dc motor would be able to run. The dc motor, which has a separately excited field, has a armature machine constant $K_m = 50$, a field constant $K_\phi = 0.02$ and an armature resistance $r_a = 0.2$ Ω. The maximum field current is 2 A, the maximum armature current is 100 A and a dc converter supplies a variable dc voltage up to 600 V.
2. Find the field current that is needed if the motor draws the maximum armature current at full speed.
3. Find the voltage that the dc converter has to supply at full speed.
4. Calculate the maximum torque that the dc motor can give. Up to what speed of the vehicle will the maximum torque be available?

FURTHER READING

Chapman S.J., *Electric Machinery Fundamentals*. 2nd edition, McGraw-Hill, Singapore, 1991.

Nasar S.A., *Handbook of Electrical Machines*. 1st edition, McGraw-Hill, New York, 1987.

Wildi T., *Electric Machines, Drives, and Power Systems*. 5th edition, Prentice Hall, Upper Saddle River, New Jersey, 2002.

CHAPTER 8

AC MACHINES

8.1 INTRODUCTION

Ac machines have been widely used for more than a hundred years. Without them, the electrical power system would have not been possible. At power stations, the mechanical power from a steam turbine is converted into electricity by synchronous machines. Induction machines then convert the electrical power back into mechanical power in factories, offices and homes. Typical loads are pumps, fans and air conditioners.

The stators of these two machines are very similar. In this chapter, we first discuss how a rotating magnetic field is generated by a stator. Two types of rotors are used in ac machines. In synchronous machines, the rotor turns with the same speed as that of the stator magnetic field. In induction machines, the rotor always turns more slowly than the stator magnetic field; therefore, these machines are sometimes called *asynchronous machines*.

8.2 THREE-PHASE AC ELECTRICAL PORT

An ac machine converts electrical power into mechanical power. It is a bidirectional device, so that the power can also flow in the opposite direction. The source or the sink of electrical power is three-phase ac. If the ac comes from the grid, then

The Principles of Electronic and Electromechanic Power Conversion: A Systems Approach, First Edition.
Braham Ferreira and Wim van der Merwe.

the frequency and voltage are fixed. Thanks to power electronics, it has become possible to vary the voltage amplitude and frequency.

In Chapter 6 and Chapter 7, we discussed power electronics and the control of dc machines. We saw that when we connect a permanent magnet dc machine to a buck and boost converter, it is possible to control the speed by adjusting the output voltage. The power electronics circuit is shown in Fig. 8.1. The two switches are opened and closed alternatively. If both of them are closed, a short circuit is created, which is not desirable. When the current flows from left to right, S_1 carries the current when it is closed and the bottom diode conducts when the switch is opened, whereas if the current to the machine is in the other direction, from right to left, then switch S_2 and the top diode become conducting.

The duty cycle is the variable that makes this possible because the output voltage is the product of the source voltage:

$$\bar{v}_o = d\,V_s. \tag{8.1}$$

In dc power supplies and dc machines, the duty cycle is kept constant, perhaps changing a little bit when the controller makes some adjustments to keep the voltage at the right value. But, because we have full control over the duty cycle it is possible to modulate it to generate any waveform and it is possible to generate ac with the two switch buck–boost converter. This two switch converter structure, when used in ac systems is given a special name; a *phase arm*.

With the one-phase arm in Fig. 8.1, it is not possible to generate ac because the output voltage is always positive. For ac, positive and negative currents are required. The solution is to connect a load between the two-phase arms, as shown in Fig. 8.2. Now, if we can make the average voltage at A larger than B, the current i will be positive, whereas if the average voltage at A is smaller than B, the current will be negative.

The next step is to add a third-phase arm making it possible to generate three-phase ac. Fig. 8.3 shows the circuit; by modulating the switching duty cycles, it is possible to generate three-phase currents i_a, i_b and i_c. So a dc/ac converter can be built by using two or three-phase arms. This is called an *inverter*.

In Fig. 8.4, a three-phase arm converter is connected to an ac electrical machine. The windings have inductance L and the motion-induced phase voltages are E_a, E_b and E_c. As the top and bottom switches are switched alternately, for example,

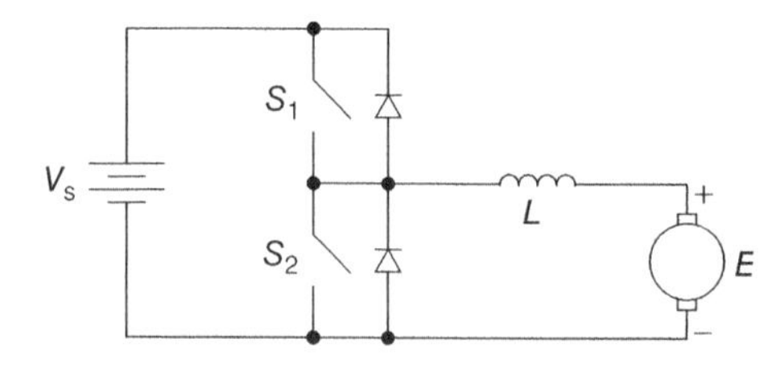

Figure 8.1 Two-quadrant dc drive based on the buck and boost converter.

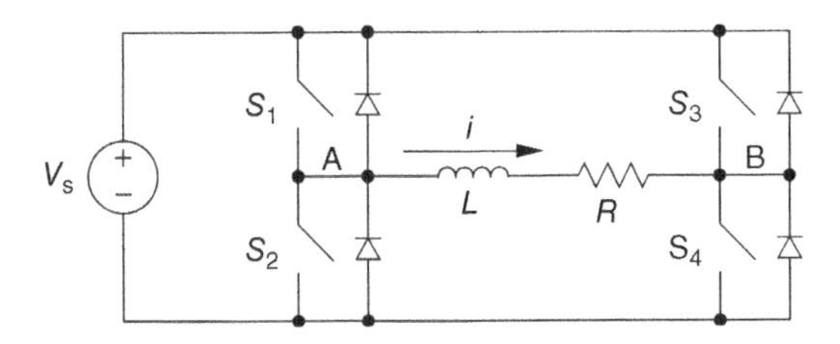

Figure 8.2 A two-phase arm converter.

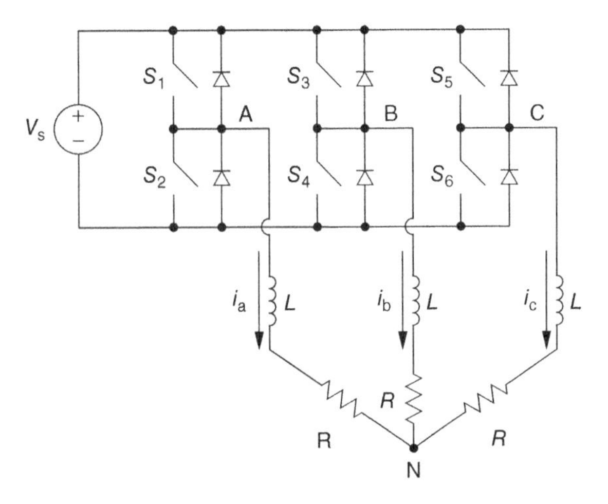

Figure 8.3 A three-phase arm converter.

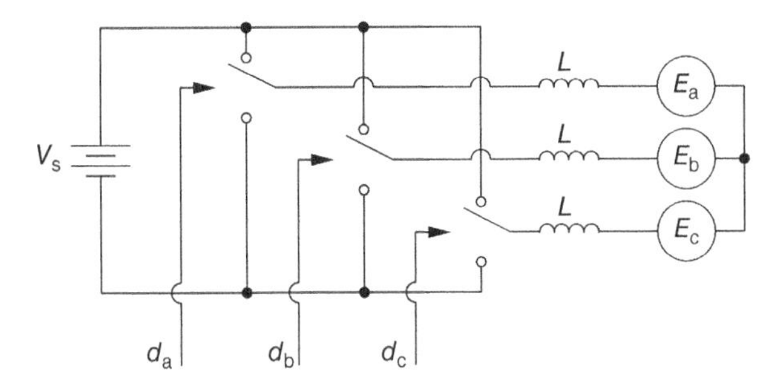

Figure 8.4 Three-phase ac inverter connected to a machine.

when S_1 is open then S_2 is closed and vice versa, for this reason we can visualise the circuit operation by replacing the phase arms with toggle switches.

We can control the voltages to become three-phase voltages that can be applied to an ac machine if we use duty cycles that are 120° phase-shifted sinusoids:

$$d_a(t) = \frac{1}{2} + \frac{m}{2}\cos(\omega t) \tag{8.2}$$

$$d_b(t) = \frac{1}{2} + \frac{m}{2}\cos\left(\omega t + \frac{2}{3}\pi\right) \tag{8.3}$$

$$d_c(t) = \frac{1}{2} + \frac{m}{2}\cos\left(\omega t - \frac{2}{3}\pi\right), \tag{8.4}$$

where m is the modulation index, a value between 0 and 1 that determines the voltage amplitude. The average value of the phase voltages, that is, ignoring the effects of the switching process, is obtained by multiplying the duty cycles by the value of the source voltage

$$\overline{v}_a(t) = d_a(t)\,V_s \qquad \overline{v}_b(t) = d_b(t)\,V_s \qquad \overline{v}_c(t) = d_c(t)\,V_s. \tag{8.5}$$

In three-phase systems, this reduces to the following expression for line-to-line voltage

$$\begin{aligned}\overline{v}_{ab}(t) &= \overline{v}_a(t) - \overline{v}_b(t)\\ &= V_s\left(\frac{1}{2} + \frac{m}{2}\cos(\omega t) - \frac{1}{2} - \frac{m}{2}\cos\left(\omega t + \frac{2}{3}\pi\right)\right)\\ &= -2V_s\frac{m}{2}\sin\left(\omega t + \frac{2}{6}\pi\right)\sin\left(-\frac{2}{6}\pi\right)\\ &= V_s m\cos\left(\omega t - \frac{1}{6}\pi\right)\frac{\sqrt{3}}{2}\\ &= \sqrt{3}V_s\frac{m}{2}\cos\left(\omega t - \frac{1}{6}\pi\right)\end{aligned} \tag{8.6}$$

So the line voltage is $\sqrt{3}$ times the phase voltage as we would expect. In Europe, the three-phase ac grid line voltage is 400 V, 50 Hz, and in the United States it is 208 V or 480 V, 60 Hz. Therefore, if one wants to connect an ac machine directly to the grid, the voltage and frequency are fixed. This is possible and many machines, in ventilation systems, are for example, directly connected to the grid. In many applications, we need to control the speed and torque of the machines. As can be seen from the above equation, we can have full control of the voltage amplitude and frequency if we use a three-phase inverter.

8.3 AC MACHINE STATOR

The ac machine stator is the same for both induction and synchronous machines. In this section, we discuss the existence of a rotating magnetic field in the stator and the basic construction of an ac machine stator.

8.3.1 Rotating Magnetic Field

Let us start the formal discussion of the ac machine by proving that a rotating magnetic field, with a constant magnitude, can be created in the air gap. What is more, this rotating magnetic field is created using only the three-phase ac currents and a well-thought through distribution of the windings around the stator.

We start with the standard definition of three-phase currents:

$$i_a(t) = I_m \cos(\omega t + 0) \text{ A} \tag{8.7}$$

$$i_b(t) = I_m \cos\left(\omega t + \frac{2}{3}\pi\right) \text{ A} \tag{8.8}$$

$$i_c(t) = I_m \cos\left(\omega t - \frac{2}{3}\pi\right) \text{ A.} \tag{8.9}$$

Consider the stator shown in Fig. 8.5, where only the a-phase winding is shown. From Chapter 4, we know that

$$B(t) \propto i(t)\,. \tag{8.10}$$

We can write an expression for the magnetic flux in the air gap due to the current i_a as

$$B_a(t) = B_m \cos(\omega t + 0) \text{ T.} \tag{8.11}$$

We can also assign a direction to this magnetic flux density according to the position of the winding. Using the definitions from Fig. 8.5, for the sake of simplicity we will not write the unit (T) in each equation while formulating our proof,

$$B_a(t) = B_m \cos(\omega t + 0)\,\hat{\mathbf{x}}. \tag{8.12}$$

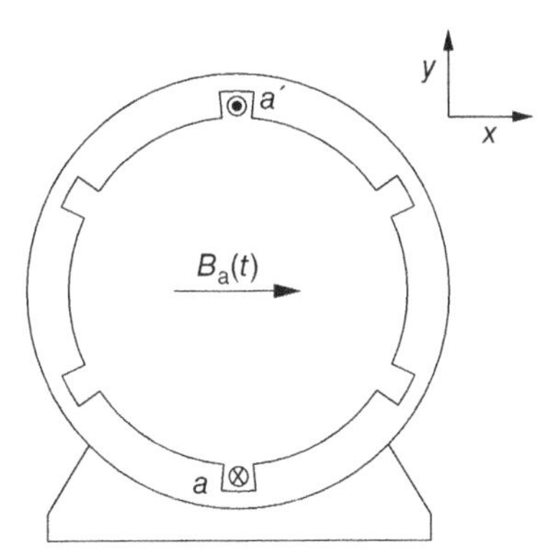

Figure 8.5 The magnetic flux created by one phase of windings—phase a.

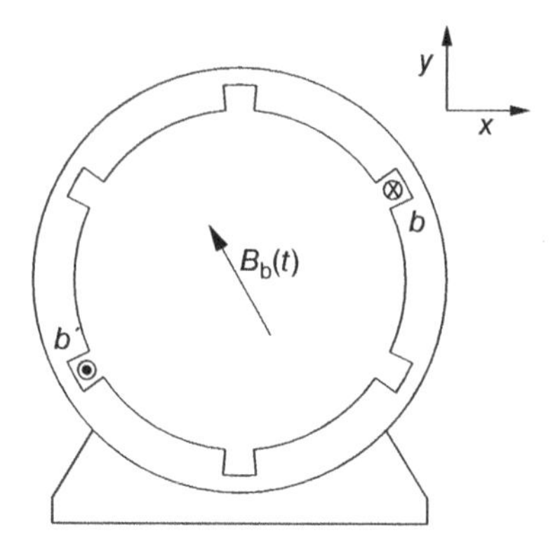

Figure 8.6 The magnetic flux created by one phase of windings—phase *b*.

Now, consider the *b*-phase winding as shown in Fig. 8.6. We can describe the magnitude and direction of the magnetic flux density created by the winding as

$$B_b(t) = -\frac{1}{2}B_m \cos\left(\omega t + \frac{2}{3}\pi\right)\hat{\mathbf{x}} + \frac{\sqrt{3}}{2}B_m \cos\left(\omega t + \frac{2}{3}\pi\right)\hat{\mathbf{y}}. \tag{8.13}$$

Now, for the *c*-phase winding in Fig. 8.7, the flux density is

$$B_c(t) = -\frac{1}{2}B_m \cos\left(\omega t - \frac{2}{3}\pi\right)\hat{\mathbf{x}} - \frac{\sqrt{3}}{2}B_m \cos\left(\omega t - \frac{2}{3}\pi\right)\hat{\mathbf{y}}. \tag{8.14}$$

The question is, however, what will the total flux density distribution be like if we have all three windings operational at the same time? As the magnetic flux equations we use here are all linear, we can use the superposition principle to find

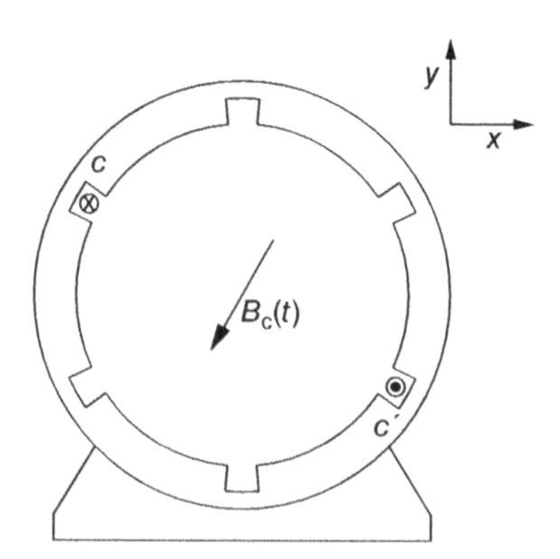

Figure 8.7 The magnetic flux created by one phase of windings—phase *c*.

the total flux density distribution inside the air gap. Therefore,

$$B_t(t) = B_a(t) + B_b(t) + B_c(t). \tag{8.15}$$

Let us first consider only the $\hat{\mathbf{x}}$-components

$$B_{t-x}(t) = \left[B_m \cos(\omega t) - \frac{1}{2}B_m \cos\left(\omega t + \frac{2}{3}\pi\right) - \frac{1}{2}B_m \cos\left(\omega t - \frac{2}{3}\pi\right)\right]\hat{\mathbf{x}}. \tag{8.16}$$

Let us use the trigonometric identities

$$\cos(\alpha + \beta) = \cos(\alpha)\cos(\beta) - \sin(\alpha)\sin(\beta) \tag{8.17}$$

$$\cos(\alpha - \beta) = \cos(\alpha)\cos(\beta) + \sin(\alpha)\sin(\beta) \tag{8.18}$$

to rewrite this expression as

$$\begin{aligned}B_{t-x}(t) = \big[&B_m \cos(\omega t)\\ &-\frac{1}{2}B_m \cos(\omega t)\cos\left(\frac{2}{3}\pi\right) - \frac{1}{2}B_m \sin(\omega t)\sin\left(\frac{2}{3}\pi\right)\\ &-\frac{1}{2}B_m \cos(\omega t)\cos\left(-120^\circ\right) + \frac{1}{2}B_m \sin(\omega t)\sin\left(\frac{2}{3}\pi\right)\Big]\hat{\mathbf{x}}\\ =&\frac{3}{2}B_m \cos(\omega t)\ \hat{\mathbf{x}}.\end{aligned} \tag{8.19}$$

We can do the same for the $\hat{\mathbf{y}}$-component of the flux

$$\begin{aligned}B_{t-y}(t) = &\left[\frac{\sqrt{3}}{2}\cos(\omega t)\cos\left(\frac{2}{3}\pi\right) - \frac{\sqrt{3}}{2}\sin(\omega t)\sin\left(\frac{2}{3}\pi\right)\right.\\ &\left.-\frac{\sqrt{3}}{2}\cos(\omega t)\cos\left(\frac{2}{3}\pi\right) - \frac{\sqrt{3}}{2}\sin(\omega t)\sin\left(\frac{2}{3}\pi\right)\right]\hat{\mathbf{y}}\\ =&-\frac{3}{2}\sin(\omega t)\ \hat{\mathbf{y}}.\end{aligned} \tag{8.20}$$

When the $\hat{\mathbf{x}}$ and the $\hat{\mathbf{y}}$-components are added, we get a complete expression for the magnetic field as

$$\boxed{B_t(t) = \frac{3}{2}B_m \cos(\omega t)\ \hat{\mathbf{x}} - \frac{3}{2}B_m \sin(\omega t)\ \hat{\mathbf{y}}.} \tag{8.21}$$

If we investigate this equation a bit more carefully, we see that it describes a vector with constant magnitude, $\frac{3}{2}B_m$ that rotates in a clockwise direction. The vector will complete f rotations in a second.

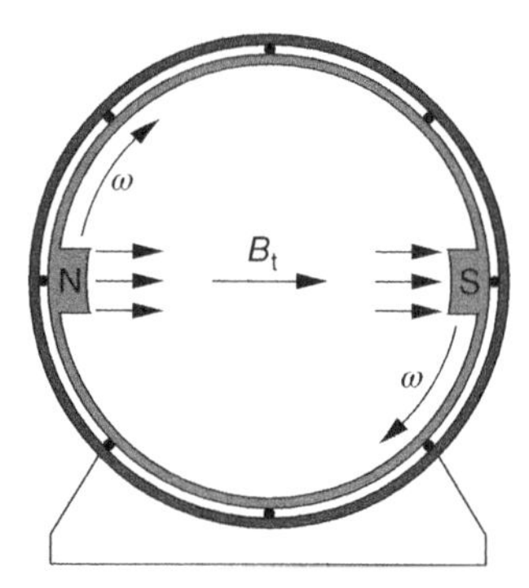

Figure 8.8 An alternate view of the rotating magnetic field using a free rotating field disc.

We can visualise this rotating magnetic field by constructing an imaginary machine, as shown in Fig. 8.8. Let us place a rotating field inside the stator. This field is free to rotate inside the stator and consists of two poles, one north pole and one south pole. If we rotate this field at $60f$ rpm in a clockwise direction, we would observe the same magnetic field in the air gap.

8.3.2 Reversing the Direction of Rotation

In the example above, we have seen that the magnetic field is rotating in a clockwise direction and that the winding sequence, in a clockwise direction, is

$$a - b' - c - a' - b - c'.$$

If we go through the proof of the magnetic field concept but interchange any two phases, we would see that the direction of rotation reverses.

Example 8.1 Consider the stator of Fig. 8.9 but with the winding sequence (in a clockwise direction)

$$a - c' - b - a' - c - b'.$$

Calculate the direction of rotation.

Solution

The three phases would create the following magnetic fields

$$B_a(t) = B_m \cos(\omega t)\ \hat{\mathbf{x}}$$

$$B_b(t) = -\frac{1}{2} B_m \cos\left(\omega t + \frac{2}{3}\pi\right)\ \hat{\mathbf{x}} - \frac{\sqrt{3}}{2} B_m \cos\left(\omega t + \frac{2}{3}\pi\right)\ \hat{\mathbf{y}}$$

$$B_c(t) = -\frac{1}{2} B_m \cos\left(\omega t - \frac{2}{3}\pi\right)\ \hat{\mathbf{x}} + \frac{\sqrt{3}}{2} B_m \cos\left(\omega t - \frac{2}{3}\pi\right)\ \hat{\mathbf{y}}.$$

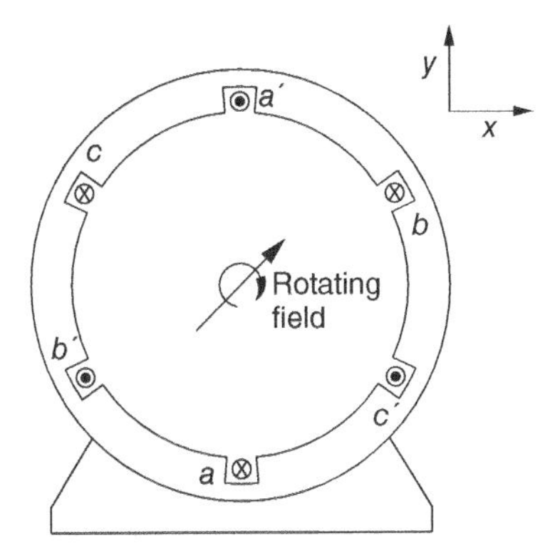

Figure 8.9 The magnetic flux created by one phase of windings—combination of phases a, b and c.

Using the trigonometric identities, we find

$$B_{m-x}(t) = \frac{3}{2}B_m \cos(\omega t)\ \hat{\mathbf{x}}$$

and

$$B_{m-y}(t) = \frac{3}{2}B_m \sin(\omega t)\ \hat{\mathbf{y}}$$

and when combined, we find

$$B_m(t) = \frac{3}{2}B_m \cos(\omega t)\ \hat{\mathbf{x}} + \frac{3}{2}B_m \sin(\omega t)\ \hat{\mathbf{y}}.$$

Therefore, the rotation would be in the anticlockwise direction.

8.3.3 Increasing the Number of Poles

In the discussion of the stator with one set of windings per phase, we saw that we can generate two magnetic poles rotating at $60f$ rpm. Consider the stator as shown in Fig. 8.9 with the winding sequence (in the clockwise direction)

$$a - b' - c - a' - b - c'.$$

What would happen if we repeated this sequence twice in the same stator? Let us create a stator with the winding sequence (in the clockwise direction)

$$a - b' - c - a' - b - c' - a - b' - c - a' - b - c'.$$

Such a stator is shown in Fig. 8.10.

If we now apply three-phase currents to this stator, we will be able to create four magnetic poles, two south poles and two north poles. Each sequence of $a - b' - c -$

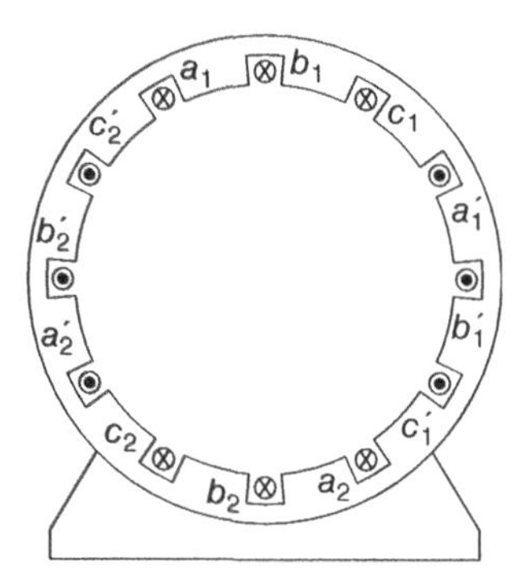

Figure 8.10 A three-phase stator with four poles.

$a' - b - c'$ will create one north and one south pole. The difference is, however, that in Fig. 8.9 the sequence is spread over 360° mechanical degrees, while in Fig. 8.10, the sequence is spread over 180° mechanical degrees. Our imaginary machine in Fig. 8.8 can be modified to become a four pole machine as illustrated Fig. 8.11. If we define an electrical degree (denoted as θ_e) such that there are 360 electrical degrees between one north pole and the closest next north pole, then we can relate the electrical degrees to the mechanical degrees (θ_m) and the number of poles (p) as

$$\boxed{\theta_e = \frac{p}{2}\theta_m.} \tag{8.22}$$

In this case, we see that each pole moves only halfway around the stator in every electrical cycle. Therefore, each pole will make one complete rotation around the stator $\frac{1}{2}f$ times per second, or stated differently, it will rotate at a speed of $30f$ rpm. We can generalise this observation to describe the rotational speed of the magnetic field, for any three-phase stator with any number of poles, as

$$\boxed{n_s = \frac{120f}{p} \text{ (rpm).}} \tag{8.23}$$

We call this speed the *synchronous speed*. It is true that once a machine is manufactured, it is (for most machines) not possible to change the number of magnetic poles, and the only way to change the synchronous speed is to change the frequency of the three-phase supply.

8.3.4 Flux Created in the Air Gap

Our discussion of the rotating air gap has so far been a very basic version of the truth. In fact, the full truth still keeps many PhD researchers and electrical machine

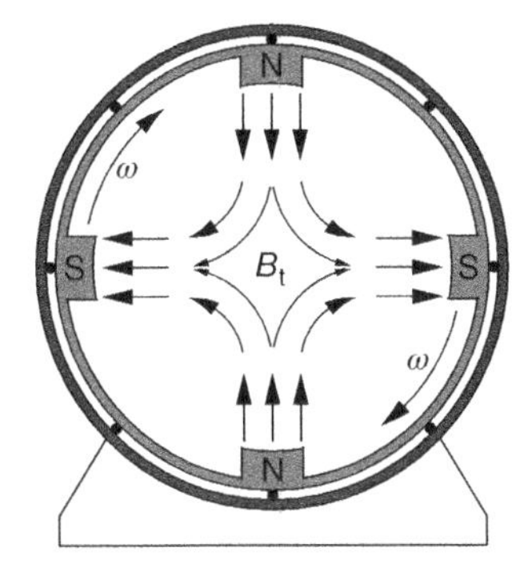

Figure 8.11 An alternative view of the four-pole rotating magnetic field using a free rotating field disc.

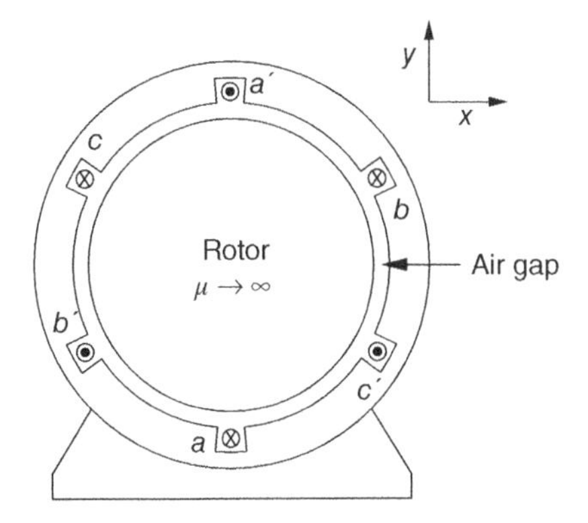

Figure 8.12 A three phase stator with an idealised rotor inside.

designers awake at night. For our purposes, let us take just a few more steps in the analysis of the rotating air gap.

The first step is to notice that, as we have seen in the analysis of the dc machine, a machine with a single turn will struggle to generate enough magnetomotive force (MMF) to really make a difference to the world. Therefore, we make allowances for our model to have N turns per slot. Secondly, we have the fact that, quite naturally if you come to think of it, an electrical machine with a stator but no rotor will not be of much use to anyone. Therefore, let us adjust the model slightly and add a simple cylindrical rotor.

Consider the machine shown in Fig. 8.12. Here, we make the simplification that the material used for the construction of the rotor, and the stator for that matter, is very magnetisable. In fact, let $\mu \to \infty$.

Now, let us revisit the discussion of the magnetic field in a toroid with an air gap described in Section 4.2.6. There we saw that the magnetic field strength in the air gap dominates the system, especially when $\mu_r \gg 1$, which is normally the case with well-designed electrical equipment. We have observed that we can make the approximation that, if H_g is the MMF in the air gap then

$$H_g \approx N\,i \tag{8.24}$$

and equivalently

$$B \approx \mu_0 \frac{N\,i}{\ell_g}, \tag{8.25}$$

where ℓ_g is the length of the air gap.

For the machine in Fig. 8.12, we can make the same approximation. Owing to the geometry of the stator and the rotor, it is more difficult to express mathematically, but the same principles as that discussed in Section 4.2.6 hold. The only difference is the fact that the flux now has to cross two air gaps. Therefore, for an electrical machine, we can make the approximation

$$\boxed{B \approx \mu_0 \frac{N\,i}{2\ell_g}} \tag{8.26}$$

and equivalently that the total MMF generated by the windings is shared equally by the two parts of the air gap.

Owing to the complexity of the system, we will not discuss the magnetic field in the air gap further using mathematical models. Instead, let us discuss a couple of finer points. Consider the finite element simulation results of a stator–rotor combination similar to that of Fig. 8.12, albeit with a much smaller air gap, shown in Fig. 8.13. Here, we can clearly see the existence of the rotating magnetic field in the air gap.

As a final comment, in the model used thus far, for example, from Fig. 8.13, we see that there is a large amount of space on the stator without any windings. Similarly, for the more astute among us, the magnetic field in the air gap changes rather rapidly. To qualify this statement, imagine that you can walk around the air gap with a magnetic flux meter. Let us take the case of Fig. 8.13(a) and we will start from the furthest right point of the air gap. As we walk in an anticlockwise direction around the air gap, we will measure the same magnetic field until we reach about 90°. This is evident from the fact that the magnetic field lines are equally spaced. When we cross the 90° mark, we suddenly measure the same magnetic field but in the opposite direction. If we graph the result, we will see a block wave.

As a rule, ac machines are designed to have a sinusoidal distribution of the magnetic field around the air gap. This is not a feat that is easily achieved or understood. However, it is mostly achieved with a clever distribution of the windings on the stator. For this reason, an ac machine stator is rarely as simple as the model shown in Fig. 8.13 but instead a complicated arrangement of windings as shown in Fig. 8.14.

For the rest of our discussion, let us assume that the machine is well designed and that the flux distribution in the air gap can best be described by the relationship

$$\boxed{B(\theta) = B_m \cos(p\,\theta + \varphi),} \tag{8.27}$$

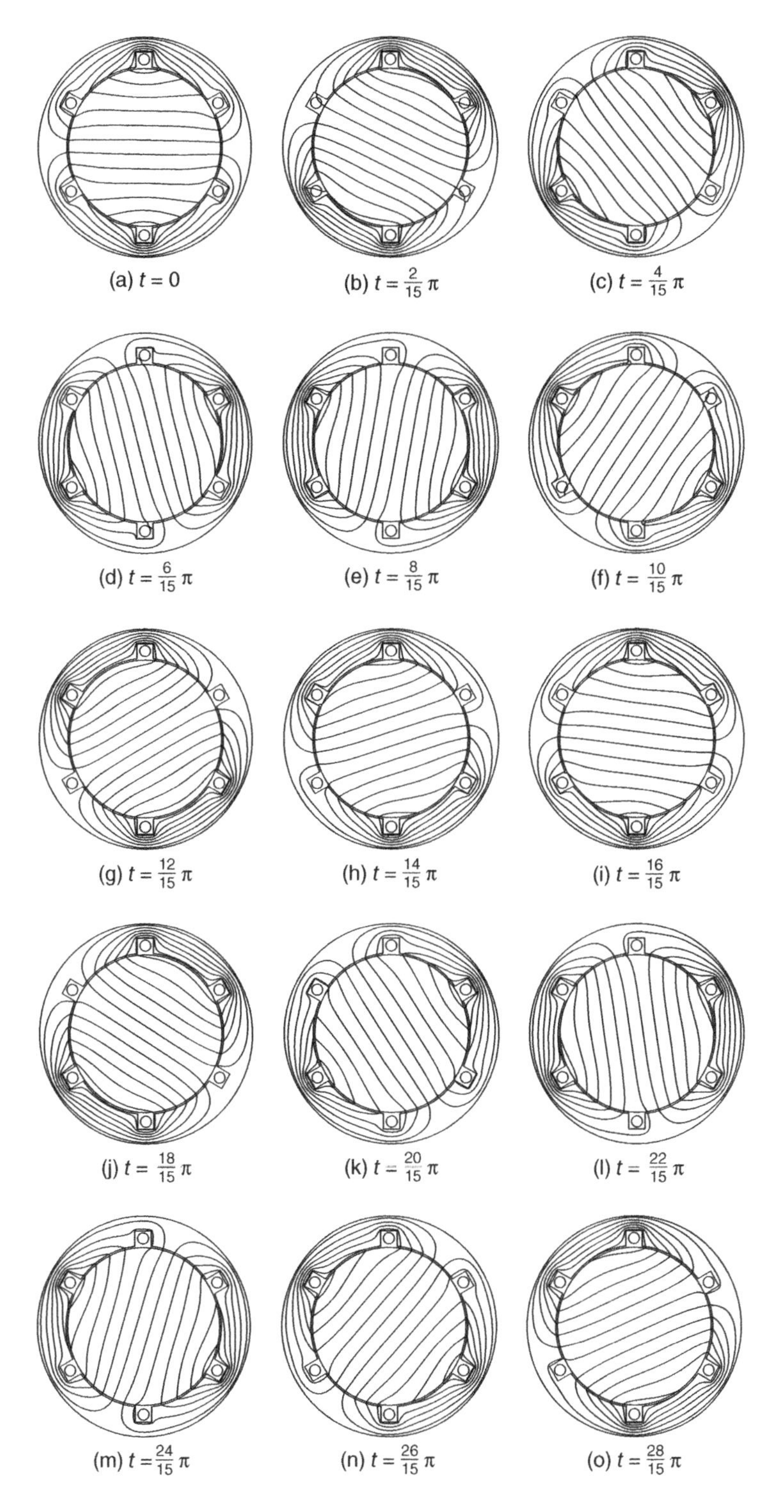

Figure 8.13 Simulation results showing the existence of the rotating magnetic field.

Figure 8.14 A photograph of a two-pole synchronous machine stator.

where P is the number of poles and φ is a constant, dependent on the machine orientation.

8.3.5 Induced Voltage in Three-Phase Stator Windings

Recall from our discussion in Chapter 4 that we have two equations which link the magnetic energy domain and the electrical energy domain

$$B(t) = \mu \frac{N\,i(t)}{\ell} \tag{8.28}$$

$$\text{and } e(t) = N \frac{d\phi(t)}{dt}. \tag{8.29}$$

We saw in the previous section how we can use the first equation to describe the existence of a rotating magnetic field in the air gap of a three-phase machine. We saw that all we need for this rotating field is sinusoidal three-phase currents and a correct spacing of the conductors around the stator. However, by using the same stator with the same spacing of the conductors, we can use the second equation to prove that a rotating magnetic field will produce a three-phase set of voltages in the stator conductors.

Consider the stator with a rotating magnetic field inside the volume. To investigate the induced voltages, let us make the following assumption: *the flux density is sinusoidally distributed along the air gap with the peaks aligned with the pole face centres*. Using the sign convention that positive flux is the flux leaving the north pole, we can express this assumption mathematically as

$$B(\theta) = B_m \cos(\theta). \tag{8.30}$$

Although this assumption is a simplification of reality, most designers of electrical machines spend a considerable amount of time and effort to get as close as possible to this ideal.

We can calculate the total flux per pole for our idealised magnetic field as

$$\phi_P = \int_{-\pi/2}^{\pi/2} B(\theta)\ \ell\ rd\theta = 2B_m\ \ell\ r, \tag{8.31}$$

where r is the radius of the stator at the air gap and ℓ is the axial length (depth into the page in Fig. 8.15) of the machine. If the winding indicated in Fig. 8.15 consists of N turns, then the total flux linked by this coil will vary as the magnetic field rotates inside the machine. The flux linkage with the coil can be expressed as

$$\lambda(\theta) = N\phi(\theta) = N\phi_P \cos(\theta). \tag{8.32}$$

If the magnetic field is rotating at a speed ω and we substitute $\theta = \omega t$, then we can calculate the induced voltage in the coil from (8.28) as

$$\begin{aligned} e(t) &= N\frac{d\phi(t)}{dt} \\ &= N\phi_P \frac{d}{dt} \cos(\omega t) \\ &= -\omega N\phi_P \sin(\omega t) = E_m \sin(\omega t). \end{aligned} \tag{8.33}$$

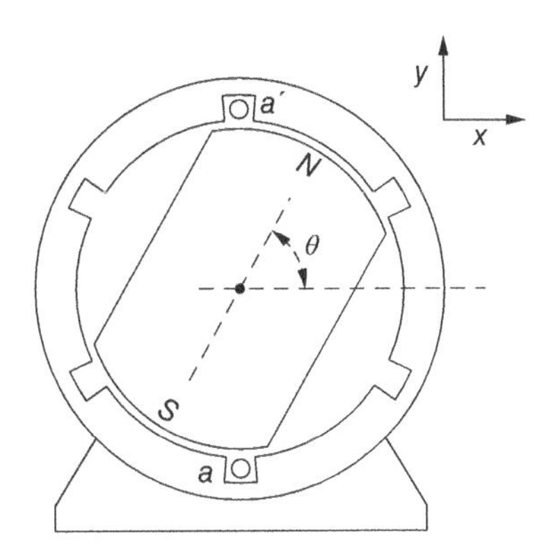

Figure 8.15 A rotating magnetic field inside a stationary coil.

Here, we have simply incorporated the negative sign into the definition of E_{m} (remember that the negative sign simply indicates which side of the coil is positive). Therefore,

$$E_m = -\omega N \phi_{\mathrm{P}} = -2\omega N \, B_m \, \ell \, r. \tag{8.34}$$

If we add the other two-phase windings into the stator, in a similar manner as shown in Fig. 8.9, the voltages induced in these windings will be phase shifted, from that in the a-phase, because of the physical distribution of the windings along the stator circumference. Therefore,

$$e_a(t) = E_m \sin(\omega t) \tag{8.35}$$

$$e_b(t) = E_m \sin\left(\omega t + \frac{2}{3}\pi\right) \tag{8.36}$$

$$e_c(t) = E_m \sin\left(\omega t - \frac{2}{3}\pi\right). \tag{8.37}$$

8.3.6 Increasing the Number of Poles

In Section 8.3.3, we saw that if the number of magnetic poles created by the stator windings increases, then the rotational speed of the field decreases. This relationship was found to be

$$n_{\mathrm{s}} = \frac{120 f}{p}. \tag{8.38}$$

Consider the stator and rotor combination shown in Fig. 8.16. In this case, we have the same stator as shown in Fig. 8.10, which is associated with a four-pole magnetic field, and a four-pole rotating magnetic field rotor. If we make the same

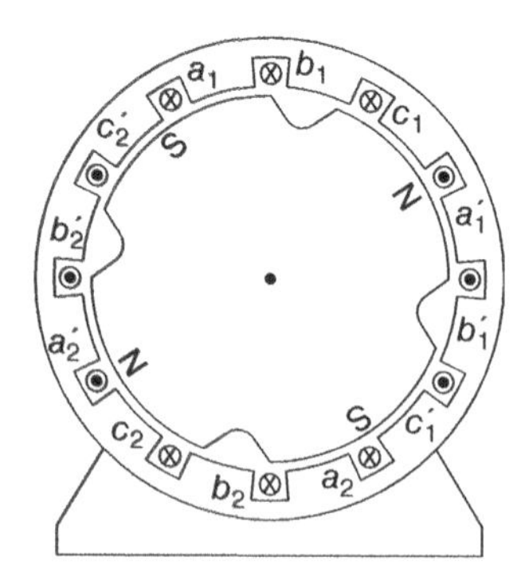

Figure 8.16 A rotating four-pole magnetic field inside a four-pole stator.

assumption as in the two-pole case, that the magnetic flux is sinusoidally distributed along the air gap, then the flux density distribution can be expressed as

$$B(\theta) = B_m \cos(2\theta) \tag{8.39}$$

and the flux per pole can be found as

$$\phi_P = \int_{-\pi/4}^{\pi/4} B_m \cos(2\theta)\ \ell\, r\, d\theta = B_m \ell\, r. \tag{8.40}$$

If we apply the same reasoning as with the two-pole case, we find the three induced voltages as

$$e_a(t) = E_m \sin(2\omega t) \tag{8.41}$$

$$e_b(t) = E_m \sin\left(2\omega t + \frac{2}{3}\pi\right) \tag{8.42}$$

$$e_c(t) = E_m \sin\left(2\omega t - \frac{2}{3}\pi\right), \tag{8.43}$$

where

$$E_m = -N\,\omega\,\phi_P = -2\omega N\, B_m\, \ell\, r. \tag{8.44}$$

From these results, we can make two important observations. The first is that the equation

$$n_s = \frac{120 f}{p} \tag{8.45}$$

holds for the induced voltages too. The second is that the peak magnitude of the induced voltage is the same for both the two-pole and the four-pole cases as the magnitude depends on the peak flux density, the geometric size of the machine and the rotational speed. From the argument, it can be seen that the difference in the flux per pole (due to the proportionally smaller pole face) does not influence the magnitude of the induced voltage as can be seen in (8.34) and (8.44).

8.3.7 Changing the Magnitude of the Induced Voltage

We can express the rms magnitude of the induced voltage as

$$E = \frac{1}{\sqrt{2}} E_m = \frac{2\omega N\, B_m\, \ell\, r}{\sqrt{2}}. \tag{8.46}$$

In general, the number of poles, the number of turns per coil and the physical size (r and ℓ) of the machine are fixed during the design and construction of the machine and cannot be varied later. This implies that once the machine is constructed, we can express the rms magnitude of the induced voltage with a generalised constant, k_e, as

$$E = k_e \, \omega \, B_m. \tag{8.47}$$

As it is not possible to change the value of k_e, the induced voltage can only be changed by either increasing the rotational speed or increasing the magnetic field on the rotor. However, there is a catch; we have established that

$$n = \frac{120f}{p} \tag{8.48}$$

therefore, we know that

$$\boxed{f \propto \omega.} \tag{8.49}$$

Therefore, if we have to keep the frequency of the induced voltage constant, something we have to do, in nearly all circumstances, the speed of rotation must be fixed. The only feasible way to change the induced voltage is therefore by changing the magnetic field on the rotor.

Example 8.2 A three-phase, six-pole machine is rotating at 850 rpm. The induced voltage is measured as 320 V (line-to-line). Calculate the frequency of the induced voltage. What will the induced voltage be if the rotational speed is increased such that the induced voltage has a frequency of 50 Hz?

Solution

At 850 rpm, the frequency of the induced voltage can be found as

$$f = \frac{np}{120} = 42.5\,\text{Hz}.$$

If the required frequency is 50 Hz, the rotational speed must increase to

$$n = \frac{120 \times 50}{6} = 1000\ \text{rpm}$$

which implies (as the relationship of (8.49) is linear) that

$$V_{\text{new}} = \frac{1000}{850} 320 = 376.5\,\text{V}$$

8.4 SYNCHRONOUS MACHINE

Synchronous machines are used widely in the modern electrical energy grid. Synchronous machines have the very desirable characteristic that the frequency of the stator currents and the rotational speed of the rotor are always related to one another by a fixed constant. This is especially beneficial in generators because the frequency of the generated ac waveform can be controlled simply by controlling the rotational speed of the rotor. This, in conjunction with the fact that any generator connected to the grid must operate at exactly the grid frequency, results in the fact that all large generators in thermal and kinetic energy power plants are synchronous generators. Figure 8.17 shows a stator of a large synchronous generator during assembly. Although this picture dates from the 1950s, the present day generator still looks, for all practical purposes, the same as it did back then.

A synchronous machine consists of a three-phase stator as discussed in Section 8.3 and a rotor that creates its own magnetic field, which is fixed with respect to the orientation of the rotor. In other words, the magnetic field created by

Figure 8.17 A large synchronous generator stator during assembly dating from the 1950s.

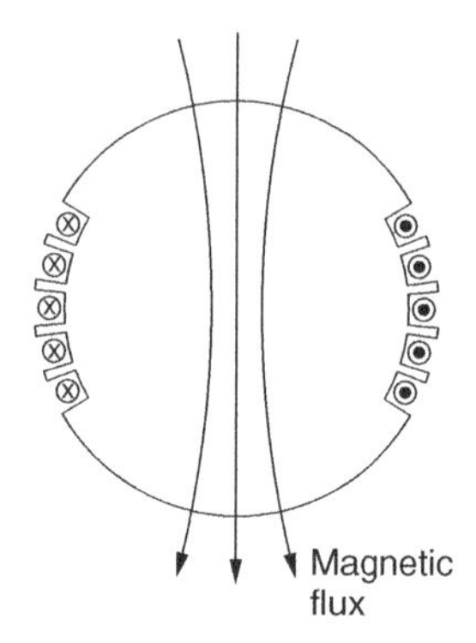

Figure 8.18 A two-pole wound rotor.

the rotor rotates *with* the rotor. A rotor with a two-pole field winding is shown in Fig. 8.18.

The synchronous machine operate with the two principles that we have discussed so far in this chapter. The stator is manufactured in such a manner that balanced three-phase currents in the stator windings will create a rotating flux vector in the air gap. Similarly, the field created by the rotor will induce balanced three-phase voltages in the stator windings if the rotor rotates in the air gap.

Now, in a synchronous machine this movement is not linear but rotational, but the principle still holds.

We can visualise the operation of the synchronous machine by imagining two rotating magnets. The rotor will rotate such that the north pole of the rotor lines up with the south pole created in the stator. As the stator magnetic field rotates, it will drag the rotor poles with it. If the magnetic fields of the rotor and stator are strong enough, we can even apply a load torque to the rotor, and the rotor will keep rotating with the stator. Of course, we can also try to accelerate the rotor, in which case the rotor will deliver power to the stator because the rotor poles will try to accelerate the stator poles. Another way to visualise the operation of the synchronous machine is by imagining two magnets. If placed close enough to one another, the north pole of one magnet will align itself with the south pole of the other magnet. If one of the magnets is moved, the other will move with it. We can even attach a weight to one magnet and, if the magnets are strong enough, when the other is moved, this magnet with the weight attached will also move. In this case, we are delivering power through a magnetic coupling! Such a system is shown in Fig. 8.19 where we can move the mass attached to magnet 2 by applying a force to magnet 1.

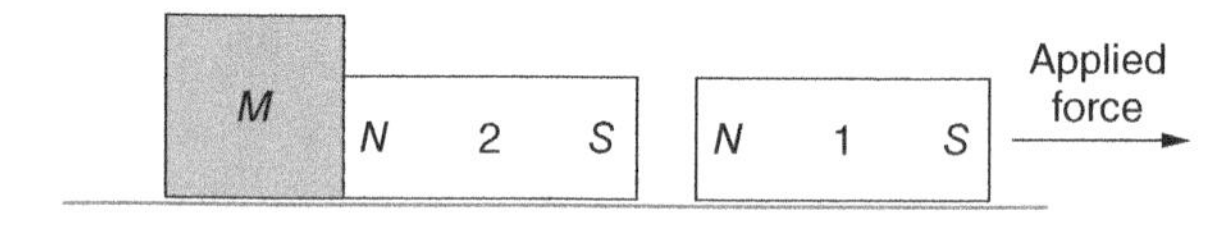

Figure 8.19 A linear view of the synchronous machine operation.

This view of synchronous machines leads us to a very important observation. Let us consider the linear case shown in Fig. 8.19; if we apply a sudden force to magnet 1 and the magnetic bond between the magnets 'breaks,' then we can no longer apply any force to magnet 2. To re-establish this bond, we first need to bring the magnet back to a position close to the other magnet and then slowly start applying force again. We know intuitively that the magnetic bond must be held intact for the system to work. If we apply this knowledge to the rotating case, it is clear that *the rotor must rotate at the same speed as the stator magnetic field lest the magnetic field bonds be broken.* This is also from where the machine gets its name, the fact that the rotor must always rotate at the synchronous speed set by the stator magnetic field.

8.4.1 The Equivalent Circuit

Having discussed the basic operation of the synchronous machine, let us start the analysis of the machine by investigating the equivalent circuit. We can make a very important simplification in the discussion of the three-phase synchronous machine, namely, that the system is a *balanced three-phase system*. From our knowledge of three-phase systems, we know that knowledge of one phase of a balanced system implies that we know all about the other two phases as well. Therefore, in the discussion of the three-phase synchronous machines, we only need to investigate one phase by looking at a single-phase equivalent of the machine.

Now, we have established that, because a synchronous machine operates by having the magnetic poles of the stator and the rotor aligned with each other, the rotor would either be rotating at synchronous speed (where it is in a position to deliver power) or it would be standing still. Let us consider the interesting case where the rotor is rotating and the machine can convert electrical energy to mechanical energy and vice versa. We know that the rotating rotor will induce an emf in the stator windings. We also know that the stator windings are made of wire wound around an iron core and therefore the stator windings will (from an electrical point of view) look like an inductor, inclusive of series wire resistance.

Taking these arguments into consideration, we can create a single-phase equivalent circuit of the machine, as shown in Fig. 8.20. Here, the voltage source V_{in} is the per-phase applied voltage, L_s and r are the winding inductance (also called the

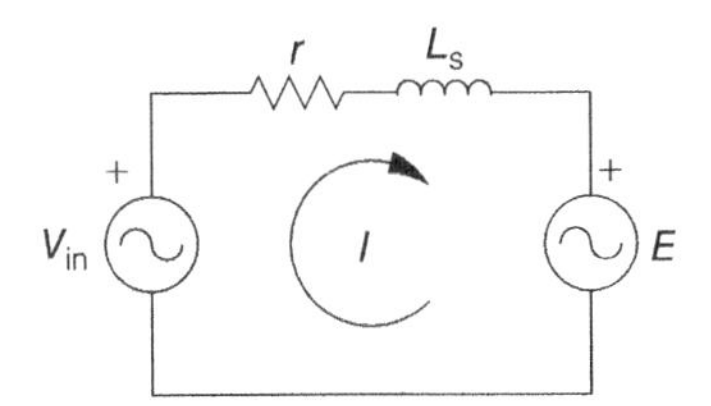

Figure 8.20 A single-phase equivalent circuit of the synchronous machine.

synchronous inductance) and wire resistance while E is the voltage induced in the stator windings. The energy dissipated in (or equivalently supplied by) the back emf (source E) is then effectively the energy that is converted by the machine.

This equivalent circuit is a simple way to understand the view of a synchronous machine operation. Let us investigate this statement with a few examples.

Example 8.3 A six-pole synchronous machine is connected to a 400 V, 50 Hz supply. The machine is Y-connected. It is known that the winding resistance is 0.1 Ω and the synchronous reactance is 4 Ω. If the machine is drawing a line current of 30 A while delivering 17 kW to a mechanical load, calculate the magnitude of the induced voltage and the power factor of the machine.

Solution

The machine is Y-connected; therefore, the per-phase input voltage is

$$V_{\text{in}-\phi} = \frac{400}{\sqrt{3}} = 230.9\ \text{V}$$

and the line current and phase current are the same. If the machine delivers 17 kW, then the machine delivers 5.6 kW per phase. We can calculate the power lost in the armature winding (per phase) as

$$P_{\text{cu}} = 90\ \text{W}.$$

The source therefore has to deliver

$$P_{\text{in}} = 5.69\ \text{kW},$$

which implies that the power factor must be

$$P_{\text{in}} = VI\text{pf} \rightarrow \text{pf} = \frac{P_{\text{in}}}{VI} = 0.82$$

however, we have no information as to whether the system has a lagging or leading power factor. If the power factor is lagging, then

$$\begin{aligned}\mathbf{E} &= \mathbf{V}_{\mathbf{in}-\phi} - \mathbf{I}\left(r + jX_{\text{s}}\right)\\ &= 230.9\angle 0^\circ - 30\angle -34.79^\circ\ (0.1 + j4)\\ &= 187\angle -31.2^\circ\ \text{V}\end{aligned}$$

Similarly, if the power factor is leading

$$\begin{aligned}\mathbf{E} &= \mathbf{V}_{\mathbf{in}-\phi} - \mathbf{I}\left(r + jX_{\text{s}}\right)\\ &= 230.9\angle 0^\circ - 30\angle 34.79^\circ\ (0.1 + j4)\\ &= 313.4\angle -18.7^\circ\ \text{V}.\end{aligned}$$

With the information supplied, either answer can be possible.

Example 8.4 A Δ-connected synchronous machine is used as a generator. The machine is connected to a 11 kV supply. The synchronous reactance is 8 Ω and the armature resistance is 0.55 Ω. The rotor field excitation is set such that the induced emf is 10.8 kV. Calculate the current and the complete value of E when the machine is delivering 4 MW at a power factor of 0.9 leading.

Solution

The machine is connected in Δ; therefore, we can work with the per-phase voltages as stated above. Now we know that since the machine operates as a generator, the current shown in Fig. 8.20 flows in the opposite direction; that is,

$$\mathbf{E} = \mathbf{V} + \mathbf{I}\left(r + jX_s\right).$$

Now, (recall that we are working with a per-phase representation of a balanced three-phase system; hence, the 3 and $\sqrt{3}$s)

$$I = \sqrt{3}\frac{P}{3\,\text{Vpf}} = 233.3\ \text{A}$$

and as the power factor is 0.9 leading

$$\mathbf{I} = 233.3\angle 25.84^\circ\ \text{A}.$$

Finally,

$$\mathbf{E} = \mathbf{V} + \mathbf{I}\left(r + jX_s\right) = 10.45\angle 9.56^\circ\ \text{kV}.$$

8.4.2 Phasor Diagram

In many instances, the information contained in the equivalent circuit can be represented in a phasor diagram form. This alternative representation yields many advantages, most of which will become evident as we continue our discussion.

Thus far, we know that

$$\mathbf{V} = \mathbf{E} + \mathbf{I}\left(r + jX_s\right) \tag{8.50}$$

now let us make the assumption that the phase shift of $\mathbf{V}$ is 0°. Then let us define

$$\mathbf{V} = V\angle 0^\circ \tag{8.51}$$

$$\mathbf{E} = E\angle \delta^\circ \tag{8.52}$$

$$\mathbf{I} = I\angle \theta^\circ \tag{8.53}$$

which has the result that

$$\text{pf} = \cos\left(\theta\right). \tag{8.54}$$

We can express all this information in a phasor diagram form, as shown in Fig. 8.21. Naturally we know, and we can see from the diagram, that the vector $\mathbf{I}r$ is in phase with the vector $\mathbf{I}$, and the vector $j\mathbf{I}X_s$ is leading the vector $\mathbf{I}$ by 90°.

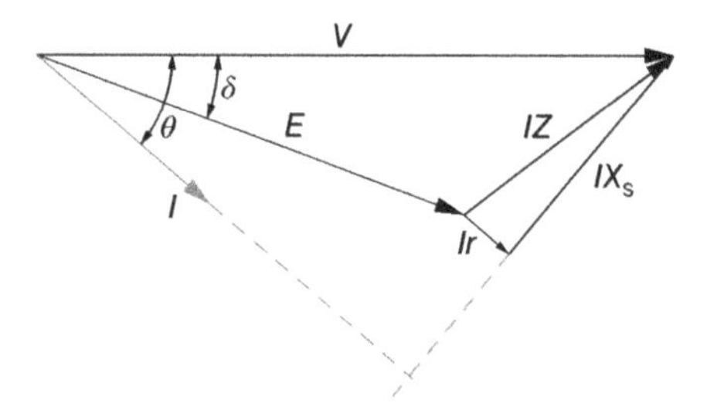

Figure 8.21 A phasor diagram of a synchronous machine.

8.4.3 Power Angle Characteristic Equation

Let us use the definitions of (8.51)–(8.53) to rewrite (8.50) into the real and imaginary components. We will use the current direction definition of Fig. 8.20:

$$V = E\cos(\delta) + Ir\cos(\theta) + IX\cos\left(90^\circ + \theta\right) \tag{8.55}$$

$$0 = E\sin(\delta) + Ir\sin(\theta) + IX\sin\left(90^\circ + \theta\right). \tag{8.56}$$

We can rewrite these using trigonometric identities as

$$V = E\cos(\delta) + Ir\cos(\theta) - IX\sin(\theta) \tag{8.57}$$

$$0 = E\sin(\delta) + Ir\sin(\theta) + IX\cos(\theta). \tag{8.58}$$

The power delivered by the source is

$$P = VI\cos(\theta), \tag{8.59}$$

which can be rewritten as

$$I\cos(\theta) = \frac{P}{\mathrm{V}}. \tag{8.60}$$

Now, substitution of (8.60) into (8.58) yields

$$\begin{aligned} 0 &= E\sin(\delta) + Ir\sin(\theta) + IX\cos(\theta) \\ &= E\sin(\delta) + Ir\cos(\theta)\tan(\theta) + IX\cos(\theta) \\ &= E\sin(\delta) + \frac{Pr}{V}\tan(\theta) + \frac{PX}{V}. \end{aligned} \tag{8.61}$$

From a first glance, this equation might not look like much, but let us investigate a bit further. In general, a synchronous machine, similar to all other machines and other energy conversion equipment, is made with the aim of keeping the efficiency as high as possible. This results in the fact that $r \ll X_s$. Now, with this fact and restricting ourselves to an area of interest where $-45° \le \theta \le 45°$ or equivalently the condition that pf $> \frac{1}{\sqrt{2}}$, we can simplify the expression to

$$\boxed{-P_{\phi} \approx \frac{E\,V}{X_s} \sin(\delta)\,.} \tag{8.62}$$

This expression is often called the *power angle* (or *torque angle*) equation or the synchronous power transfer characteristic equation.

The angle between the induced voltage and the terminal voltage, δ, is an important characteristic of the synchronous machine operation. It is called the *power angle* or the *torque angle*. Recall that we derived (8.62) by investigating the power delivered by the source. If we draw a graph of the power delivered as a function of the torque angle, as shown in Fig. 8.22, we can clearly see that the source would be delivering power when $-\pi < \delta < 0$ and it would be absorbing power when $0 < \delta < \pi$. Therefore, when **E** lags **V** the machine operates as a motor where the source delivers energy to the machine where it is converted to mechanical energy. Similarly, when **E** leads **V** the machine operates as a generator and the source absorbs energy delivered from the machine. This energy is converted inside the machine from mechanical to electrical energy.

Example 8.5 A four-pole, Y-connected, synchronous machine is connected to a 400 V, 50 Hz supply. It is a permanent magnet rotor machine and it is known that the induced voltage is 240 V (per phase). If the machine delivers 125 Nm and the synchronous reactance is 7 Ω, calculate the input current. Neglect the armature resistance.

Solution

As the machine is Y-connected, the per-phase input voltage is

$$V = \frac{400}{\sqrt{3}} = 230.9\,\text{V}.$$

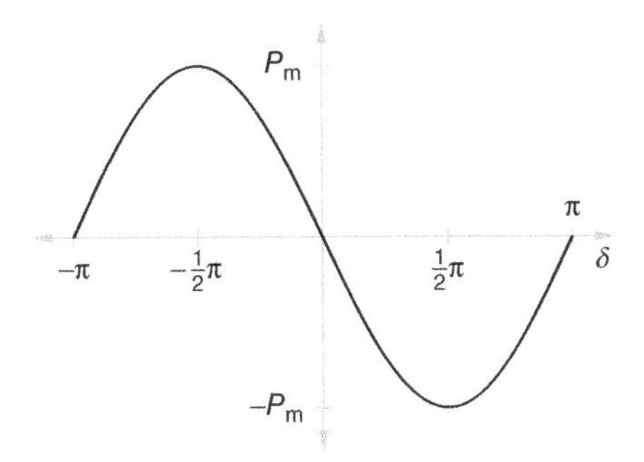

Figure 8.22 The power transferred by the machine as a function of the torque angle.

The synchronous speed is

$$n_s = \frac{120f}{p} = 1\,500 \text{ rpm}$$

and the machine therefore delivers

$$P = Tn_s\frac{2\pi}{60} = 19.64\,\text{kW}$$

so the torque angle of the machine is (remember we are working with a single-phase equivalent circuit),

$$\delta = \sin^{-1}\left(\frac{PX_s}{3VE}\right) = 55.75°.$$

Now,

$$j\mathbf{IX} = \mathbf{V} - \mathbf{E} = 230.9\angle 0° - 240\angle -55.75° = 220.33\angle 64.2°\text{ V}$$

and therefore,

$$\mathbf{I} = \frac{220.33\angle 64.2°}{j7} = 31.48\angle - 25.8°\text{ A}.$$

8.4.4 Controlling the Power Factor

If we investigate the torque angle characteristics of the synchronous machine, (8.62), a couple of interesting facts emerge. Firstly, recall from the discussion in Chapter 7 that it is the mechanical part of the energy conversion process that determines the power delivery of the system. We can show this argument in the following way. The synchronous machine always rotates at the synchronous speed. The characteristics of the mechanical load in the case of motor operation (or the mechanical source in the case of generator operation) will determine the amount of energy that is transferred.

From (8.62), we see that there is a maximum power point (absolute value)

$$P_{max} = 3\frac{EV}{X_s} \tag{8.63}$$

that occurs when $\delta = \pm 90°$ depending on whether the machine is operating as a motor or as a generator. Recall from Section 8.4 that the machine operates through the magnetic bond between the stator magnetic field and the rotor magnetic field. For the ideal synchronous machine under discussion here, this bond 'breaks' at 180 electrical degrees. However, we see that the amount of torque that can be generated by the machine decreases after 90 electrical degrees. Therefore, if the mechanical power required (or delivered) from the machine exceeds the maximum power that the machine can convert, (8.63), then we know that the machine cannot deliver. The end result would be that the magnetic bonds will break; if the machine is operating as a motor the rotor will decelerate to standstill, and if operating as a generator, the rotor will accelerate to some arbitrary high speed where either someone will realise something is wrong and end the misery, or the system will suffer catastrophic failure.

The second observation we can make from the torque angle equation, (8.62), is that we have some measure of control over the torque angle, although the mechanical circuit load sets the power. In most applications, the voltage applied to the synchronous machine cannot be varied. This is true because in most cases the machine is directly connected to the grid and the terminal voltage is therefore fixed. That said, it is true that in some applications the machine is operated with a drive, as is described later where the terminal voltage can be changed by the drive. However, even in this case, the drive unit will keep the input voltage at the rated level.

We know now that for most applications we do not have control over the power which is delivered, the terminal voltage or, for obvious reasons, the synchronous reactance. However, if the machine is a wound rotor machine where the magnetic field on the rotor is created by current-carrying windings on the rotor, as shown in Fig. 8.18, we have a measure of control over **E**. If we increase the field current, the flux per pole on the rotor will increase and therefore the magnitude of the induced voltage will increase. Remember that as the rotor speed is constant at the synchronous speed, the only method for adjusting the magnitude of the induced voltage is by varying the rotor flux. Let us define two terms: if $E > V$, the machine is said to be *over-excited* and when $E < V$, the machine is *under-excited*.

Now, according to (8.62), if the magnitude of the induced voltage changes but the power, terminal voltage and the synchronous reactance remain unchanged, then the value of δ must change. This brings us to an interesting conclusion, as (neglecting the effect of the stator resistance)

$$j\mathbf{I}X_s = \mathbf{V} - \mathbf{E} = V - E\angle\delta^\circ. \tag{8.64}$$

Let us investigate the operation of the synchronous machine as either over- or under-excited and in either motor or generator mode by means of an example.

Example 8.6 A Y-connected synchronous machine delivers 20 kW from a 400 V supply. The power factor is 0.65 lagging. If the synchronous reactance is 5 Ω, what change should be made to the rotor excitation current to change the power factor to 0.9 lagging?

Solution

At a power factor of 0.65 lagging, the input current is

$$|\mathbf{I}| = \frac{P}{\sqrt{3}V_{LL}\mathrm{pf}} = 44.4\,A$$

and as pf = 0.65 lagging

$$\mathbf{I} = 44.4\angle -49.46^\circ \text{ A}.$$

Now, the existing excitation of the machine is

$$\mathbf{E} = \mathbf{V} - j\mathbf{I}X_s = 157.1\angle - 66.7^\circ \text{ V}.$$

If we change the power factor to 0.9 lagging, then

$$|\mathbf{I}| = \frac{P}{\sqrt{3}V_{LL}\text{pf}} = 32.08 \text{ A}$$

$$\therefore \mathbf{I} = 32.08\angle -25.84^\circ \text{ A}$$

and the new excitation becomes $\mathbf{E} = \mathbf{V} - j\mathbf{I}X_s$

$$= 216.67\angle - 41.9^\circ \text{ V}.$$

Example 8.7 A large factory uses a 3.3 kV, Y-connected synchronous machine to drive a fan as part of the process. The fan load is 55 kW and the rest of the process, apart from the fan, takes 250 kW at a power factor of 0.75 lagging. Specify the power rating of the machine if the machine is to be used to correct the power factor of the process to 0.92 lagging. *Take the nameplate rating of the synchronous machine as at a power factor of 0.85.*

Solution

The required reactive power to be delivered by the synchronous machine is

$$\frac{250}{0.75}\angle 45.4^\circ - \frac{250}{0.92}\angle 23.1^\circ = -114 \text{ kVAR}.$$

The machine therefore needs to deliver

$$|S| = \sqrt{P^2 + Q^2} = 126.6 \text{ kVA},$$

which translates to a line current of

$$|\mathbf{I}| = \frac{|S|}{\sqrt{3}|\mathbf{V}|} = 22.14 \text{ A},$$

which at a power factor of 0.85 would translate to a machine rating of

$$P = \sqrt{3}VI\,\text{pf} = 107.6 \text{ kW}.$$

8.5 INDUCTION MACHINE

Induction machines share a very important component and characteristic with synchronous machines: the stator and the rotating magnetic field in the air gap. The main difference between them is the construction of the rotor. In synchronous machines, the rotor creates its own magnetic field independently of the magnetic field of the stator. On the contrary, the induction machine rotor does not create its own magnetic field. Instead the magnetic field of the stator induces currents in the rotor. The logical consequence of currents circulating in the rotor is the creation of a rotor magnetic field. The rotor of the induction machine cannot therefore create its own magnetic field.

We start the discussion of the induction machine by investigating the rotor and the induced currents in the rotor. The rotor can be clearly seen in the photograph shown in Fig. 8.23. The stator is cut open and the rotor shaft is supported by two bearings. Mounted on the right side is a fan that blows air across the machine to keep it cool.

8.5.1 Induced Currents in the Induction Machine Rotor

What will happen if we place a coil on a rotor inside the ac machine stator? This situation is shown in Fig. 8.24. Let us first consider the magnetic field created by the stator. In Section 8.3, we found that the stator, when operational with three-phase currents, will create a rotating sinusoidally distributed magnetic field in the air gap. This magnetic field will rotate at a synchronous speed given by

$$n_s = \frac{120f}{p}. \tag{8.65}$$

Figure 8.23 A cross section through an induction machine.

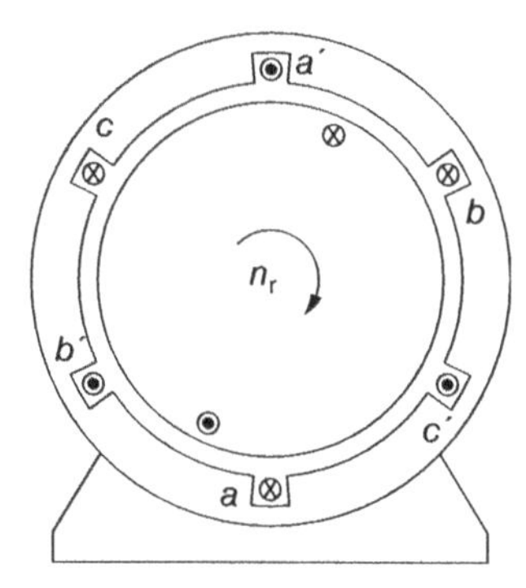

Figure 8.24 A coil placed on a rotor inside an ac machine stator.

Let us, for simplicity, consider a two-pole machine here. The results of the two-pole machine are directly translatable to other pole numbers.

Now let us place a freely rotating rotor in the air gap. For ease, we will assume that both the stator and the rotor are readily magnetised; in fact we will assume that $\mu_r \rightarrow \infty$. The rotational speed of the rotor, in the same direction as the stator magnetic field rotation, is denoted as n_r.

Now let us place a single coil in the rotor. The coil is a simple coil with no end points. It is important to note that the coil is embedded in the rotor; therefore, as the rotor turns, the coil orientation with respect to the stator will also change.

8.5.1.1 *The Case Where $n_r = 0$* What will happen when the stator is energised and the rotor is standing still? We know that the energised stator will create a rotating magnetic field. As the magnetic field rotates, the amount of flux coupled to the coil in the rotor will change. To visualise this process, consider the rotor shown in Fig. 8.24 in the magnetic field procession as shown in Fig. 8.13. It is clear that as the magnetic field rotates the amount of flux captured by the coil on the rotor will change. Remember that

$$\phi = \int B(A)\, dA, \tag{8.66}$$

which can, simply be written as $\phi = BA$.

Now, because the flux is changing, we know that according to Faraday's law,

$$e = N\frac{d\phi}{dt} \tag{8.67}$$

a voltage will be induced in the coil. For a simple coil, $N = 1$. This might sound a little strange at first as it does not make sense that a voltage can be induced in a closed coil. However, every coil will have some inductance and resistance. Therefore, the induced voltage can be present in the coil when there is a current flowing in the coil. Let L_r and r_r be the coil inductance and resistance, respectively. Then, we can express the coil current,[1] i_r, as

$$i_r = \frac{e}{\sqrt{(2\pi f_r L_r)^2 + r_r^2}} = \frac{d\phi}{dt}\frac{1}{\sqrt{(2\pi f_r L_r)^2 + r_r^2}}. \tag{8.68}$$

Here, we still have one unknown, namely, the frequency f_r (the rotor current/induced voltage frequency). We can get this frequency by investigating the rotating field and the change of the coupled flux in the rotor coil. As the field rotates at a speed, in rotations per minute, of

$$n_s = \frac{120f}{p} \tag{8.69}$$

we know that if we stood somewhere on the stator that a north pole would come past us n_s times per minute. The frequency of this change is then simply the number of times per second that a north pole will come past. Let us call this frequency f_s, which is then

$$f_s = \frac{2f}{p} \tag{8.70}$$

and for the two-pole machine this is simply the applied line frequency.

Now, if the rotor stands still, the change in the coupled flux will also occur at the same frequency. Therefore, *when the rotor is standing still*, the rotor frequency can be found as

$$f_r = f_s = f. \tag{8.71}$$

8.5.1.2 *The Case Where $n_r = n_s$* Now, what will happen when $n_r = n_s$? In this case, the rotor is turning just as fast as the stator rotating magnetic field. Therefore, the coil in the rotor would not perceive any change in flux. Again, this can be visualised by using the rotor shown in Fig. 8.24 and the images shown in Fig. 8.13 together. Just remember to let the rotor rotate with the same speed as the rotating field. Naturally, because there is no changing flux, the rotor frequency is also zero.

8.5.1.3 *The Case Where $0 < n_r < n_s$* Now, we get to the interesting part. What would happen if the rotor is not standing still but nor is it rotating at synchronous speed? Well, because the rotor is not turning at synchronous speed, we know that the coil on the rotor will see a changing flux. The size of this changing flux and the rotor frequency are both naturally dependent on the relative speed difference between the rotating flux and the rotor.

[1] Here we are only considering the absolute value of the current. Technically, if we were to keep with the notation used thus far, we would have to speak of $\mathbf{I}_r$ rather than i_r. However, this requires a certain reference angle which, in the rotating reference frame, can be a bit tricky. We will therefore just use the term i_r, which should be read as $|\mathbf{I}_r|$ if you feel strongly about the fact. That said, we will later revert to the more formal definition.

This is in fact something most well-travelled business (and leisure for that matter) travellers are quite acquainted with. If we take the very medieval viewpoint that the earth is the centre of the universe, then it would appear that the sun rotates around the earth with a synchronous speed of one rotation per twenty four hours. This change of coordinates system is something we engineers are quite well versed in; maybe the medieval scientists had a point after all, but let that be. Now, if you are standing still on the earth's surface you would see the changing sunlight levels as described by the synchronous speed of the sun's rotation. As you start moving in a westerly direction, the sunlight levels will start changing less rapidly. The perceived day will then become slightly longer. This is of course not really noticeable when walking, running or even driving; but have you ever taken a direct flight from Beijing to Amsterdam? The perceived day can become very long indeed. In fact, if one can find an aircraft fast enough, it should, in theory at least, be possible to fulfil the dream of having continuous sunlight, twenty four hours a day.

We see therefore that the faster we move in a westerly direction, the longer the day becomes and the lower the frequency of change and the rate of change (in sunlight intensity). We see the same with the flux change perceived by the coil on the rotor. When standing still, it sees the full rate of change in magnetic flux density as prescribed by the stator and the rotor frequency is the same as the stator frequency. As the rotor starts to rotate, the frequency of change starts to decrease. Because the time period in which the change occurs lengthens, the change also occurs more gradually. As a result, the induced voltage in the coil is lower and likewise the current is lower.

8.5.1.4 Slip We have seen that the induced currents in the rotor, both in magnitude and frequency, are dependent on the relative speed between the rotating field and the rotor. For the analysis of induction machines, it is beneficial to define a new variable to describe this difference in the rotational speeds. We will call this variable the slip of the machine, denoted by s. We can probably describe the difference in speed in many ways, but the most convenient is to make the amount of slip a dimensionless entity where the value 1 implies that the difference is at a maximum and the value 0 implies that the difference is zero. Formally, the slip is defined as

$$\boxed{s = \frac{n_\mathrm{s} - n}{n_\mathrm{s}}.} \tag{8.72}$$

From here on, we will refer to the rotor speed (previously denoted as n_r for the sake of clarity) simply as the machine speed denoted as n.

It is now possible to express the machine speed as

$$\boxed{n = (1 - s)\, n_\mathrm{s}} \tag{8.73}$$

the induced voltage as

$$\boxed{E = s\, E_m,} \tag{8.74}$$

where E_m is the voltage induced when the rotor is standing still, and the rotor frequency as

$$\boxed{f_\mathrm{r} = sf.} \tag{8.75}$$

At this point, it might be tempting to think that there is also a linear relationship between the current in the rotor and the slip. However, because of the fact that both e and f_r change in (8.68), this relationship is not linear. We will derive expressions for this current later.

It is worthwhile noting from (8.72) that slip is conveniently dimensioned and often expressed as a percentage.

Example 8.8 A six-pole 50 Hz induction machine runs at 4% slip at a certain load. At this point, calculate the

1. synchronous speed
2. rotor speed
3. and the frequency of the rotor currents.

Solution

1. The synchronous speed is

$$n_\mathrm{s} = \frac{120f}{p} = 1\,000 \text{ rpm}.$$

2. The rotor speed is

$$n = (1 - s)\,n_\mathrm{s} = 960 \text{ rpm}.$$

3. The frequency of the rotor currents is

$$f_\mathrm{r} = sf = 2\,\mathrm{Hz}.$$

8.5.1.5 The Magnetic Field Created by the Rotor Currents According to Ampère's law, every current will create a corresponding magnetic field strength. This is also true for the currents flowing in the rotor. In fact, because of the fact that the rotor and stator are manufactured using readily magnetisable materials, the corresponding magnetic field is substantial. The question is therefore not whether there is a magnetic field associated with the rotor currents but rather what it will look like when viewed from the stator.

Let us view the rotor from a stationary position on the stator. Let us assume that the motor is operating with a slip s. Now the rotor speed will be $s\,n_\mathrm{s}$ and the frequency of the rotor currents will be sf. At this point, we will have to increase the complexity of the model; let us allow for many different coils on the rotor, not only

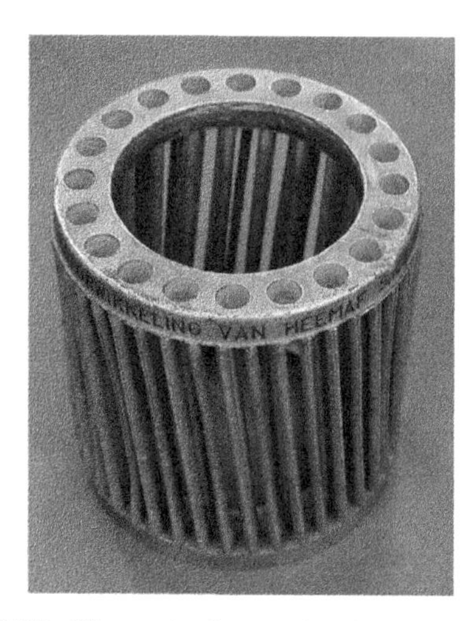

Figure 8.25 The squirrel cage of an induction machine.

the one shown in Fig. 8.24. In fact, most induction machine rotors have the basic shape shown in Fig. 8.25. This rotor is called the *squirrel cage rotor*, for obvious reasons.

If we allow for a sufficient number of coils on the rotor, it is possible to approximate the currents in the rotor windings as a three-phase system. The rotor currents will then also create a magnetic field that rotates about the rotor at a 'synchronous speed' of

$$n_s' = \frac{120 f_r}{p} = s\, n_s. \tag{8.76}$$

This leads us to an interesting conclusion. As the rotor is turning at n and the magnetic field induced by the rotor currents rotates at the speed n_s' around the rotor, the total speed of the magnetic field created by the rotor (with respect to a stationary observer) is

$$n + n_s' = (1 - s)\, n_s + s\, n_s = n_s. \tag{8.77}$$

This is interesting—the magnetic field created by the rotor currents therefore rotates at the same speed as the stator magnetic field.

Actually, this makes sense when we remember the operation of the synchronous machine where the magnetic field created by stator 'locks' to the magnetic field of the rotor. This bond between the magnetic fields is the method of power conversion. The main difference in the operation of the synchronous machine and the induction machine is the fact that while the synchronous machine rotor creates its own magnetic field (and therefore rotates at synchronous speed), the induction machine rotor magnetic field is created by the currents induced in the rotor windings. Consequently, the rotor cannot rotate at synchronous speed because the induced currents in the rotor will be zero and no magnetic field will be created by the rotor.

8.5.2 Development of an Equivalent Circuit

In this section, we develop an equivalent circuit of the induction machine. When an induction machine does not rotate but stands still, it can be viewed as a three-phase transformer. The squirrel cage behaves similarly to a single-turn secondary winding that is short-circuited and it has a certain value of reactance and resistance that will prevent the current from becoming too large. In some induction machines, for example the ones that are used in turbines, the rotor uses regular three-phase windings, similar to the stator windings that were discussed in the previous section.

One can visualise the operation as a three-phase transformer that can rotate. The primary windings are the stator windings and the secondary windings are placed on the rotor, as is shown in Fig. 8.26. The windings on the rotor is, however, short-circuited; the impedance Z_r, as shown in the figure, is the inductance and resistance of the short-circuited coils on the rotor. A complication that we have to deal with is that the frequencies on the stator and rotor are different. The primary frequency has the same value of the line frequency, but the secondary frequency depends on how fast the machine turns and is therefore a function of the slip. The basic premise of this section, as with the equivalent circuit of the synchronous machine, is the assumption that the system is a balanced three-phase system. This allows us to use a single-phase equivalent circuit.

The first step in developing an equivalent circuit is to investigate what is happening in the rotor. We can summarise the conclusions of the above discussions with the equivalent circuit shown in Fig. 8.27. From this circuit, it is clear that the magnitude of the induced voltage will increase linearly with the slip. However, the current does not increase linearly with the slip as the impedance of the circuit also increases with the increasing frequency of the rotor currents.

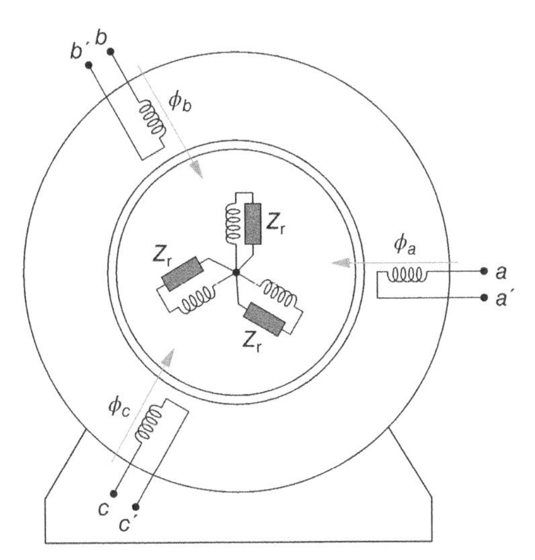

Figure 8.26 Three-phase windings on an induction machine.

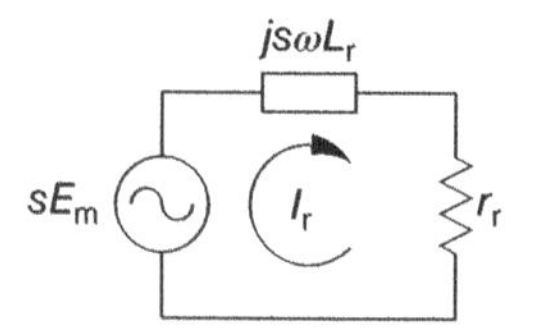

Figure 8.27 The basic rotor equivalent circuit.

Now, let us make the mind shift that we are interested in the rotor current and not the rotor voltages. We can rewrite the expression for the rotor current, repeated for convenience

$$|\mathbf{I}_r| = \frac{|s\mathbf{E}|}{\sqrt{(j2\pi sfL_r)^2 + r_r^2}} \tag{8.78}$$

as

$$|\mathbf{I}_r| = \frac{|E|}{\sqrt{(j2\pi fL_r)^2 + \left(\frac{r_r}{s}\right)^2}}. \tag{8.79}$$

It is important to note that, from the viewpoint of the magnitude of the rotor current, these two expressions are equivalent. We can redraw the equivalent circuit shown in Fig. 8.27 using this definition as shown in Fig. 8.28.

During our discussion of the induced currents in the rotor, in Section 8.5.1, we have seen that the stator creates a magnetic field and that this magnetic field in turn induces a voltage and therefore currents in the rotor. Furthermore, we have seen that the induced current in the rotor creates its own magnetic field that interacts with the stator magnetic field. We have seen this behaviour before.

Seen from a global perspective, the induction machine operates with the same basic operation as the transformer. However, in this case, there is no well-defined magnetic core but the coupling medium is air. Let us approximate the operation of the induction machine then with that of an air-coupled transformer. As we have seen in Chapter 4, it is possible to represent the operation of the circuit using the ideal transformer model. Consider the equivalent circuit shown in Fig. 8.29; here, we see the coupling of the stator to the rotor via an ideal transformer model.

Now, it is important that just as there are similarities between the operation of the induction machine and the transformer, there are also major differences, the

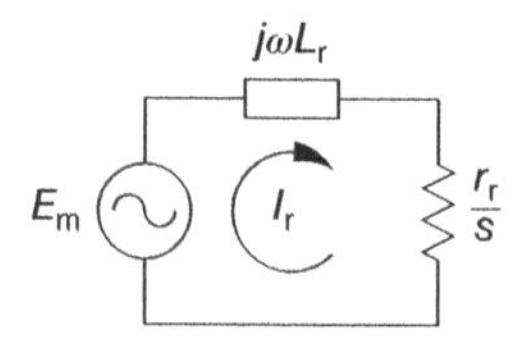

Figure 8.28 Rewriting the basic rotor equivalent circuit.

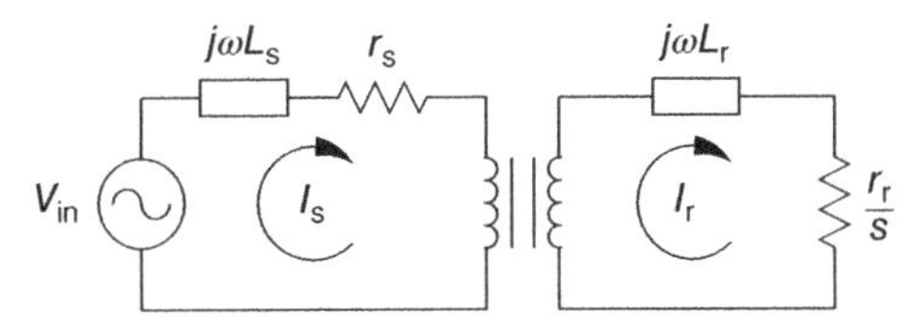

Figure 8.29 The complete induction machine-equivalent circuit using an ideal transformer model.

most important being the rotation of the rotor. This rotation has two very important influences on the system. Firstly, the frequency of the rotor and that of the stator are not the same. Secondly, the voltage induced in the rotor varies with the slip. Luckily, we had already considered these discrepancies when we redrew the circuit shown in Fig. 8.27 as Fig. 8.28.

Now, the circuit of Fig. 8.29 might provide a false sense of confidence. However, when we investigate the circuit more closely, we see that we did not include any information on the induced voltage or even the supposed turns ratio of the ideal transformer. This is because it is very difficult to calculate these values analytically. Although it is, in principle, possible to measure the inductance and resistance of a simple rotor coil as shown in Fig. 8.24, which coil should be measured in the more complex rotor shown in Fig 8.25?

This is not a simple problem to solve. However, we can circumvent the problem by simply referring the rotor circuit to the stator. Now, it is true that we do not have accurate information on the real value of the current in the rotor, or for that matter the rotor resistance or inductance, but we know what their value would be when seen through the eyes of the stator. In other words, we know that there is a 'transformer'-like behaviour in the circuit, but it is much easier to simply work with the referred values of the rotor and to measure them directly from the stator.

We can draw this referred rotor circuit as shown in Fig. 8.30. To simplify the discussion, we denote the inductive impedances as X where $X = 2\pi fL$. Furthermore, a certain amount of magnetic energy is stored in the stator, rotor and air gap, consistent with the transformer-like behaviour. We will represent this stored energy, in the same way as we did with the transformer-equivalent circuit, by means of a magnetising inductance. Finally, as we have referred the rotor circuit to the stator side, we should probably denote the referred value as r_r' rather than r_r. However, as

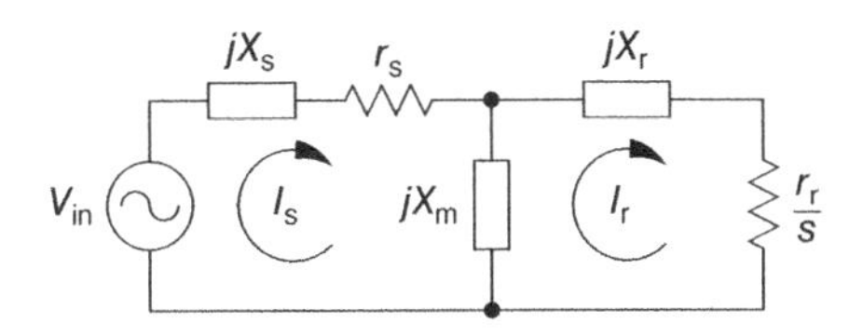

Figure 8.30 The complete induction machine-equivalent circuit.

we never knew the 'real' value and we are only ever going to work with the referred value, we will keep with the simple r_r to keep our notation clear.

Our final act in the development of the induction machine-equivalent circuit is splitting the rotor resistance. In the circuit shown in Fig. 8.30, we can readily calculate the stator winding losses as

$$P_{cu_s} = |\mathbf{I}_s|^2 r_s \tag{8.80}$$

however, we cannot easily do the same with the rotor winding losses. Luckily, we can solve this problem quite easily using a little mathematical trick shot. As

$$\frac{r_r}{s} = r_r + r_r\frac{1-s}{s} \tag{8.81}$$

we can redraw the circuit as shown in Fig. 8.31. Now, it is easy to calculate the rotor winding losses.

Example 8.9 A 400 V, four pole, 50 Hz induction machine runs at 1 425 rpm. The machine is Y-connected and the following information is available

$X_s = 1.3\,\Omega$ $\quad$ $r_s = 0.5\,\Omega$

$X_r = 1\,\Omega$ $\quad$ $r_r = 0.35\,\Omega$ both referred to the stator

$X_m = 350\,\Omega$

Calculate the input power, the stator winding loss and the rotor winding loss.

Solution

The machine is operating at a slip of

$$s = \frac{n_s - n}{n_s} = 5\%.$$

The total impedance seen by the source is

$$Z_t = r_s + jX_s + \left(\frac{1}{jX_m} + \frac{1}{\frac{r_r}{s} + jX_r}\right)^{-1}$$

$$= 7.85\angle 18.1^\circ\,\Omega.$$

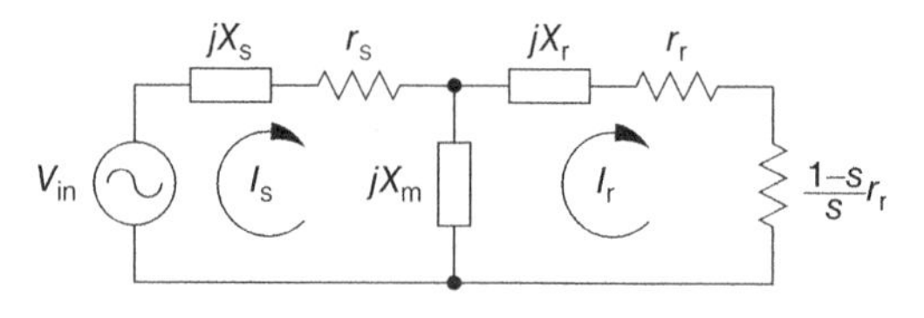

Figure 8.31 The redrawn induction machine-equivalent circuit.

The input current is

$$\mathbf{I}_{\rm s} = \frac{400}{\sqrt{3}Z_{\rm t}} = 29.4\angle - 18.1^\circ\ {\rm A}$$

and the input power is therefore,

$$P_{\rm in} = 3{\rm Re}\left\{\mathbf{V}_\phi \mathbf{I}_\phi^*\right\} = 3\frac{400}{\sqrt{3}}{\rm Re}\left\{\mathbf{I}_{\rm s}^*\right\} = 19.4\,{\rm kW}.$$

The rotor current is

$$\mathbf{I}_{\rm r} = \frac{\frac{400}{\sqrt{3}} - \mathbf{I}_{\rm s}\left(r_{\rm s} + jX_{\rm s}\right)}{\frac{r_{\rm r}}{s} + jX_{\rm r}} = 28.1\angle{-8.55^\circ}\ {\rm A}.$$

Therefore, the stator winding loss is

$$P_{{\rm cu}_{\rm s}} = 3|\mathbf{I}_{\rm s}|^2 r_{\rm s} = 1.3\,{\rm kW}$$

and that of the rotor

$$P_{{\rm cu}_{\rm r}} = 3|\mathbf{I}_{\rm r}|^2 r_{\rm r} = 829\,{\rm W}.$$

8.5.3 Measurement of the Induction Machine Parameters

It is all good and well to use the referred parameters (to the stator) in the equivalent circuit. This cheap shot really does make life easier. However, how do we determine the values that we use in the equivalent circuit? As said earlier, in principle, one can probably measure the parameters directly while manufacturing the rotor, but then how do you refer it to the stator if you do not have the required information with regard to the effective turns ratio of the transformer used to model the energy exchange between stator and rotor?

The answer is to measure it directly from the stator side. The trick here is to realise the fact that

$$\lim_{s \to 0} \frac{r_{\rm r}}{s} = \infty. \tag{8.82}$$

Therefore, the rotor current approaches zero as the rotor speed approaches the synchronous speed. We can mimic this situation by letting the machine run at no load. What exactly will become clear later in our discussion of the induction machine?

We can make a very good approximation of the parameters of the machine by carrying out two tests: the locked rotor test and the no-load test.

8.5.3.1 *The Locked Rotor Test* If the rotor is locked, that is, standing still, then the slip is 100%. Under these conditions

$$\frac{r_r}{s} = r_r. \tag{8.83}$$

Now, if we make the assumption, which is valid for every well-designed induction machine, that

$$|r_s + jX_s + r_r + jX_r| \ll X_m \tag{8.84}$$

we can simplify the equivalent circuit of Fig. 8.31 to that of Fig. 8.32.

For the locked rotor test, the rotor is prevented from rotating by some external means, normally some form of clamp, and then the machine is excited with *a lower than normal voltage*. The applied voltage is normally adjusted such that the machine takes rated current. Now, we can calculate the following, where P_{in} is the power supplied to the machine (per phase), I_ϕ is the phase current and V_ϕ is the phase voltage:

$$r_s + r_s = \frac{P_{in}}{I_\phi^2} \tag{8.85}$$

$$X_s + X_r = \frac{\sqrt{V_\phi^2 I_\phi^2 - P_{in}^2}}{I_{in}^2}. \tag{8.86}$$

Normally, it is trivial to split the $r_s + r_r$ term into its components as the stator resistance can be measured directly. However, it is not so simple to split the $X_s + X_r$ term. In most cases, where an accurate model is required, the values can only be found with significant effort. However, for most cases, the assumption that $X_s = X_r$ will yield an induction machine model that is accurate enough for most calculations.

8.5.3.2 *The No-Load Test* When the motor is running under no load, the assumption is that $s \approx 0$. Under this condition, as there is no current in the rotor, the rotor circuit can be neglected. Furthermore, again using the assumption of (8.84), we can draw the equivalent circuit of the machine under no-load conditions as shown in Fig. 8.33.

It is now trivial to calculate the machine-magnetising inductance when the current and voltage are known.

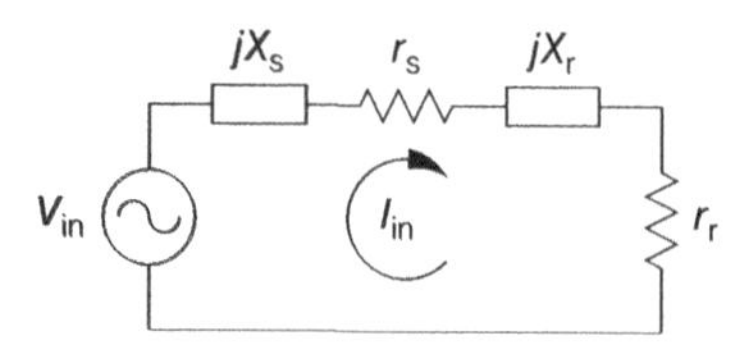

Figure 8.32 The induction machine-equivalent circuit under the locked rotor test.

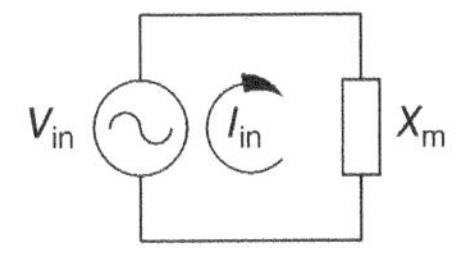

Figure 8.33 The induction machine-equivalent circuit under the no-load test.

8.5.4 Performance Calculations

The equivalent circuit shown in Fig. 8.31 contains all the information needed to analyse the operation of the induction machine. The only question that needs to be answered is the function of the resistance $r_r\frac{1-s}{s}$.

To answer this question, let us refer to Example 8.9. In this case, the input power was calculated to be 19.4 kW and the total losses in the machine were 2.1 kW. The remainder of the power (about 17.3 kW) is dissipated in the slip-dependent resistance. This relatively large power dissipation is actually the mechanical energy delivered by the machine. The resistance is a function of slip as the amount of slip determines the magnitude of the current in the rotor windings (Section 8.5.1) and therefore the size of the magnetic field created by the rotor. The power delivered by the machine depends on the interlocking of the magnetic fields of the rotor and stator and is therefore also a function of the slip.

In our discussion thus far, we have however neglected two further loss mechanisms: the magnetic core losses, and the windage and friction losses. The magnetic core losses are because of the fact that all magnetic materials offer some opposition to the creation of a magnetic field, while a small air gap in the machine (normally in the range of 2–7 mm) causes a significant amount of air friction opposing the rotation of the rotor. The complete modelling of these losses is fairly complex and in most cases, it is sufficient to consider them as a constant.

The power flow diagram of the induction machine is shown in Fig. 8.34. We can relate this diagram to the circuit shown in Fig. 8.31 on a per-phase basis. The input power is

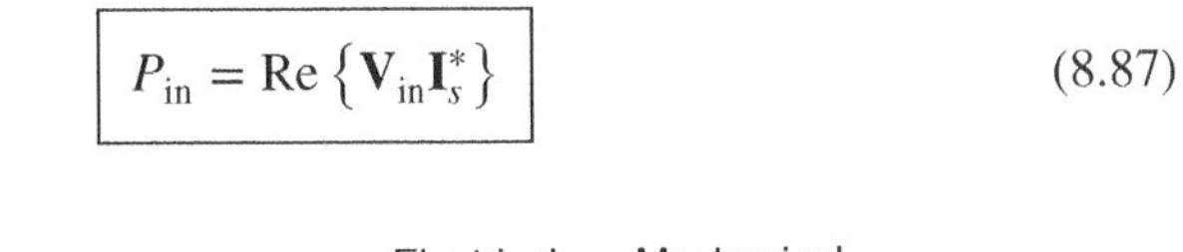

$$P_{in} = \mathrm{Re}\left\{\mathbf{V}_{in}\mathbf{I}_s^*\right\} \tag{8.87}$$

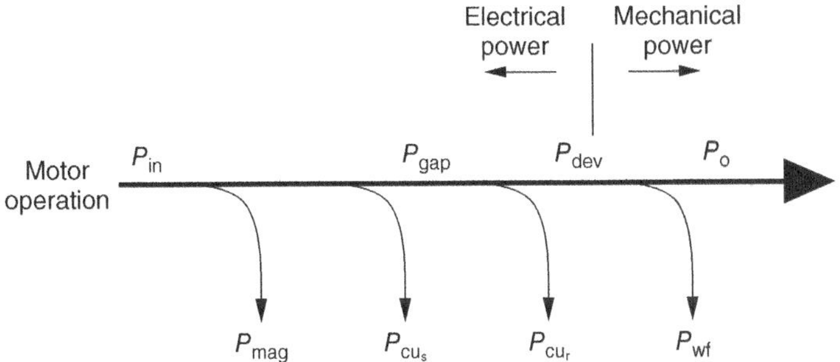

Figure 8.34 The power flow diagram of the induction machine.

and the stator copper losses are

$$P_{cu_s} = |\mathbf{I}_s|^2 r_s. \tag{8.88}$$

From here, we can easily calculate the power crossing the air gap as

$$P_{gap} = P_{in} - P_{cu_s} - P_{mag}. \tag{8.89}$$

Now interestingly, according to Fig. 8.30, the air gap power is simply the power dissipated in $\frac{r_r}{s}$; we now have another method of calculating the air gap power:

$$P_{gap} = |\mathbf{I}_r|^2 \frac{r_r}{s} \tag{8.90}$$

The rotor copper losses are

$$P_{cu_r} = |\mathbf{I}_r|^2 r_r \tag{8.91}$$

and the mechanical power delivered by the machine is

$$P_{dev} = P_{gap} - P_{cu_r} \tag{8.92}$$

which is simply the power that appears across the resistor $r_r \frac{1-s}{s}$ as

$$P_{dev} = |\mathbf{I}_r|^2 r_r \frac{1-s}{s}. \tag{8.93}$$

Combining these expressions for the developed power gives us another way of expressing the power as

$$P_{dev} = (1 - s) P_{gap}. \tag{8.94}$$

Finally, the output power is

$$P_o = P_{dev} - P_{wf}. \tag{8.95}$$

The efficiency of the machine is

$$\eta = \frac{P_o}{P_{in}}. \tag{8.96}$$

Once the output power that ends the developed power is known, it is a simple task to calculate the torque developed and the output torque of the machine. Let

$$\omega = 2\pi \frac{n}{60} \tag{8.97}$$

then

$$\boxed{T_o = \frac{P_o}{\omega}} \tag{8.98}$$

and

$$\boxed{T_{dev} = \frac{P_{dev}}{\omega}.} \tag{8.99}$$

Now using

$$\omega_s = 2\pi \frac{n_s}{60} \tag{8.100}$$

and therefore,

$$\omega = (1 - s)\,\omega_s \tag{8.101}$$

it is also true that

$$\boxed{T_{dev} = \frac{P_{gap}}{\omega_s}.} \tag{8.102}$$

Example 8.10 A Y-connected, 400 V, four-pole, 50 Hz induction machine has the following parameters:

$X_s = 1.1\,\Omega$ $\qquad r_s = 0.3\,\Omega$

$X_r = 0.8\,\Omega$ $\qquad r_r = 0.2\,\Omega$ both referred to the stator

$X_m = 250\,\Omega$

The total magnetising losses are 250 W and the windage and friction losses are 420 W. Calculate the input power, input power factor, output torque and efficiency if the slip is 3%.

Solution

The total impedance seen by the source is

$$Z_t = r_s + jX_s + \left(\frac{1}{jX_m} + \frac{1}{\frac{r_r}{s} + jX_r}\right)^{-1}$$
$$= 7.22\angle 16.7^\circ\ \Omega.$$

The input current is

$$\mathbf{I}_s = 31.97\angle - 16.7^\circ \text{ A}$$

and the input power (per phase) is therefore,

$$P_{in} = \text{Re}\left\{\mathbf{V}_\phi \mathbf{I}_\phi^*\right\} = \frac{400}{\sqrt{3}}\text{Re}\left\{\mathbf{I}_s^*\right\} = 7.07 \text{ kW}$$

at a power factor

$$\text{pf} = \cos(-16.7) = 0.96 \text{ lagging}.$$

The (per phase) power crossing the air gap is

$$P_{gap} = P_{in} - \frac{1}{3}P_{mag} - |\mathbf{I}_r|^2 r_s = 6.68 \text{ kW}$$

and the mechanical power developed by the machine (per phase) is

$$P_{dev} = (1 - s)\, P_{gap} = 6.48 \text{ kW}.$$

Therefore, the output power (per phase) is

$$P_o = P_{dev} - \frac{1}{3}P_{wf} = 6.34 \text{ kW}$$

and therefore,

$$\eta = \frac{P_o}{P_{in}} = 89.7\%.$$

The output torque of the machine is

$$T_o = 3\frac{P_o}{\omega} = 124.8 \text{ Nm}.$$

Example 8.11 A 50 Hz, six-pole, Y-connected, 400 V induction machine delivers 300 Nm to a load at 960 rpm. If we neglect the magnetising losses estimate the stator resistance if it is known that the efficiency at this point is 88% and the input power factor is 0.92 lagging.

Solution

The synchronous speed is

$$n_s = \frac{120f}{p} = 1\,000 \text{ rpm}$$

and the slip therefore,

$$s = \frac{n_s - n}{n_s} = 4\%.$$

The machine delivers

$$P_o = T\omega = 30.16 \text{ kW}$$

which means that the input power to the machine is

$$P_{in} = \frac{P_o}{\eta} = 34.27 \text{ kW}.$$

The power crossing the air gap is

$$P_{gap} = \frac{P_o}{1 - s} = 31.42 \text{ kW}.$$

Now, the magnitude of the input current is

$$|\mathbf{I}_s| = \frac{400}{\sqrt{3}V\text{pf}} = 57.8 \text{ A}$$

and as the copper losses in the stator are

$$P_{cu_s} = P_{in} - P_{gap} = 3 \text{ kW}$$

the stator winding resistance is estimated as

$$r_s = \frac{P_{cu_s}}{3|\mathbf{I}_s|^2} = 38.4 \text{ m}\Omega.$$

8.5.5 Induction Motor as a Component in a System

Induction motors are the work horse for converting the electrical power from the ac grid into mechanical power. They are used for pumping water, powering fans and air conditioners and for doing numerous tasks in factories.

8.5.5.1 Torque–Speed Characteristics A typical torque–speed characteristic of an induction machine is shown in Fig. 8.35, with the different characteristic points named. Although it is typically true that a machine can deliver more torque than the rated torque, this point of the breakdown torque falls outside the desired operating range. Machines are designed with a definite margin between the rated torque (the machine is taking rated current) and the breakdown torque, because if the load exceeds the breakdown torque the operating point will slide down to the left of the curve towards standstill. This is clearly an undesirable situation.

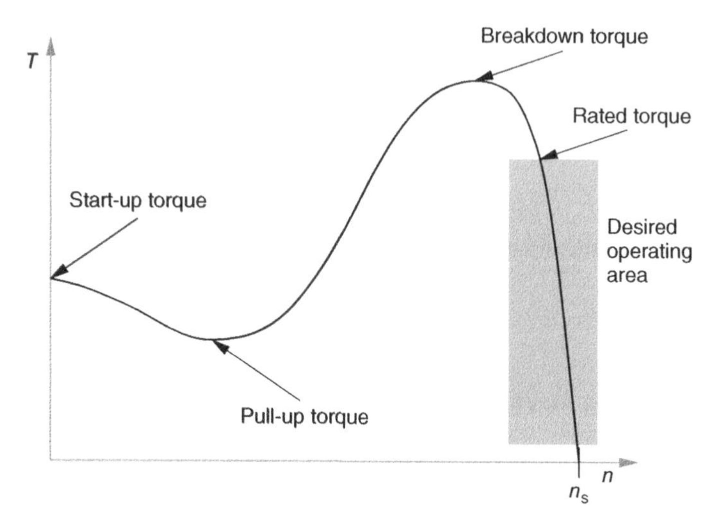

Figure 8.35 The induction machine torque–speed graph.

8.5.5.2 Variable Speed Drives If the three-phase ac is generated by a power electronics inverter, then it becomes possible to vary the frequency in order to achieve speed control with ac machines. Each stator electrical frequency ω_s has its own speed–torque curve, and if we plot them on the same graph, a family of curves is obtained as is illustrated in Fig. 8.36. Naturally, we have to divide ω_s by the number of pole pairs p to get the correct mechanical frequency ω_m.

In Chapter 7, we learned that with dc machines the armature current was proportional to the torque, and the induced voltage was proportional to the speed of the machine. This is very handy because it makes it possible to control torque and speed independently. By varying the field current, it is possible to adjust the constant that determines the relationship between torque and current, and speed and voltage.

What do we do when we need to control the speed and torque of an ac machine? Unfortunately, the behaviour of ac machines is more complicated than that of dc machines, but thanks to modern microprocessors it is possible to run in real-time

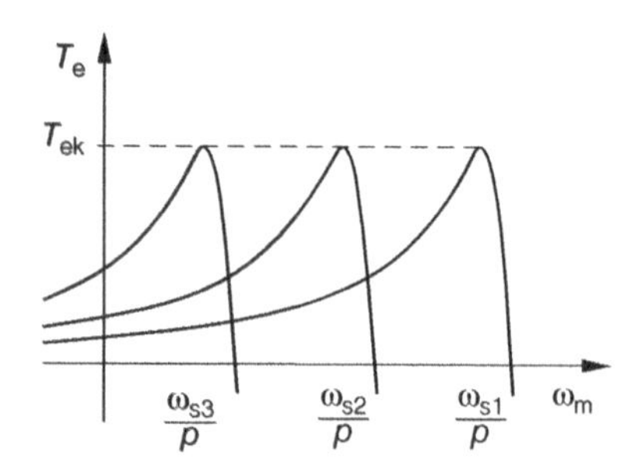

Figure 8.36 Induction motor torque–speed curves as a function of frequency.

models that convert the voltages and currents of an ac machine into the voltage and currents of a dc machine. This is called *vector control* whereby an ac induction or synchronous motor is controlled to behave similarly to a separately excited dc motor. Vector control generates a three-phase PWM machine voltage output that is derived from a complex voltage vector and is used to control a complex current vector derived from the motor's three-phase motor stator current input. Vector control is complicated and beyond the scope of this book.

Vector control of ac motors was pioneered by Hasse and Blaschke, starting in 1968 and during the early 1970s. It took about two decades before microprocessors became powerful and cheap enough to be used to implement vector control. Now vector-controlled drives are used in many industry applications, for example, in the printing industry. Although most manufacturers of ac machine drives are continually improving the control algorithms—for example, the ACS800 drive shown in Fig. 8.37 operates with ABBs trademark control method DTC (direct torque control—these methods use the same fundamental principles as vector control described here.

8.6 HIGHLIGHTS

- An inverter generates variable frequency, variable amplitude ac. It comprises three-phase arms that are controlled by the duty cycles that are 120° phase-shifted sinusoids

$$d_a(t) = \frac{1}{2} + \frac{m}{2}\cos(\omega t)$$
$$d_b(t) = \frac{1}{2} + \frac{m}{2}\cos\left(\omega t + \frac{2}{3}\pi\right)$$
$$d_c(t) = \frac{1}{2} + \frac{m}{2}\cos\left(\omega t - \frac{2}{3}\pi\right).$$

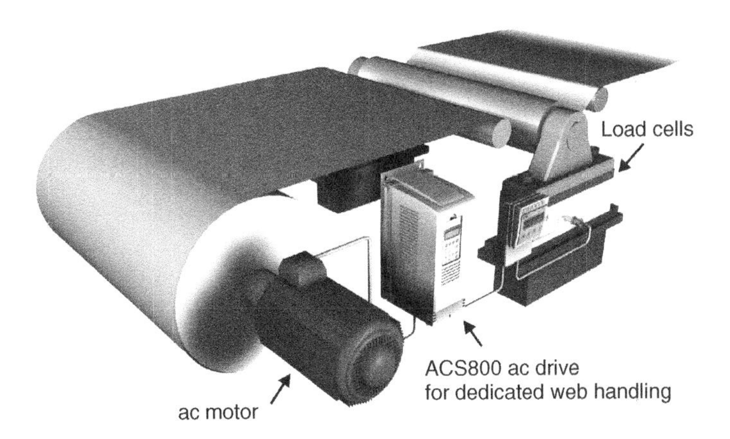

Figure 8.37 An ac drive used with an induction machine in a paper mill application.

- For an electrical machine, we can make the approximation

$$B \approx \mu_0 \frac{N\,i}{2\ell_{\mathrm{g}}}$$

because the total MMF generated by the windings is shared equally between the two parts of the air gap.
- The three-phase currents

$$\begin{aligned} i_a(t) &= I_m \cos(\omega t + 0)\ \mathrm{A} \\ i_b(t) &= I_m \cos\left(\omega t + \frac{2}{3}\pi\right)\ \mathrm{A} \\ i_c(t) &= I_m \cos\left(\omega t - \frac{2}{3}\pi\right)\ \mathrm{A}. \end{aligned}$$

give a rotating magnetic field in the stator

$$B_t(t) = \frac{3}{2} B_m \cos(\omega t)\ \hat{\mathbf{x}} - \frac{3}{2} B_m \sin(\omega t)\ \hat{\mathbf{y}}$$

in the clockwise direction, corresponding to a phase sequence

$$a - b' - c - a' - b - c'.$$

If the winding sequence is reversed

$$a - c' - b - a' - c - b'$$

the field rotates in an anticlockwise direction

$$B_m(t) = \frac{3}{2} B_m \cos(\omega t)\ \hat{\mathbf{x}} + \frac{3}{2} B_m \sin(\omega t)\ \hat{\mathbf{y}}$$

- The relationship of the electrical degrees to the mechanical degrees (θ_{m}) and the number of poles (p) is

$$\theta_{\mathrm{e}} = \frac{p}{2} \theta_{\mathrm{m}}.$$

- According to Faraday's law, the induced voltage is

$$\begin{aligned} e(t) &= N \frac{d\phi(t)}{dt} \\ &= -\omega N \phi_{\mathrm{P}}\ \sin(\omega t) = E_m \sin(\omega t) \end{aligned}$$

and the voltages induced in the windings are phase shifted due to the physical distribution of the windings along the stator circumference, resulting in

$$e_a(t) = E_m \sin(\omega t)$$
$$e_b(t) = E_m \sin\left(\omega t + \frac{2}{3}\pi\right)$$
$$e_c(t) = E_m \sin\left(\omega t - \frac{2}{3}\pi\right).$$

- The equivalent circuit of a synchronous machine can be solved with the phasor equations

$$\mathbf{E} = \mathbf{V} + \mathbf{I}\left(r + jX_\mathrm{s}\right)$$

  and the expression for the electromechanical power is

$$-P_\phi \approx \frac{EV}{X_\mathrm{s}} \sin(\delta).$$

- Slip of an induction machine is defined with respect to the stator synchronous speed as

$$s = \frac{n_\mathrm{s} - n}{n_\mathrm{s}}$$

  and the machine speed as

$$n = (1 - s)\, n_\mathrm{s}$$

  the induced voltage as

$$E = s\, E_m,$$

  where E_m is the voltage induced when the rotor is standing still, and finally the rotor frequency is given by

$$f_\mathrm{r} = sf.$$

- The power and losses equations of the induction machine can be solved by the following equations:
  The input power is

$$P_\mathrm{in} = \mathrm{Re}\left\{\mathbf{V}_\mathrm{in}\mathbf{I}_\mathrm{s}^*\right\}.$$

  The stator copper losses are

$$P_{\mathrm{cu}_\mathrm{s}} = |\mathbf{I}_\mathrm{s}|^2 r_\mathrm{s}.$$

The power crossing the air gap is

$$P_{\text{gap}} = P_{\text{in}} - P_{\text{cu}_s} - P_{\text{mag}}$$

or the air gap power can be calculated directly

$$P_{\text{gap}} = |\mathbf{I}_r|^2 \frac{r_r}{s}.$$

The rotor copper losses are

$$P_{\text{cu}_r} = |\mathbf{I}_r|^2 r_r.$$

Mechanical power delivered by the machine is

$$P_{\text{dev}} = P_{\text{gap}} - P_{\text{cu}_r}.$$

Mechanical power is the power that appears across the resistor $r_r \frac{1-s}{s}$ as

$$P_{\text{dev}} = |\mathbf{I}_r|^2 r_r \frac{1-s}{s}.$$

Output power is

$$P_o = P_{\text{shaft}} = P_{\text{dev}} - P_{\text{wf}}.$$

- Speed control is possible when variable frequency, variable voltage and three-phase voltage are generated with an inverter and applied to an ac machine. Vector control is the means where an ac induction or synchronous motor is controlled to behave like a separately excited dc motor.

PROBLEMS

8.1 How is it possible that the induced voltage could be larger or smaller than the grid voltage when a synchronous machine operates as a motor?

8.2 Imagine that you are fly sitting on the rotor of an induction machine. At what speed would the stator and the magnetic field rotate with respect to you?

8.3 A three-phase inverter is supplied with 720 V dc. The output is connected to a two-pole permanent magnet synchronous motor. The method of control is to keep the product of applied ac voltage times the frequency constant so that $V_{\text{line}} f_{\text{electrical}} = 3 \times 10^4$ where V_{line} is the rms value of the line voltage. This ensures that the magnetic material used in the machine will not saturate.

(Can you explain why? Hint: the flux is the integral of voltage according to Faraday's law.) The maximum modulation index is 0.9.

1. Find the speed of the motor for $m = 0.9$.
2. Find the ac voltage applied to the motor when it runs at its maximum speed = 10 000 rpm.
3. Find the depth of modulation at 100 rpm.

8.4 A three-phase wye connected to a 3300 V, 500 kVA synchronous generator has $X_s = 10\,\Omega$ and $R_w = 0.1\,\Omega$. The machine operates at rated load and voltage at a power factor of 0.867 lagging. Find the induced voltage per phase and the torque angle.

8.5 A wye-connected eight-pole synchronous motor has a reactance of 10 Ω and a negligible resistance. It is connected to a 400 V, 50 Hz three-phase line. The load requires a torque of 150 Nm. The line current is 15 A leading the phase voltage. Assume that all the losses can be neglected.

1. Find the power angle and induced voltage.
2. Find the line current when the load is removed, ignoring losses.

8.6 A 10 hp, 400 V wye-connected synchronous motor has a reactance of 4 Ω and a negligible resistance . The power factor is 0.8 leading and the rotational loss is 230 W and the field loss is 50 W.

1. Find the armature current.
2. Find the motor efficiency.
3. Find the power angle.

8.7 A six-pole induction motor that is used to drive one wheel set in a tram has a 60 kW input power rating and is 85% efficient. If the inverter supply is 400 V at 40 Hz, compute the motor speed and torque at a slip of 0.03.

8.8 A 74.6 kW wye-connected six-pole induction motor is connected to a 440 V, 50 Hz grid. The following parameters apply: $R_s = 0.06\,\Omega$, $R_r = 0.08\,\Omega$, $X_s = 0.3\,\Omega$, $X_r = 0.3\,\Omega$, $X_m = 5\,\Omega$, $s = 0.02\,\Omega$. The no-load power input is 3 240 W at a current of 45 A.

1. Find the line current and the input power.
2. Find the developed torque and shaft torque.

8.9 A 50 Hz, two-pole, wye-connected induction motor is connected to a three-phase 400 V (line-to-line). The following parameters apply: $R_s = 0.5\,\Omega$, $R_r = 0.2\,\Omega$, $X_s = 10\,\Omega$, $X_r = 0.5\,\Omega$, $X_m = 40\,\Omega$. The machine is running at 2 950 rpm and the total rotational and stray-load losses are 400 W.

1. Find the slip and the input current
2. Find the input power and the mechanical power developed.
3. Find the shaft torque and efficiency.

FURTHER READING

Chapman S.J., *Electric Machinery Fundamentals*. 3rd edition, McGraw-Hill, Boston, 1999.

Fitzgerald A.E., Kingsley C. jnr and Umans S.D., *Electric Machinery*. 6th edition, McGraw-Hill, Boston, 2003.

Nasar S.A., *Electric Energy Systems*. 1st edition, Prentice Hall, Upper Saddle River, 1996.

Sen P.C., *Principles of Electric Machines and Power Electronics*. 2nd edition, John Wiley and Sons, New York, 1997.

INDEX

The Principles of Electronic and Electromechanic Power Conversion: A Systems Approach, First Edition.
Braham Ferreira and Wim van der Merwe.

Printed and bound by CPI Group (UK) Ltd, Croydon, CR0 4YY

16/04/2025

14658595-0005